Chemical Aspects of Photodynamic Therapy

ADVANCED CHEMISTRY TEXTS

A series edited by DAVID PHILLIPS, *Imperial College, London, UK*, PAUL O'BRIEN, *Imperial College, London, UK* and STAN ROBERTS, *University of Liverpool, UK*

Volume 1
Chemical Aspects of Photodynamic Therapy
Raymond Bonnett

This book is part of a series. The publisher will accept continuation orders which may be cancelled at any time and which provide for automatic billing and shipping of each title in the series upon publication. Please write for details.

Chemical Aspects of Photodynamic Therapy

Raymond Bonnett

Queen Mary and Westfield College
University of London, UK

Gordon and Breach Science Publishers

Australia • Canada • France • Germany • India • Japan
Luxembourg • Malaysia • The Netherlands • Russia
Singapore • Switzerland

Amsteldijk 166
1st Floor
1079 LH Amsterdam
The Netherlands

British Library Cataloguing in Publication Data

Bonnett, Raymond
Chemical aspects of photodynamic therapy. – (Advanced chemistry texts ; v. 1)
1. – Photochemotherapy
I. Title
615.8′31

ISBN: 90-5699-248-1
ISSN: 1029-3654

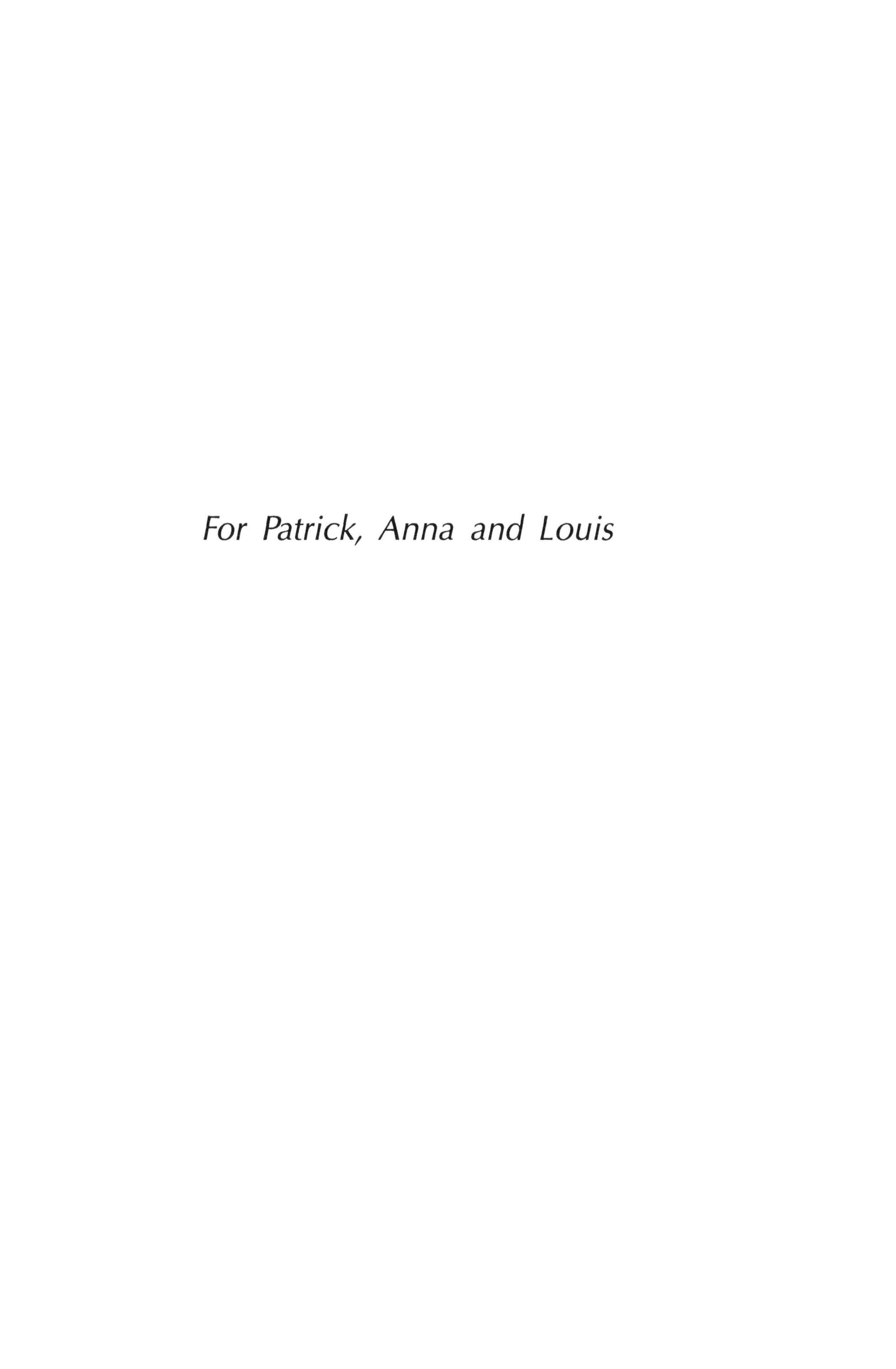

For Patrick, Anna and Louis

Contents

Preface xiii

Acronyms xv

Chapter 1 Introduction **1**

1.1 Definition of Area of Study 1
1.2 Historical Aspects 3
1.2.1 Vitamin D and Rickets 4
1.2.2 Phototherapy of Psoriasis 6
1.2.3 Neonatal Hyperbilirubinemia 7
1.2.4 Cancer 9
Bibliography 13

Chapter 2 Physical Matters **15**

2.1 Light 15
2.2 Light Sources 19
2.2.1 Sunlight 19
2.2.2 Incandescent Lamps 19
2.2.3 Arc Lamps 19
2.2.4 Light-emitting Diodes 20
2.2.5 Lasers 21
2.3 Light Absorption 23
2.3.1 The Beer-Lambert Law 23
2.3.2 Excitation 24
2.3.2.1 *Multiplicity* 24
2.3.2.2 *Probability* 25
2.4 Emission 27
2.5 Jablonski Diagram 28
2.6 Quantum Efficiency 30
2.7 Intermolecular Electronic Excitation Transfer 31
2.8 Singlet Oxygen Quantum Yields 32
Bibliography 37

Chapter 3 Singlet Oxygen **39**

3.1 General 39
3.2 Generation of Singlet Oxygen 42
3.2.1 Chemical Methods 42
3.2.2 Physical Methods 42

3.3 Chemical Reactions 47
3.3.1 Alkenes — the "ene" Reaction 47
3.3.2 Electron-rich Alkenes 49
3.3.3 Conjugated Dienes — Diels-Alder Addition 52
3.3.4 Miscellaneous Reactions 54
Bibliography 56

Chapter 4 Photodynamic Action 57

4.1 Definition 57
4.2 Discovery and Development 57
4.3 Photodynamic Agents — Structural Types 64
4.3.1 General Considerations 64
4.3.2 Structural Types 64
4.4 Mechanism of Photodynamic Action 70
4.4.1 The Type I Mechanism — Electron Transfer 70
4.4.2 The Type II Mechanism — Energy Transfer 74
4.4.3 Distinguishing between Type I and Type II Photooxygenation Processes 76
4.4.3.1 *Tests for radical intermediates (Type I mechanism)* 80
4.4.3.2 *Tests for singlet oxygen (Type II mechanism)* 81
4.4.4 The Overall Mechanistic Picture 83
Bibliography 87

Chapter 5 Some Other Examples of Photomedicine 89

5.1 Introduction 89
5.2 Sunscreens 89
5.2.1 Sunlight and the Skin 89
5.2.2 Artificial Sunscreens 92
5.3 Skin Carcinogenesis 93
5.4 Phototherapy of Rickets (Osteomalacia) 94
5.5 Psoriasis 100
5.5.1 Photosensitisers for PUVA 100
5.5.2 Mechanism 102
5.6 Phototherapy of Neonatal Hyperbilirubinemia 103
5.6.1 Photofragmentation 104
5.6.2 Photoisomerisation / Photosolubilisation 107
5.6.2.1 *Photochemical configurational isomerisation (Photobilirubins I)* 107
5.6.2.2 *Photocyclisation (Photobilirubins II, lumirubins)* 109
Bibliography 113

Chapter 6 The Chemistry of Haematoporphyrin Derivative (HpD) 115

6.1 Introduction 115
6.2 Chemistry of HpD Stage I 117
6.3 Chemistry of HpD Stage II 118
6.4 Clinical Development 126
Bibliography 128

Chapter 7 Second Generation Photosensitisers 129

7.1 First Generation Photosensitisers 129
7.1.1 Advantages 129
7.1.2 Disadvantages 129
7.2 Design Criteria for Second Generation Photosensitisers 130
7.2.1 Dark Toxicity 130
7.2.2 Composition 130
7.2.3 Synthesis 130
7.2.4 Solution Behaviour 130
7.2.4.1 *Tetraphenylporphyrin sulphonic acids* 131
7.2.4.2 *Sulphonated metallophthalocyanines* 132
7.2.4.3 *3-(1-Alkyloxyethyl)-3-devinylpyropheophorbide a* 135
7.2.4.4 *Spacing of hydrophilic substituents on the hydrophobic framework* 136
7.2.5 Delivery Systems 137
7.2.6 Photophysical Properties 143
7.2.7 Red Absorption 144
Bibliography 147

Chapter 8 Porphyrin Photosensitisers 149

8.1 Introduction 149
8.2 Porphyrin Sources 149
8.3 Porphyrins Derived from Haemoglobin 156
8.4 Synthetic Porphyrins 159
8.4.1 General Approaches 159
8.4.2 Specific Examples 166
8.5 Endogenous Porphyrin: δ-ALA as a Pro-drug 173
Bibliography 174

Chapter 9 Chlorins and Bacteriochlorins 177

9.1 Structural Definition 177
9.2 Absorption Spectra 178
9.3 Naturally Derived Chlorins and Bacteriochlorins 183
9.4 Chlorins Derived from Protohaem 186
9.4.1 From Deuteroporphyrin (**88**) 187
9.4.2 From Protoporphyrin (**61**) 188
9.4.3 From Photoprotoporphyrin (**132**) 189

9.5 Chlorins and Bacteriochlorins by Total Synthesis 190
9.5.1 By Reduction 190
9.5.2 By Other β,β′-Addition Reactions 193
9.5.3 By Meso-β Cyclisation 193
Bibliography 198

Chapter 10 Phthalocyanines and Naphthalocyanines 199

10.1 Structural Considerations 199
10.2 Spectra 201
10.3 5-Azaporphyrins 203
10.4 Benzoporphyrins 203
10.5 Phthalocyanines 203
10.5.1 General 203
10.5.2 Synthesis 211
10.5.3 PDT Photosensitisers 219
10.5.3.1 *Zinc(II) phthalocyanine* 219
10.5.3.2 *Sulphonic acids* 220
10.5.3.3 *Hydroxyphthalocyanines* 220
10.5.3.4 *Axial ligand variation* 221
10.6 Naphthalocyanines 222
Bibliography 223

Chapter 11 Other Photosensitisers 225

11.1 Cyanine Dyes 225
11.1.1 Merocyanine 540 225
11.1.2 Indocyanine Green 226
11.2 Hypericin 226
11.3 Phenothiazines 226
11.4 Porphycenes 227
11.5 Squaraines 229
11.6 Texaphyrins 229
11.7 Xanthenes 231
11.8 Conjugated Photosensitisers 233
Bibliography 235

Chapter 12 Photobleaching 237

12.1 Definitions 237
12.2 Potential Value of Photobleaching in PDT 238
12.3 Kinetic Studies 238
12.4 Product Studies 241
12.4.1 Photomodification 241
12.4.1.1 *Protoporphyrin* 241
12.4.1.2 *m-THPP* 243
12.4.1.3 *m-THPC* 244

12.4.1.4 *Purpurins* 244
12.4.1.5 *Haematoporphyrin* 245
12.4.2 Photobleaching 246
12.4.2.1 *m-THPC and m-THPBC* 246
12.4.2.2 *Other photosensitisers related to porphyrins* 246
12.5 Biological Systems 248
12.5.1 *In vitro* Studies 248
12.5.2 *In vivo* Studies 248
Bibliography 256

Chapter 13 Biological Aspects 257

13.1 Biological Assay 257
13.1.1 Preliminary Physical Tests 257
13.1.2 *In vitro* Bioassays 258
13.1.3 Intermediate Bioassays 259
13.1.4 *In vivo* Bioassays 260
13.1.4.1 *Tumour photonecrotic activity* 260
13.1.4.2 *Tumour selectivity* 261
13.2 Localisation 262
13.2.1 Protein Binding 264
13.2.2 Intracellular Localisation 264
13.2.3 Localisation *in vivo* 265
13.3 Photosensitiser Excretion 268
13.4 Mechanisms of Tumour Destruction 269
Bibliography 270

Chapter 14 Clinical and Commercial Developments: Prospects for the Future 271

14.1 Introduction 271
14.2 Photofrin^R (also referred to as porfimer sodium) 272
14.3 Second Generation Drugs at the Clinical Stage 273
14.3.1 δ-ALA (**104**) 273
14.3.2 m-THPC (**77**) 274
14.3.3 Tin Etiopurpurin — SnEt2 275
14.3.4 Mono-L-aspartylchlorin e_6 — MACE (**125**) 275
14.3.5 3-(1-Hexyloxyethyl)-3-devinylpyropheophorbide *a* (**75**) 275
14.3.6 Sulphonated Aluminium Phthalocyanine 275
14.3.7 Benzoporphyrin Derivative (**131**) 275
14.3.8 Lutetium Texaphyrin (**187**) 276
14.3.9 Porphycenes 276
14.4 Comparative Data 276
14.5 Future Prospects 277
14.5.1 Synthesis of New Sensitisers 277
14.5.2 Light Sources 277

14.5.3 New Clinical Directions in PDT 278
14.5.3.1 *Psoriasis* 278
14.5.3.2 *Arthritis* 278
14.5.3.3 *Age-related Macular Degeneration (AMD)* 278
14.5.3.4 *Other Conditions and Treatments* 278
14.5.3.5 *Microbicidal Activity* 278
Bibliography 289

Index 291

Preface

This book is written to fill a need I have felt over the past 15 years or so for a book on photodynamic therapy (PDT) for third year undergraduate chemistry students, and for organic chemistry research students. The book might, indeed, find a wider audience, since PDT, after a very slow start, is expanding rapidly. Although what is written now will inevitably be overtaken in a few years, nevertheless the basics of the various aspects involved will not change, and it is these that I would like to get across to my students.

In its very essence, the subject is interdisciplinary, but I have tried to present the areas less familiar to the chemist in a way which is easily assimilable. In doing so I may have oversimplified here and there. So be it. But any errors I have made in doing this are my responsibility, and I would be glad if they can be drawn to my attention.

I am grateful to those who have helped me write this book. My wife, Shirley, for help and forebearance; to my son, Paul, for his sustained help with the intricacies of the computer; to colleagues in various laboratories, both at QM and elsewhere, for providing information and (with the publishers) giving permission for reproducing figures, in particular amongst this group Professor Hans-Dieter Brauer (Frankfurt), Dr Tom Dougherty (Buffalo), Professor Josép Ribo and Dr Asuncion Vallés (Barcelona), and Professor Jeremy Sanders (Cambridge); to Dr Roger Evans (Biology, QM), Dr Robin Whelpton (Pharmacology, QM) and Dr Charles Stewart (Scotia Pharmaceuticals) for advice on biological and clinical matters; to Dr M.N. Lilly (Chemistry, QM) for advice about Chapter 2; to staff in the QM Library, and especially Susan Richards, for unfailing help; to Gabriel Martinez for help with some of the figures; and to Professor Albert Gossauer (Fribourg) and Professor Thierry Patrice (Nantes) for invitations to give a lecture series and a plenary lecture, respectively, during the preparations for which this book took shape.

I would also like to acknowledge the comradeship over many years that I enjoyed with the late Morris Berenbaum and his students (E.B. Chevretton, S. Akande and M. Ruston), and with my own students and associates working on this project (Birgul Djelal, Aslam Galia, Antonia Gomez, Stella Ioannou, Andrei Kozyrev, Gabriel Martinez, David Moffat, Alexander Nizhnik, Kawulia Okolo, Tony Salgado, Jason Siu, Phinda Songca, Asun Vallés, Franz Weirrani, Rosemary White and Una-Jane Winfield). They each contributed in their individual ways to taking the subject forward, and increasing our understanding of the complexities of PDT.

Thanks are warmly accorded to Mrs Lesley Lambert (QM) for invaluable help with the preparation of the manuscript and to Gordon and Breach for expert and rapid production of the book.

A final word about sources. It turns out that, although PDT papers can be found in a variety of well-known journals, of late they have tended to congregate in conference proceedings, which are not very accessible, but also in two journals which are much easier to find: *Photochemistry and Photobiology*, which is the periodical of the American Society of Photobiology, and *Journal of Photochemistry and Photobiology, B: Biology*, which is the periodical of the European Society of Photobiology. The web sites of the pharmaceutical companies involved (Panels in Chapter 14) generally provide up-to-date information on progress at the cutting edge.

Ray Bonnett
Chemistry Department
Queen Mary, University of London
Mile End Road
London E1 4NS, UK

Acronyms

ALA, δ-ALA, 5-ALA	*delta*-Aminolaevulinic acid
$AlPcS_{2a}$	Aluminium phthalocyanine substituted with 2 sulphonic acid groups on adjacent benzenoid rings
$AlPcS_{2o}$	Aluminium phthalocyanine substituted with 2 sulphonic acid groups on opposite benzenoid rings
$AlPcS_n$	Aluminium phthalocyanine substituted with n sulphonic acid groups
AMD	Age-related macular degeneration
BCC	Basal cell carcinoma
BHT	Butylated hydroxytoluene; 2,6-di-*t*-butyl-4-methylphenol
BPDMA	Benzoporphyrin derivative, mono acid, ring A
CAM	Chorioallantoic membrane
CON	Conrotatory
CPC	Cetylpyridinium chloride
DABCO	1,4-Diazabicyclo[2.2.2]octane
DBU	1,8-Diazabicyclo[5.4.0]undec-7-ene
DCC	Dicyclohexylcarbodiimide
DDQ	2,3-Dichloro-4,5-dicyano-1,4-benzoquinone
DLI	Drug-light interval
DME	Dimethyl ester
DMPO	5,5-Dimethyl-1-pyrroline-1-oxide
DMSO	Dimethyl sulphoxide
DNA	Deoxyribonucleic acid
DOPA	3,4-Dihydroxyphenylalanine
DPA	9,10-Diphenylanthracene
DPPC	Dipalmitoylphosphatidylcholine
ESR	Electron spin resonance
FANFT	2-Formamido-4-(5-nitro-2-furanyl)thiazole
FBS	Foetal bovine serum
HDL	High density lipoprotein
HIV	Human immunodeficiency virus
HOMO	Highest occupied molecular orbital
HpD	Haematoporphyrin derivative
HPLC	High pressure liquid chromatography
LDL	Low density lipoprotein
LED	Light-emitting diode
LUMO	Lowest unoccupied molecular orbital
m-THPBC	5,10,15,20 -Tetrakis(*m*-hydroxyphenyl)bacteriochlorin

m-THPC	5,10,15,20 -Tetrakis(*m*-hydroxyphenyl)chlorin
m-THPP	5,10,15,20 -Tetrakis(*m*-hydroxyphenyl)porphyrin
MACE	Mono-L-aspartyl chlorin e_6
MB	Methylene blue
MOP	Methoxypsoralen
MS	Mass spectrometry
MTS	Multicellular tumour spheroids
MTT	3-(4,5-Dimethylthiazol-2-yl)-2,5-diphenyl-2*H*-tetrazolium bromide
NMR	Nuclear magnetic resonance
NPc	Naphthalocyanine
OOPS	Dioleylphosphatidylcholine
PBG	Porphobilinogen
PBS	Phosphate-bufferred saline
Pc	Phthalocyanine
PD	Photodynamic
PDT	Photodynamic therapy
POPC	Palmitoyloleylphosphatidylcholine
PUVA	Psoralen + UVA
RB	Rose bengal
RNA	Ribonucleic acid
RT	Room temperature
SDS	Sodium dodecyl sulphate
SnEt2	Dichlorotin(IV) complex of a purpurin derivative formulated as **78**
THF	Tetrahydrofuran
TLC	Thin layer chromatography
TPC	Tetraphenylchlorin
TPP	5,10,15,20 -Tetraphenylporphyrin
$TPPS_{2a}$	TPP substituted with 2 sulphonic acid groups on adjacent phenyl rings
$TPPS_{2o}$	TPP substituted with 2 sulphonic acid groups on opposite phenyl rings
$TPPS_n$	TPP substituted with n sulphonic acid groups on the phenyl rings
UDPGT	Uridine diphosphoglucuronosyl transferase
VLDL	Very low density lipoprotein
YAG	Yttrium-aluminium-garnet

Chapter 1

INTRODUCTION

1.1 DEFINITION OF AREA OF STUDY

This book is about the photochemical processes which are important, or of potential importance, in therapeutic medicine. Although the topic thus falls within the general area of photobiology, the approach adopted here is a molecular one: chemistry in an interdisciplinary setting.

The basic laws which apply to photobiology are necessarily the same as those which apply to photochemistry, that is:

Law 1

The *Grotthus-Draper Law* which states that only light which is absorbed by a system can cause a photochemical change. Thus, it is not productive to irradiate a colourless solution with visible light: if no photons are absorbed, no photochemical reaction can occur.

Law 2

The *Stark-Einstein Law of Photochemical Equivalence.* This can be expressed in various ways, but here we shall say that in the initial excitation step of a photochemical reaction, the absorption of one photon excites one substrate molecule. Thus the quantum yield (which expresses the yield of a subsequent event or product in terms of the number of photons absorbed : see section 2.6), cannot be greater than 1. This statement becomes untrue in special circustances; for example, if the photochemical step initiates a dark chain reaction, as in the photochemical chlorination of methane:

$$Cl_2 \xrightarrow{hv} 2Cl\bullet$$
$$Cl\bullet + CH_4 \longrightarrow HCl + CH_3\bullet$$
$$\bullet CH_3 + Cl_2 \longrightarrow CH_3Cl + Cl\bullet \quad \phi_{MeCl} \approx 10^5$$

Although the same basic laws apply, photobiological processes are far more complex than photochemical ones. Whereas photochemists generally study homogeneous systems (gas phase or solution), in photobiology the system is almost always heterogeneous, being divided into various phases and compartments, as is the nature of living things.

However, it is not necessary to search far in order to find a principle, or razor, which simplifies this complex topic by neatly dividing it into two parts, albeit unequal ones. This razor is based on biological evolution. Thus photobiological processes can be broadly divided into two groups:

Group 1. Evolved, biologically vital processes, which occur in specialised systems which are highly organised at the macroscopic and molecular levels, and which have evolved to follow essentially only one main pathway, the biologically desirable one. Photosynthesis and vision, which are key processes in the biosphere, are examples of this group. For the photosynthetic bacterium, *Rhodopseudomonas spheroides*, the photochemical step and the course of the dark reactions which immediately follow it are shown in Figure 1.1.

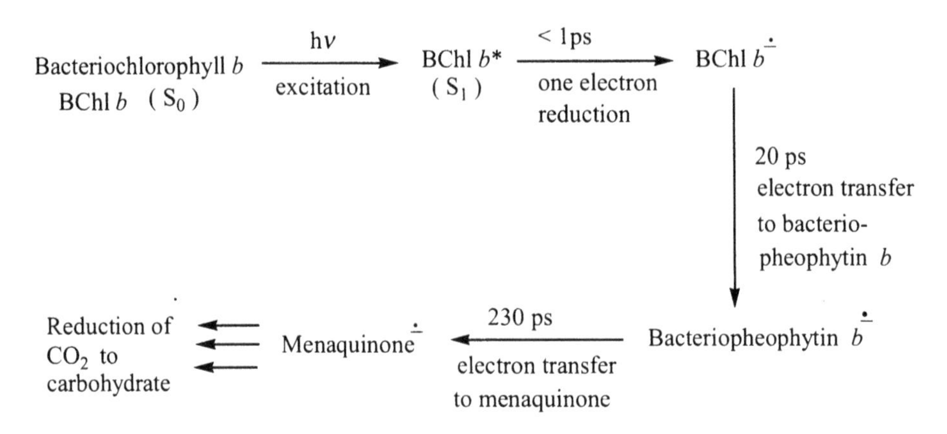

FIGURE 1.1 A PHOTOBIOLOGICAL PROCESS IN AN EVOLVED HIGHLY ORDERED SYSTEM (GROUP 1). S_0 REFERS TO THE GROUND SINGLET STATE, AND S_1 TO THE FIRST EXCITED STATE.

Group 2. These photobiological processes are at the other extreme, and are adventitious, biologically peripheral, processes which lack macroscopic order in relation to the photochemical event. A variety of reactive biomolecules may be involved, and hence, a single chemical pathway is not to be expected.

It will emerge that many of the processes to be considered in phototherapy, and, in particular, in the photodynamic effect (Chapter 4), fall into this category. Figure 1.2 serves to illustrate the diversity of pathways in the photodynamic effect in contrast to what is observed in Group 1 processes (Figure 1.1).

Having indicated what is to be considered in the present book, it is proper to state clearly what is not to be covered. Firstly, *X*-radiation is not covered, because although *X*-rays are simply another part of the electromagnetic spectrum, they are of very high energy and tend to damage everything that they encounter by ionising processes. The photochemical processes we shall be discussing here involve visible and near visible light, and are much more selective and much more elegant. Secondly we are not looking at the effects and applications of high power lasers, even though they may fall in the wavelength range under consideration here (about 250–900 nm). Such lasers are essentially thermal lances and are excellent for cutting, ablating and sealing tissue, but not for carrying out controlled photochemical processes within it.

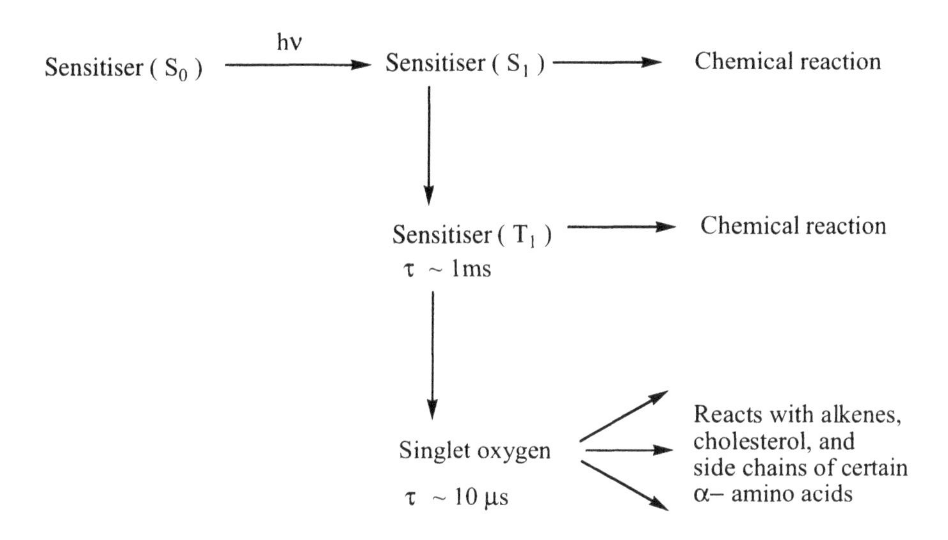

FIGURE 1.2 AN ADVENTITIOUS MULTIPATHWAY PHOTOBIOLOGICAL PROCESS (GROUP 2). T_1 REFERS TO THE FIRST EXCITED TRIPLET STATE, AND τ TO THE LIFETIME OF THE SPECIES.

1.2 HISTORICAL ASPECTS

Phototherapy is the use of visible or near-visible light in the treatment of disease. Essentially it involves electronic transitions leading to photochemical reactions, which may be followed by dark reactions of the initial products. Phototherapy falls into two broad categories:

Direct. No drug is administered, and, thus, according to the Grotthus-Draper Law, if an effect is to be observed, light must be absorbed by molecules already present in the organism.

Indirect. In this case an additional substance, a sensitiser, is administered before irradiation. This sort of treatment is sometimes referred to as *photochemotherapy.* The therapeutic process is initiated by light being absorbed by the sensitiser. Many of the processes used or being developed in phototherapy today fall into the indirect, sensitised category, but phototherapy was introduced in the closing years of the nineteenth century with examples which did not employ a sensitiser ie they fell into the "direct" category.

Of course, the benefits of sunlight must have been appreciated by human beings from the beginning, and, indeed, there are indications in the ancient authors of medicinal effects (heliotherapy). Thus Herodotus (6C BC) is recorded as noticing the beneficial effect of sunlight on bone growth, and Hippocrates (460–375 BC) advocated the use of heliotherapy for various human maladies . But the introduction of phototherapy backed up by the scientific method is due to Niels Rydberg Finsen, and, with justification, he is generally regarded as the father of the subject. He was born in the Faroe Islands in 1860 and worked in Copenhagen. He published mainly in Danish, French and German, but wrote a book which was translated into English ("Phototherapy", Arnold, London, 1901).

Finsen carried out experiments on the effect of light of different colours on animals. Related to this study he advocated the use of red light (ie light devoid of ultraviolet radiation, or as Finsen called it "the chemical rays") for the amelioration of suppuration and scarring in cases of smallpox. His most noted finding, however, was that sunlight, or, better, light from a carbon arc suitably fitted to reduce infrared intensity (water filter, quartz optics), could be used to treat and cure lupus vulgaris, a tubercular condition of the skin which was then prevalent in Nordic countries, particularly in the winter. It caused reddish brown plaques to develop, especially on the face, and is fortunately now rare. His countrymen evidently appreciated his efforts: a Medical Light Institute named after him was set up in Copenhagen, and he was awarded the Nobel Prize for Physiology-Medicine in 1903. This award came just in time: he had always been a frail person and he died the next year.

There followed a period of considerable enthusiasm for phototherapy. Queen Alexandra, a fellow Dane, was instrumental in technology transfer to London, where the technique was introduced into the London Hospital (in Whitechapel) which had a flourishing Light Department in the early 1900s. Figure 1.3 shows bronze bas-reliefs on the plinth of a statue of Queen Alexandra, which stands in the grounds of the London Hospital (now Royal London Hospital), of which she was President. These bas-reliefs record her activities and actually portray her making a visit to the Light Department, with phototherapy in progress. The original light source is now preserved in the Science Museum in South Kensington, London.

Curiously, even with royal patronage, phototherapy did not at that stage gather strength, in contrast to *X*-radiation therapy, which was emerging at about the same time. Probably the only contribution to medicine which has survived from this early period is the germicidal effect of ultraviolet light, which had been appreciated by Finsen, and which, of course, still finds everyday application. The reasons for this falling away of interest are not altogether clear, but probably include a diminution of the prevalence of lupus vulgaris, the main target disease; a certain amount of exaggeration about the benefits of the treatment; and a healthy scepticism towards the approach as a whole. After all, one might argue, visible light is all around us all day — why is it necessary to have more to cure anything?

After a period of stagnation between the World Wars, phototherapy has seen a revival over recent decades. This is illustrated here for four areas: vitamin D synthesis, psoriasis treatment, the phototherapy of jaundice in the newborn, and tumour phototherapy. These are very diverse areas, and phototherapy appears to have developed in each one without reference to the others.

1.2.1 Vitamin D and Rickets

Vitamin D refers to a family of substances which are responsible for calcium and phosphate absorption and metabolism in bone deposition and maintenance. The deficiency disease is known clinically as rickets or osteomalacia: the former term is often used for the condition in childhood, the latter for the disease of adults (often the elderly).

FIGURE 1.3 BRONZE *BAS-RELIEFS* ON THE BASE OF A STATUE OF QUEEN ALEXANDRA IN THE GROUNDS OF THE ROYAL LONDON HOSPITAL, WHITECHAPEL. IN THE SECOND PHOTOGRAPH, QUEEN ALEXANDRA IS SEEN VISITING THE LIGHT DEPARTMENT: THE LIGHT SOURCE, A CARBON ARC, IS IN THE CANISTER AT THE TOP RIGHT CORNER, AND LIGHT TRAINS CAN BE CLEARLY SEEN EMERGING FROM IT. REMARKABLY FOR THIS DATE (1904) THE NURSES ARE WEARING PROTECTIVE GLASSES. KING EDWARD IS STANDING WELL BACK.

The condition is associated with bone malformation and characteristically with the bowing of the long bones in childhood. Rickets, which had been described in English children in 1645 by Whistler, became prevalent under the smoke-laden skies of industrial cities of England at the time of the industrial revolution, and became known as the "English disease". By the 1920s, folklore, clinical observation, and scientific demonstration had in succession established that the disease could be prevented or treated in two ways: by incorporating fish oil into the diet, or by exposing the skin to sunlight. Convincing observations on phototherapy were made by Huldschinsky in 1919. Subsequent experiments with animals fed on a diet deficient in calcium and phosphate showed that irradiation of the food supply had a similar curative effect to that observed following the direct irradiation of the ricketic animals with ultraviolet light. At that stage the structures of the vitamin D family of compounds were not known, and it took many years of endeavour (Windaus, Heilbron, Havinga) before the structures of the compounds and the detail of the photochemical processes were uncovered (section 5.4).

It seems likely that in primitive man, essentially unclothed and having to hunt for food, exposure to sunlight may well have been a very significant source of vitamin D. Under modern conditions it appears that dietary sources may often outweigh the photochemical production. Nevertheless the latter remains important. In the temperate zones, the amount of ultraviolet light available in the relevant 295 nm region is sufficient in the summer months to cause an increase in the concentration of vitamin D_3 in the plasma, but it is often insufficient in the winter months. Thus ultraviolet irradiation of school children living at high latitudes (eg Siberia) has been carried out during winter months as a routine preventative measure, the children being equipped with dark glasses to prevent retinal damage. In 1978, Jung and his colleagues working at the Dunn Nutritional Unit in Cambridge redirected attention to the benefit observed in treating osteomalacia in an adult who showed poor absorption of vitamin D administered orally or intramuscularly. Whole body treatment with short exposures of ultraviolet light had a marked beneficial effect, leading to a rapid rise in the concentration of 25-hydroxyvitamin D_3 in the plasma (section 5.4). Benefit has also been found in such phototherapy in elderly patients.

However, there is a downside to this. The substrates for the photochemical processes have absorption maxima (ca. 280nm–300nm) at wavelengths close to where the purine and pyrimidine bases of DNA absorb, and the potential carcinogenic effect of radiation in this region (UV-B and UV-C) needs to be kept in mind (section 5.3).

1.2.2 Phototherapy of Psoriasis

The treatment of various skin diseases with plant extracts now known to contain furocoumarins (psoralens) was practised in ancient civilisations. Thus in India, extracts of *Psoralea corylifolia* were given orally, followed by exposure to sunlight, in order to treat vitiligo, a disorder of the skin associated with loss of pigmentation. By the 1930s it had been shown that various plants of the *Rutaceae* and *Umbelliferae* families contained linear furocoumarins (eg psoralen) and angular furocoumarins (eg angelicin), and the structures (see section 5.5) and syntheses of these compounds were investigated (Thoms, Späth). The compounds had interesting physiological properties: they included compounds which were highly toxic to fish, but not to mammals; and they caused phototoxicity if ingested.

As early as 1953, Lerner, Denton and Fitzpatrick had reported promising studies on the treatment of vitiligo with 8-methoxypsoralen (abbreviated to 8-MOP in the trade). In 1974 Parrish and his colleagues revealed that oral administration of 8-MOP, followed by exposure to ultraviolet light in the 320–400 nm range, could be used to treat psoriasis, and this observation started another field of phototherapy. In this example a photosensitiser was needed, so that this is indirect, sensitised phototherapy or photochemotherapy. Since the 320–400 nm region is known as the UV-A (section 2.1) the treatment was christened PUVA = psoralen + UVA.

Psoriasis is a strange disease: the origin is unknown, and it cannot be cured as yet. It is an autoimmune disorder which manifests itself in the inadequately controlled growth of the epidermal layer, which develops raised reddish plaques covered by tiny scales of dead skin. These lesions can appear anywhere, but often occur at the axes (knees, elbows, crotch). And they are itchy. The Greeks have a word for the disease: it is *psoriasis* from *psorian*, to have the itch.The disease erupts and then quietens down again. It is not life threatening, but it can be exceedingly uncomfortable, and socially inconvenient. About 2% of the population of the United Kingdom suffer from it, and there is a genetic link in its distribution. In current practice, control is afforded in various ways — by topical application (coal tar, corticosteroids), by systemic administration of antimetabolites (eg methotrexate) and by PUVA. The latter is used only in moderately severe cases, and carries a carcinogenic risk. Interestingly the use of coal tar in combination with ultraviolet light in the treatment of psoriasis was described by Goeckerman in 1925, and has been in use for many years. It may be an example of the photodynamic effect (Chapter 4), with polycyclic aromatic hydrocarbons acting as the sensitisers.

In recent years, PUVA has been examined as a treatment for a number of other dermatological conditions, for example, mycosis fungoides (a rather rare neoplastic disease of the lymphoid cells in the upper dermis) and atopic eczema. Nevertheless psoriasis remains the most common target.

1.2.3 Neonatal Hyperbilirubinemia

Bilirubin is a linear tetrapyrrole which is formed as a product of haem catabolism, that is the breakdown of the haem of haemoglobin. The red blood cells do not last forever: after a vigorous working life (in Man, about 3 months) they are worn out, are beginning to leak, and are then destroyed and replaced. The haem system is oxidatively opened to give carbon monoxide and biliverdin (a blue green substance) and the latter is reduced to bilirubin (orange yellow, the chromophore of jaundice). Bilirubin is virtually insoluble in water, and has to be solubilised by reaction with a water-soluble unit before it can be secreted into the bile and passed out of the body. This solubilisation occurs in the liver, the main pathway being an enzymatic incorporation of D-glucuronic acid to give bilirubin diglucuronide. The enzyme involved is uridine diphosphoglucuronyl transferase (UDPGT). In biochemical parlance, this type of reaction is called conjugation, and the product is an example of a bilirubin conjugate (see Panel 5.2).

In newly-born infants this conjugation is sometimes impaired, and bilirubin itself is retained in the circulation. This happens especially in prematurely-born infants, where the enzymatic system has not developed to full efficiency, and in haemolytic disease, where the enzyme system is overloaded. The condition is called neonatal hyperbilirubinemia: the skin of the infant becomes yellow (jaundiced) due to bilirubin.

Because of its lipophilic properties bilirubin is able to penetrate membranes and in particular to get into developing nervous tissue, including the brain, where it exerts a toxic effect which, at this stage of development, can be permanently damaging. In extreme cases the infant dies, and the brain is stained yellow (kernicterus).

When bilirubin concentrations in plasma exceed a threshold of about 10 mg/100 ml the classical treatment for this condition is exchange transfusion, in which the contaminated blood of the infant is entirely replaced by new blood. As the prematurely-born baby comes to what would have been normal term, the UDPGT activity develops, and bilirubin conjugation and excretion become normal. The jaundice disappears.

Phototherapy of neonatal hyperbilirubinemia came about in a remarkable way. During the summer of 1956, which must have been particularly fine in Eastern England, the sister (Miss J. Ward) in charge of a pediatric unit at Rochford General Hospital (near Southend in Essex) noticed in the course of her duties that where the skin of a jaundiced infant was exposed to sunlight it became noticeably less yellow than skin which had been covered. Richard Cremer, the pediatrician, took these observations seriously, and decided to carry out some scientific experiments to test them. With the help of biochemical colleagues the concentration of bilirubin in plasma was measured, and it emerged that bilirubin levels were indeed reduced after exposure of the jaundiced infants to sunlight. However, sunlight is a sporadic and unreliable feature of the English scene, and the next step was to irradiate with a controllable source. Various lamps were tried (including a street lamp borrowed from Southend Municipal Corporation) until a satisfactory system was developed (Figure 1.4). Again it was found that plasma bilirubin levels fell after irradiation: this was the first scientific demonstration of the phototherapy of neonatal hyperbilirubinemia. It was published in the *Lancet* in 1958.

Sad to say, the world was not quite ready for this result: it seems that some people, particularly in the USA, refused to believe that something as ordinary as visible light could have such a beneficial and specific effect. Cremer and his colleagues never received the recognition that they clearly deserved for their discovery.

Gradually, however, the world woke up, and the methodology began to seem less bizarre, especially after 1968, when Lucey, Ferriero and Hewitt published confirmatory studies in the USA. The phototherapy of neonatal hyperbilirubinemia of the prematurely born using white light or blue-enriched white light (fluorescent lamps) is today a standard procedure in maternity and pediatric hospitals worldwide. It

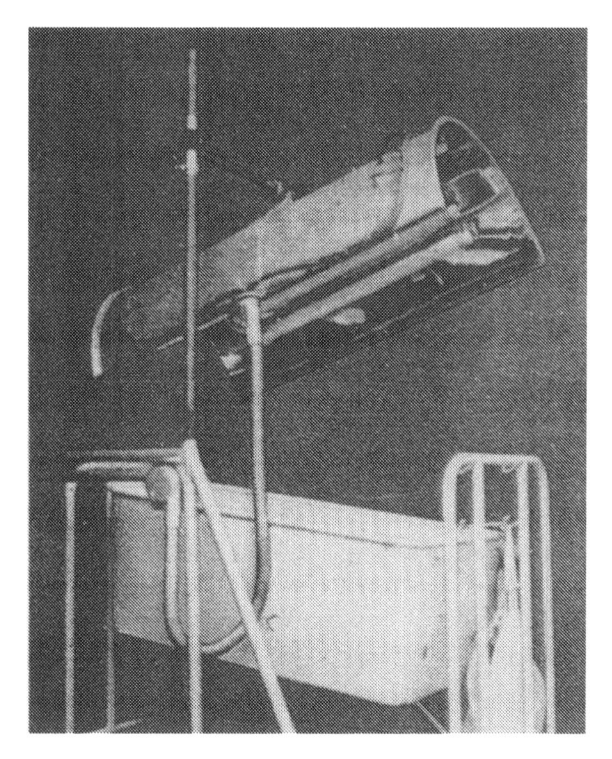

FIGURE 1.4 ORIGINAL APPARATUS USED FOR THE PHOTOTHERAPY OF NEONATAL HYPERBILIRUBINEMIA IN ROCHFORD HOSPITAL, ESSEX. (REPRODUCED WITH PERMISSION FROM R.J.CREMER, P.W.PERRYMAN, AND D.H.RICHARDS, "INFLUENCE OF LIGHT ON THE HYPERBILIRUBINEMIA OF INFANTS", *LANCET*, 1958 (1), 1094–1097. COPYRIGHT THE LANCET LTD).

appears to suffer from no significant drawbacks, and is to be regarded as a most successful example of phototherapy.

1.2.4 Cancer

The phototherapy of cancer is essentially photodynamic therapy, requiring visible light, a photosensitiser and oxygen. The photodynamic effect is the subject of Chapter 4: here we examine the development of ideas with particular reference to cancer therapy. In 1903, not long after the discovery of the photodynamic effect, Jesionek and von Tappeiner (Munich) reported experiments in which tumours had been treated with eosin topically, and then exposed to visible light. Two years later they reported an extension of this work in which various sensitisers, and two light sources (sunlight, arc lamp) were employed. Again the application of the sensitiser was at or near the surface (brushing, local injection) but the results (mainly on patients with basal cell carcinoma) appeared to be promising.

Nonetheless, nothing substantial was done to follow this up, although there were several indications that some dyestuffs could preferentially stain tumour tissue. Policard (Lyons) in 1924 observed what appears to have been a natural porphyrin fluorescence in experimental tumours, indicating an association between porphyrins and cancer tissue.

The most significant finding in these in-between years seems to have been made by Auler and Banzer, who wrote:

> "A small number of tumour-bearing animals were injected with Photodyn (= haematoporphyrin) and at the same time irradiated with a powerful quartz lamp. Besides known irritation symptoms (light rash, oedema, uneasiness) the animals showed necrosis, fluorescence, and cellular softening of the tumours. For one animal, which was given 4 irradiations, there remained only a small scab-like tumour mass, which showed strong fluorescence. The presence of porphyrin was identified spectroscopically. Analogous experiments have begun in humans" (*translated from Z.Krebsforsch.*, 1942, **53**, 65–68).

However it was 1942, in Berlin; World War II was at its height, and there are no subsequent reports of those analogous experiments.

The next developments came in the USA. In 1948 Figge, Wieland and Manganiello (Baltimore) reported that haematoporphyrin and its zinc complex localised in mouse tumours, (and also in embryonic and regenerating wound tissue) but this report does not seem to have had much impact. A decade or so later it was different. Lipson and his colleagues introduced a preparation called "Haematoporphyrin Derivative" (HpD), the chemistry of which will be discussed in Chapter 6. This preparation was shown in 1961 to be localised with some degree of selectivity in tumour tissue, with sufficient selectivity indeed that, making use of the red fluorescence of the porphyrin when irradiated in the ultraviolet (366 nm), the tumour could be visualised by its red emission. For a decade HpD was investigated as a diagnostic agent for cancer, and this remains an important application. It was eventually realised, however, that, by changing the irradiation conditions, the photodynamic effect could be brought into play and the tumour tissue, identified by its fluorescence, could be photodegraded. The initial observation of this sort was made by Diamond and his colleagues (San Francisco) who in 1972 reported the photodegradation of glioma implants in the rat after parenteral administration of haematoporphyrin. These workers also introduced the term photodynamic therapy. In 1974 Dougherty showed that fluorescein given intraperitoneally reduced the growth rate of mammary tumour implants in irradiated animals. In the same year Thomson, Emmett and Fox (Cincinnati) described the photodestruction of mouse epithelial tumours after oral administration of acridine orange.

However it was haematoporphyrin derivative (HpD) that led to rapid development. Experiments *in vivo* by Dougherty (Buffalo, USA) and Berenbaum (London, UK) and their colleagues showed clearly the effectiveness of the photodynamic therapy of cancer in animal models and in humans. The first photodynamic experiments in man were published by Kelly and Snell working in London (1976). These authors concluded "that haematoporphyrin derivative could be used as an aid to the diagnosis and treatment of carcinoma of the bladder." (*J. Urol.,* 1976, **115**, 150–151).

The concept of the treatment is presented in schematic form in Figure 1.5.

HpD and various commercial preparations derived from it (eg Photofrin) have been used in clinical experimentation for some years. Several thousand patients have been treated by photodynamic therapy using HpD. The first regulatory

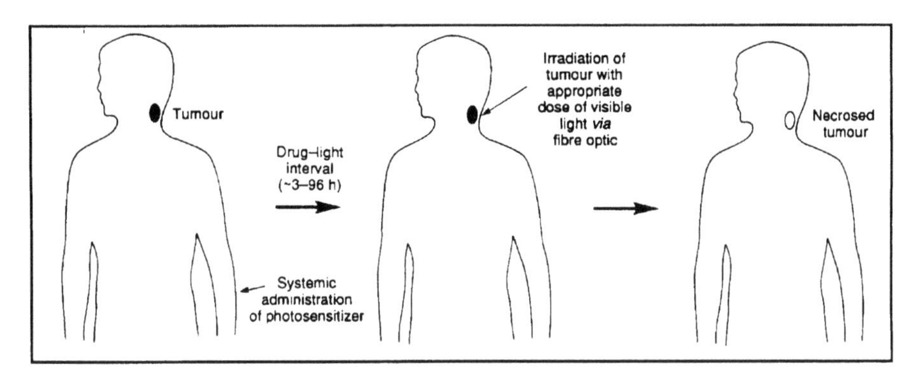

FIGURE 1.5 SCHEMATIC DIAGRAM TO SHOW THE PRINCIPLE OF PHOTODYNAMIC THERAPY OF CANCER. (REPRODUCED WITH PERMISSION FROM R.BONNETT, *CHEM.SOC.REV.*, 1995, **24**, 19).

approval for a photodynamic drug was accorded to Photofrin by the Canadian authorities in 1993. This approval was for the treatment of recurrent superficial papillary bladder cancer. Subsequent approvals, generally for limited applications, have been made in other countries (section 6.3). At the same time there has been increasing interest in developing more powerful and more specific tumour sensitisers, the so-called second generation sensitisers, and this topic is a major feature of the rest of this book.

Panel 1.1 Terms Used in Describing Light Treatment

PHOTOTHERAPY: The use of ultraviolet, visible, or near infrared light in the treatment of disease. In cases where no photosensitiser is administered the light is absorbed by chromophores present in the tissue, e.g. phototherapy of neonatal hyperbilirubinemia; photogeneration of vitamin D_3.

PHOTOCHEMOTHERAPY: The use of ultraviolet, visible or near infrared light together with an administered photosensitiser (the photochemotherapeutic agent) in the treatment of disease. Light is absorbed by the photosensitiser, eg PUVA therapy of psoriasis and other skin diseases.

PHOTODYNAMIC THERAPY (PDT): The treatment of disease by the use of visible or near infrared light together with an administered photosensitiser and in the presence of molecular oxygen at ambient levels. Light is absorbed by the photosensitiser, which then serves to activate oxygen in some way. The activated oxygen (principally singlet oxygen) then causes damage to the living system, eg PDT of cancer; photodynamic destruction of microbes.

Phototherapy is thus a general term used to describe all cases where light is used therapeutically: photochemotherapy is a subset of it, and photodynamic therapy is a division of that, as shown in the following Venn diagram.

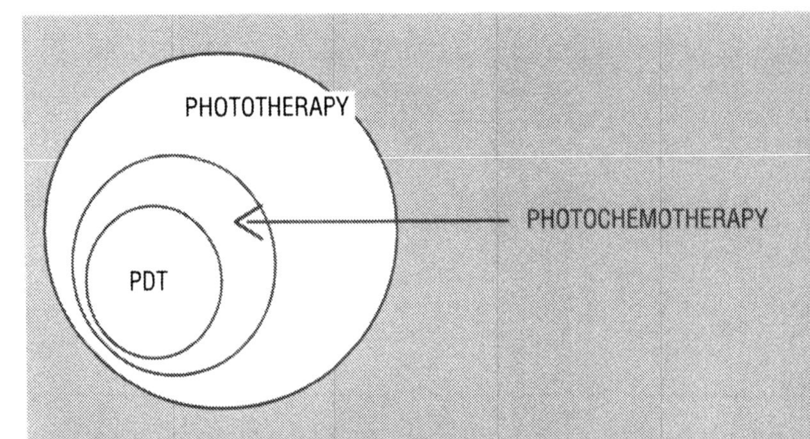

PHOTOPHERESIS: This refers to extracorporeal photochemotherapy. Thus the blood (or blood components) are removed from the body, treated by photochemotherapy *in vitro*, and then put back again. This procedure has been used to treat cutaneous T cell lymphoma (following oral 8-MOP) with beneficial effect.

Panel 1.2 Terms Used in Drug Administration

Topical: applied to the surface, for example in a lotion or a cream.
Parenteral: administered in any way, other than by mouth.
Oral: administered by mouth.
Intravenous (i.v.): by injection into a vein.
Intradermal: by injection into the skin.
Intramuscular: by injection into a muscle.
Intraperitoneal (i.p.): by injection into the peritoneal cavity.
Liposomal: solubilised (eg for injection) by inclusion in liposomes, which are artificial vesicles prepared from phospholipids.
Cremophor: a neutral detergent material prepared from castor oil (major component: glycerides of ricinoleic acid) by reaction with ethylene oxide. The generalised structure of this substance is shown at (**84**) in Panel 7.2. It has the ability to form stable emulsions of insoluble photosensitisers in water or buffer solutions, and so facilitate injection.

Panel 1.3 Terms Used in Describing Tumours

Tumour: Abnormal excessive growth on any part of body: may be benign or malignant.
Cancer: Malignant tumour, including sarcoma and carcinoma.
-oma: suffix denoting a (malignant) tumour eg glioma, tumour of the nervous system; hepatoma, tumour of the liver; cutaneous T-cell lymphoma, malignancy of certain lymphocytes which infiltrate the skin (eg mycosis fungoides).

Sarcoma: cancer of connective tissue eg fibrosarcoma, of fibrous tissue; osteosarcoma, of bone.
Carcinoma: cancer that occurs in the epithelium, the tissue that covers the external surface of the body, and lines hollow structures (except the blood and lymph vesicular systems) eg basal cell carcinoma; squamous (scaly) cell carcinoma; adenocarcinoma (an epithelial cancer originating in glandular structures).
Metastasis: spread of cancer from original site to other parts of the body.
Dysplasia: abnormal change in structure of tissue; may be an early indication of malignancy.

BIBLIOGRAPHY

N.R. Finsen, *Phototherapy*, Arnold, London 1901. Translated from the German version, which is possibly easier to find. Considerable historical interest.

J.D. Regan and J.A. Parrish (Eds.) *The Science of Photomedicine*, Plenum, New York and London, 1982. Broad coverage, with historical background in many areas.

R. Bonnett, 'Photodynamic therapy in historical perspective', *Rev. Contemp. Pharmacother.*, 1999, **10**, 1–19. Historical introduction to PDT, with many references.

G. Jori and C. Perria (Eds) *Photodiagnostic and Phototherapeutic Techniques in Medicine*, Documento Editoriale, Milan, 1995.

F. Bicknell and F. Prescott, *The Vitamins in Medicine*, Heinemann, London, 1942, pp.442–503. Early history and clinical observations on the D vitamins, and a warning on the careful control of uv radiation — *"We have seen twenty-five people who had to go to bed with severe sunburn because of over-exposure at the christening party of a new lamp".*

J.A. Parrish, T.B. Fitzpatrick, L. Tanenbaum and M.A. Pathak, 'Photochemotherapy of psoriasis with oral methoxsalen and longwave ultraviolet light', *N.Engl. J. Med.*, 1974, **291**, 1207–1222. Although not the first paper on the psoralens, this one had the greatest impact and introduced the term photochemotherapy.

R.J. Cremer, P.W. Perryman and D.H. Richards, 'Influence of light on the hyperbilirubinemia of infants', *Lancet*, 1958, (i), 1094–1097. The discovery of the phototherapy of jaundice of the newborn using visible light. A seminal paper.

G. Bock and S. Harnett (eds) *Photosensitising Compounds: Their Chemistry, Biology and Clinical Use*, Ciba Foundation Symposium 146, Wiley, London, 1989. Broad interdisciplinary coverage, but largely concerned with photodynamic therapy of cancer.

Dorland's Medical Dictionary, Saunders, Philadelphia. Latest edition. Excellent coverage, with pronunciation.

Black's Medical Dictionary, Black, London. Latest edition. Not so detailed, but clearly presented.

CHAPTER 2

PHYSICAL MATTERS

In an interdisciplinary subject such as photodynamic therapy we cannot go much further without setting down the basic physical chemistry that is needed. To some, indeed, this will be the essence of the subject. This chapter is really designed for those readers who are not in that position.

2.1 LIGHT

In order to come to an appreciation of the nature of electromagnetic radiation, of which visible light is a small part, it is necessary to be able to think of it as a form of energy which can be expressed both as a wave and as a particle (quantum, photon). The wave approach allows a treatment of light propagation, refraction, diffraction and polarisation, whereas the particle approach is suited for an understanding of absorption and emission phenomena.

The place of visible light in the electromagnetic spectrum is shown schematically in Figure 2.1. Electromagnetic radiation can be interpreted in terms of a transverse plane wave of wavelength λ associated with electric and magnetic fields, the vectors of which are in planes perpendicular to one another and to the direction of propagation (Figure 2.2). The wavelength of visible light falls in the range 400–750 nm: 400 nm we see as indigo, and 750 nm we see as red. At higher energies the near

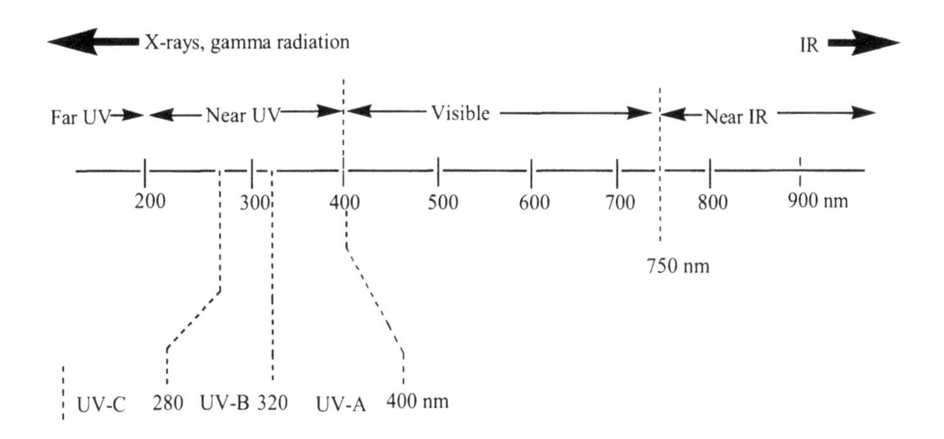

FIGURE 2.1 THE ELECTROMAGNETIC SPECTRUM, SHOWING PARTICULARLY THE NEAR-UV, VISIBLE, AND NEAR-IR REGIONS IN WHICH OCCUR THE PRINCIPAL ELECTRONIC TRANSITIONS OF INTEREST IN PHOTOTHERAPY.

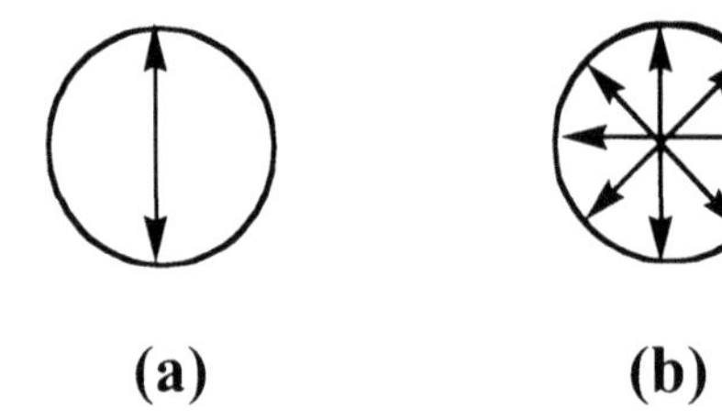

FIGURE 2.2 REPRESENTATION OF ELECTROMAGNETIC RADIATION IN TERMS OF SINUSOIDAL ELECTRIC AND MAGNETIC FIELDS, THE VECTORS OF WHICH ARE AT RIGHT ANGLES TO ONE ANOTHER AND TO THE DIRECTION OF PROPAGATION. FOR SIMPLICITY, PLANE POLARISED LIGHT IS SHOWN. THE CIRCLES REPRESENT THE TRANSVERSE WAVE IN (A) POLARISED AND (B) UNPOLARISED LIGHT, LOOKING DOWN THE AXIS OF PROPAGATION.

ultra-violet region extends to 200 nm, beyond which vacuum spectrometers are required because of absorption by air. Photobiologists divide the near ultraviolet region into three parts, thus:-

UV-A	320–400 nm
UV-B	280–320 nm
UV-C	< 280 nm

This division may seem arbitrary in physical terms, and so it is. But in terms of photobiology it is useful in relation to photodamage to living tissue. Energy per mole of photons (called an *einstein*) increases from A to C. Thus whereas UV-A is relatively harmless, by the point at which UV-B is reached a sunburn reaction occurs in humans, and this is the region that suncreams are designed to cope with (section 5.2). The DNA bases and certain α-amino acid side chains (eg tryptophan and tyrosine) have broad absorption maxima in the UV-C/UV-B region. Irradiation here does not penetrate far into tissue, but it can cause skin cancer (section 5.3). In the near infrared (~ 750–2000 nm) we find mainly vibrational and rotational transitions, but some extended organic chromophores (eg naphthalocyanines) show electronic transitions as far out as this.

The wavelength (λ) of light may be expressed in various units. Here we shall use nanometres (nm, 10^{-9} m, 10^{-7} cm). The Angstrom unit (Å, 10^{-10} m) is also employed. Because of the relationship:

$$\nu = \frac{c}{\lambda}$$

where c is the velocity of light (3.00×10^{10} cm s^{-1}) and ν is the frequency (waves or cycles per second), it is also possible to refer to the light in terms of its *frequency*

Panel 2.1 Some conversions between units

Wavelength

1 nm = 10^{-9}m = 10^{-7}cm

1Å = 10^{-10}m

The nanometre (nm) is often referred to as the millimicron (mμ) in the older literature.

Energy 1 cal = 4.184 Joules

It is convenient to refer to energy in terms of the Avogadro number of photons, and so it is expressed per mole.

1 kJ mol^{-1}	=	1.036×10^{-2} eV
	=	83.61 cm^{-1}
	=	0.239 kcal mol^{-1}
1 eV	=	8065 cm^{-1}
	=	23.06 kcal mol^{-1}
	=	96.48 kJ mol^{-1}
1 kcal mol^{-1}	=	4.335×10^{-2} eV
	=	349.8 cm^{-1}
	=	4.184 kJ mol^{-1}
1 cm^{-1}	=	1.240×10^{-4} eV
	=	2.859×10^{-3} kcal mol^{-1}
	=	1.196×10^{-2} kJ mol^{-1}

Energy conversions for wavelength *expressed in nm*

$$E = \frac{1.196 \times 10^5}{\lambda} \text{ kJ mol}^{-1}$$

$$= \frac{1.196 \times 10^5}{\lambda \times 96.48} \text{ eV}$$

$$= \frac{10^7}{\lambda} \text{ cm}^{-1}$$

(units s^{-1}, or waves s^{-1} = Hertz, Hz). A variant is to employ *wave numbers*, another frequency unit (represented by $\bar{\nu}$). This is reciprocal wavelength ($1/\lambda$), the units being length^{-1}, and generally cm^{-1}: hence it specifies the number of waves per centimetre.

It is a matter of custom and convenience which of these units is used, and they are readily interconverted. For example, the wavelength of 400 nm is equivalent to

$$\nu = \frac{3\times10^{10}\,\text{cm s}^{-1}}{400\,\text{nm}} = \frac{3\times10^{10}\,\text{cm s}^{-1}}{4\times10^{-5}\,\text{cm}} = 0.75\times10^{15}\,\text{s}^{-1}$$

that is 0.75×10^{15} waves per second in frequency units; and

$$\bar{\nu} = \frac{1}{400 \text{ nm}} = \frac{1}{4 \times 10^{-5} \text{cm}} = 25{,}000 \text{ cm}^{-1}$$

or 25000 waves for each centimetre, in wave number units.

Clearly we are talking about light as a wave. Another way of talking about wavelength is in terms of energy, visualised as the energy of the corresponding photon. This is a switch to the quantum approach, and illustrates the intermingling of the wave and particle concepts. Energy is given by the equation:

$$E = h\nu = \frac{hc}{\lambda}$$

where E is the energy of the transition and h is Planck's constant (6.626×10^{-34} J s). Continuing with the example of light at 400 nm, and expressing the result on a molar basis, then the energy is

$$\frac{\text{hc}}{\lambda} \times \text{Avogadro's number} = \frac{6.626 \times 10^{-34} \text{ J s} \times 3.00 \times 10^{10} \text{ cm s}^{-1}}{4 \times 10^{-5} \text{ cm}} \times 6.022 \times 10^{23} \text{ mol}^{-1}$$

$$= 299000 \text{J mol}^{-1} = 299 \text{kJ mol}^{-1} \text{ (to 3 sig. fig.)}$$

Simplifying, a direct conversion of λ (nm) to E (kJ mol^{-1}) is given by:

$$E = \frac{1.196 \times 10^{5} \text{ kJ mol}^{-1}}{\lambda \text{ (in nm)}}$$

The energy of the Avogadro number of photons at a given wavelength is sometimes referred to as the *einstein.* Thus an einstein of light of wavelength 400 nm is 299 kJ.

Several other energy units are employed, particularly kcal and electron volts: some workers prefer to use frequency units (ν or $\bar{\nu}$), which are linear with energy. When speaking of peak widths (eg width at half height) and peak separations (eg the separation of Bands I and III in porphyrin spectra; the Stokes shift) it makes sense to use frequency units (usually cm^{-1}) so that if energetic regularities exist, they can emerge clearly. With these exceptions, we shall use kJ mol^{-1} for energy and nm for wavelength in this book.

The total *amount* of light energy employed in a given photoprocess and the *rate* at which it is applied to the system are important parameters in both photochemistry and photomedicine. In chemistry it is the amount of light (number of quanta) *absorbed* which is important in deriving quantities such as quantum yield (section 2.6). In photobiology and photomedicine it is generally the amount of light *arriving* at the surface of a given object (referred to as light dose, precisely analogous to drug dose) that is measured, usually with a commercial light meter. (Sometimes the light is *fractionated,* that is, given as a series of short exposures with short dark intervals

— a few seconds, perhaps). The quantities recorded are usually *fluence* and *fluence rate,* but other terms are also employed (see Panel 2.5).

Fluence is the total radiant energy traversing a small transparent imaginary spherical target containing the point under consideration, divided by the cross sectional area of the target. Units: J cm^{-2}.

Fluence rate (irradiance, power density) is the rate of fluence.Units: W cm^{-2}. So fluence is the total energy arriving at a unit surface area, while fluence rate is the corresponding power, or rate of energy. (Power unit: 1 Watt = 1 Joule per second). Thus, in using m-THPC as a photosensitiser in cancer treatment, the fluence ('light dose') at 652 nm may typically be 20 J cm^{-2} and the fluence rate is 100 mW cm^{-2} (= 0.1 J s^{-1} cm^{-2}). The time of treatment is therefore 20/0.1 = 200 s, or just over three minutes.

2.2 LIGHT SOURCES

2.2.1 Sunlight

It is evident from everyday observation that solar energy maintains life on Earth, and it is commonly thought to have been a key factor in the generation of that life. Being so important and so widespread, photosynthesis is the most obvious sunlight-dependant process in the biosphere, but is is by no means the only one: in the plant world we have phototropism and photoperiodism; in the animal world, vision and seasonal affective disorder. Photobiology is a rich subject, and much of it is related to sunlight. The topic of sunlight with respect to sunburn and skin cancer will be outlined in sections 5.2 and 5.3; but for the phototherapeutic purposes to be considered here, sunlight is neither convenient nor reliable enough for routine use.

2.2.2 Incandescent Lamps

These depend on a filament (eg tungsten) which is heated in an evacuated glass envelope by an electric current. Such lamps are inexpensive and have been used in PDT. For example, in the treatment of basal cell carcinoma using δ-aminolaevulinic acid as a pro-drug (section 8.5), Kennedy and Pottier employed a projector lamp as the light source. In the phototherapy of neonatal hyperbilirubinemia fluorescent tubes have commonly been used (sections 1.2.3 and 5.6) but Ohmeda have recently developed a woven fibre optic pad (the Biliblanket^{R}) which wraps around the infant and which is supplied with light from a 100w quartz halogen bulb. Infrared and ultraviolet light are filtered out with filters and a dichroic reflector to give illumination in the required 400–550 nm region (bilirubin λ_{max} 450 nm).

2.2.3 Arc Lamps

The mercury arc lamp is the workhorse of organic photochemistry, and has some applications in photomedicine. There are three main types. The low-pressure mercury arc operates at room temperature and about 10^{-3} mm pressure: the main emission is a single line at 253.7 nm. This is well into the UV-C region, and germicidal lamps are of this type. Medium-pressure mercury lamps operate at about 1 atmosphere: the emission contains a number of lines, of which 366 nm and 546 nm are the principal ones. When such a source is filtered by Wood's glass, the 366

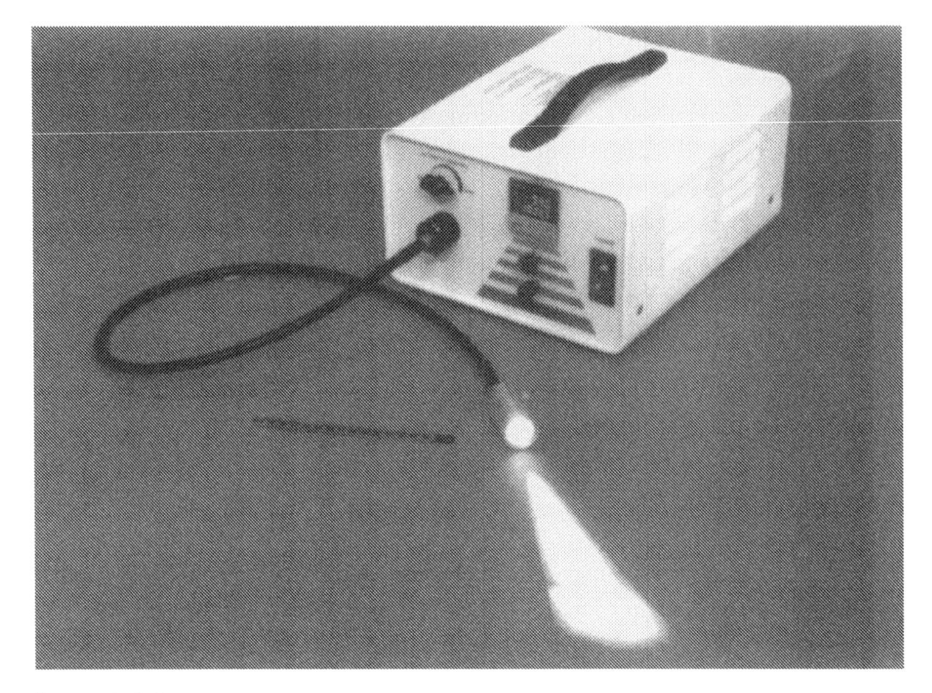

FIGURE 2.3 PATERSON LIGHT SOURCE BASED ON A SHORT ARC XENON LAMP COUPLED WITH A LIGHT GUIDE, THE TIP OF WHICH CAN BE DESIGNED TO DELIVER LIGHT TO SUIT THE APPLICATION IN HAND. COURTESY OF DR COLIN WHITEHURST, LASER ONCOLOGY GROUP, PATERSON INSTITUTE, MANCHESTER.

nm line remains, and it is this filtered ultraviolet light which is commonly used to observe the red fluorescence of porphyrins and other dyestuffs.

The high-pressure mercury arc operates at ~100 atmospheres and is an intense source: the emission is practically a continuum. Such lamps must be cooled, and even then have a short life. More convenient to use are point-source arcs using xenon or xenon-mercury: these are also intense sources, operating at about 20 atmospheres, but do not require cooling. Thus the Paterson lamp (Figure 2.3) is a low-cost portable xenon short arc lamp (300 W) which plugs into the mains and which has no special cooling demands. It is provided with rapidly interchangeable wavelength filters, and the output can be effectively coupled with a liquid light guide.

Where it is desired to generate a narrow wavelength band from lamps (such as the xenon lamp just referred to, or the incandescent lamp), this can be done by using a monochromator or by using commercial filters. Water cuts out much of the infrared: 2mm of Pyrex glass effectively removes ultraviolet below ~ 300 nm. In those cases where light has to be delivered to an internal site, the light is carried through a guide such as a fibre optic or a plastic tube full of liquid.(A laser beam (see below) is carried most efficiently in this way). The fibre tip is specially shaped to deliver the light as required, for example, to send it in one direction, or to diffuse it generally.

2.2.4 Light-emitting Diodes

Light-emitting diodes (LEDs) are III-IV semiconductor-based devices driven by an electric current. The emission is not coherent (this is not a laser source), and is low power, so not much heat is produced. Wavelength can be adjusted by changing the semiconductor, and the devices are small, but they can be bunched together to fit a particular structure. They are finding increasing application. Thus , the LED system recently announced by Diomed consists of a close-packed array of LEDs in a water-cooled head which is designed to be kept in contact with the area of treatment. The irradiation wavelength is 635 nm or 652 nm, with a fluence rate of up to 200 mW cm^{-2}.

2.2.5 Lasers

Laser is an acronym for "light amplification by stimulated emission of radiation". "Stimulated" emission means the release of a photon by an electronically excited species S* on interaction with a photon, thus:-

$$S^* + h\nu \rightarrow S + 2h\nu$$

The photon released by this process has exactly the same properties as the stimulating photon.

Under normal conditions where [S]>>[S*} this process is negligible, but where a population inversion exists ([S*]>[S], achievable by excitation with a powerful flash or a laser) this amplification does occur. The importance of this result can be judged from the award in 1964 of the Nobel Prize in Physics to its discoverers [Townes (USA), and Basov and Prokhorov (USSR)].

Laser emission has special properties some of which are important in photomedicine. The beam emerging from the laser is

(i) *intense.* This leads to uses in laser surgery (eg the carbon dioxide laser emitting at 10600 nm) for cutting, ablating, and sealing. High intensity is not needed for PDT, however.
(ii) *coherent.* This means all the waves, besides being of the same wavelength, are in phase (cf Figure 2.2).
(iii) *deliverable in some cases in pulses of short duration.* Pulses down to the femtosecond (10^{-15}s) range have allowed the study of initial events in photochemical and photobiological processes.
(iv) *monochromatic.* This is a very useful characteristic, especially in tunable lasers where, for example, the wavelength can be set to the maximum of an intense absorption band of a photosensitiser.
(v) *a parallel beam.* Further collimation is not required, and the beam can be conducted into and along a fibre optic, for example in an endoscope, with little loss of intensity.

Some commonly used lasers are listed in Panel 2.2.

Many lasers emit light at too long a wavelength for photochemical use, but the technique of *frequency doubling* (second harmonic generation) can be employed to get into the useful UV-VIS region. Essentially, these are two-photon absorption processes, sequential not simultaneous, and require materials with non-linear optical properties. Thus, the emission from a ruby laser (694 nm) on passage through a

Panel 2.2 Selected Examples of Lasers

	Active Species	Tunable	λ emission (nm)
Solid State Lasers			
1. Ruby	Cr^{3+}	No	694.3
2. Alexandrite	Cr^{3+}	Yes (720–800 nm)	680.4 + tunable
3. Titanium-Sapphire	Ti^{3+}	Yes (670–1100 nm)	tunable
4. Neodymium-YAG	Nd^{3+}	No	1060
Gas Lasers			
5. Carbon dioxide laser (CO_2:N_2:He)	CO_2	No	10600 (9600)
6. Helium-neon	Ne	No	632.8 (1152, 3391)
7. Argon ion	Ar^+	No	488, 514.5 + 8 others
8. Krypton ion	Kr^+	No	647.1 + 8 others
Dye Lasers			
9.	various dyes	Yes, ~ 40 nm for each dye, then change dye.	~ 365-930
	eg Rhodamine 6G(1)	Yes	570–620
Diode Lasers			
10.	semiconductor *n-p* type junction	Yes, by varying composition, controlling temperature	
	eg $Pb_{1-x}Sn_xSe$ $PbS_{1-x}Se_x$	Yes	2800–30000 nm

crystal of ammonium dihydrogen phosphate is converted to an ultraviolet laser beam of λ 347 nm with about 10% efficiency. Again, the neodymium-YAG laser is of such power in pulsed operation that second, third and fourth harmonics can be generated at 533, 355 and 266 nm, respectively.

For many purposes (eg assessing new photosensitising drugs) it is highly desirable to have a tunable laser source. Some solid state lasers show some tunability (Panel 2.2), but organic dye lasers provide the most useful facility. They are optically pumped (eg by a flash lamp or a gas laser) and can be tuned through the fluorescence range of the dye. Rhodamine 6G (**1**) is an example of such a dye: its fluorescence emission occurs over the range 570–620 nm and the dye laser is tunable over this region. A wide range of laser dyes is available covering the spectrum from about 450 to 950 nm.

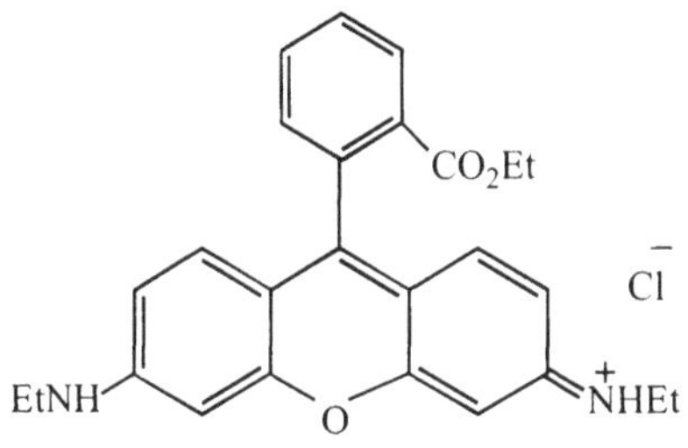

1 Rhodamine 6G

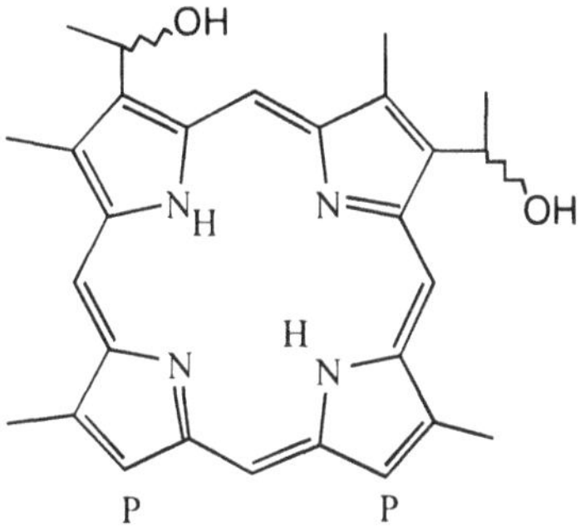

2 Haematoporphyrin
$P = CH_2CH_2CO_2H$

One of the problems with conventional lasers is that they are *expensive*: they also need expert technical supervision and maintenance. The diode lasers which are now emerging are smaller and much cheaper, and look promising. They are based on semiconductors ($n - p$ junctions activated electrically) and although most examples emit in the infrared, the range is rapidly being extended to the red region of the visible, which is needed for PDT applications. Thus, Diomed have produced diode lasers adapted for δ-ALA (AlGaInP, 635 nm), mTHPC (InGaAlP, 652 nm), and lutetium texaphyrin (AlGaAs, 730 nm).

2.3 LIGHT ABSORPTION

2.3.1 The Beer-Lambert Law

The intensity of a light absorption band of a given substance in a stated solvent follows two laws. Lambert's law states that the fraction of light absorbed is independent of the incident radiant power; while Beer's Law relates that the amount of light absorbed is directly proportional to the concentration of the solution. Put together (Beer-Lambert law) in mathematical form this becomes:

$$\log_{10} \frac{Po}{P} = A = \varepsilon \, \mathrm{cl} = \log_{10} \frac{I_0}{I}$$

where P_o and P are the incident and transmitted radiant powers (in common parlance called intensities, and represented by I_o and I), A is the absorbance, ε is the molar absorption coefficient (also called by some the molar or molecular extinction coefficient), c is the concentration (mol l^{-1}) and l is the path length (in centimetres, so this is conveniently unity when the common 10mm cuvettes are used). The units of ε are therefore l mol^{-1} cm^{-1} (although they are often not stated). [Chemists usually use the *decadic* molar absorption coefficient (ie logarithms to the base 10 as above). Physicists tend to use natural (*Naperian*) logarithms].

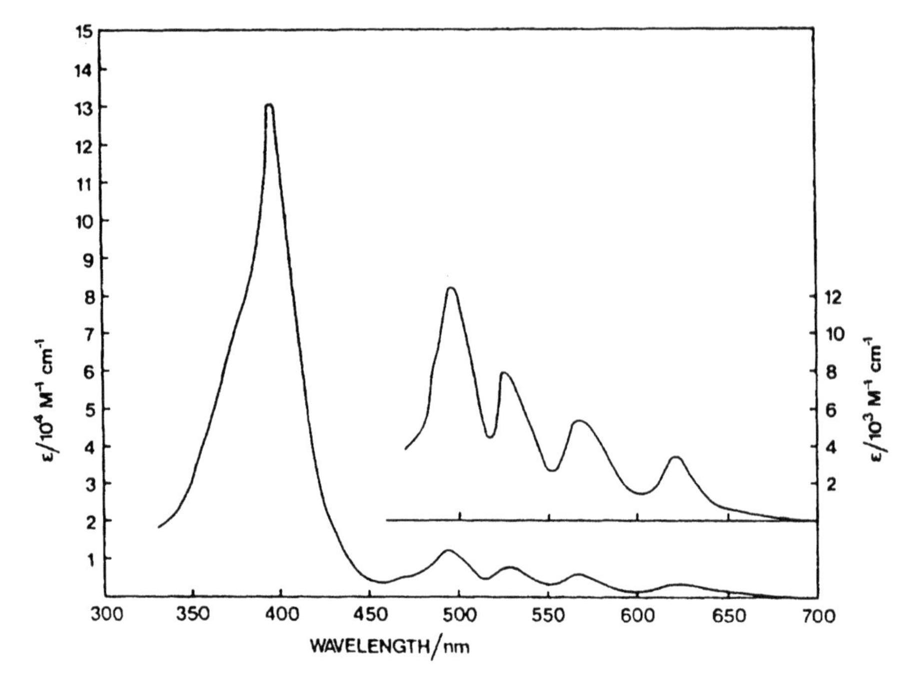

FIGURE 2.4 ELECTRONIC SPECTRUM OF HAEMATOPORPHYRIN (**2**) IN ACETONE. THE BANDS IN THE 500–600 NM REGION ARE REFERRED TO AS BANDS I–IV, STARTING AT THE LONGEST WAVELENGTH. (REPRODUCED FROM R. BONNETT, C. LAMBERT, E.J. LAND, P.A. SCOURIDES, R.S. SINCLAIR AND T.G. TRUSCOTT, *PHOTOCHEM. PHOTOBIOL.*, 1983, **38**, 1 WITH PERMISSION OF THE PUBLISHER).

Beer's Law usually holds well for dilute solution, but it breaks down in concentrated solution or under limiting solubility conditions when aggregation occurs (Panel 12.1).

Figure 2.4 shows the electronic spectrum (ie a plot of ε against λ) of haematoporphyrin (2) in acetone by way of illustration. This has a typical porphyrin spectrum — an intense absorption at about 400 nm (called the Soret band, or the *B* band) and four bands (the *Q* bands) in the 500–650 nm region.

2.3.2 Excitation

2.3.2.1 Multiplicity

Light absorption in the ultraviolet and visible region involves electronic transitions: a photon of a given energy interacts with the substrate molecule in its ground state (S_0) to cause an electron to be promoted from a bonding or non-bonding orbital to a higher energy orbital, normally an unoccupied antibonding orbital in organic molecules. The molecule is energised from its ground state to an electronically excited state.

Such excitation can formally result in two electronic configurations, which differ in multiplicity. Multiplicity is defined as $2S + 1$, where S is the total spin of the system ($\Sigma \pm {}^1/_2$). The configurations in question are excited singlet and triplet states, and are shown in Figure 2.5.

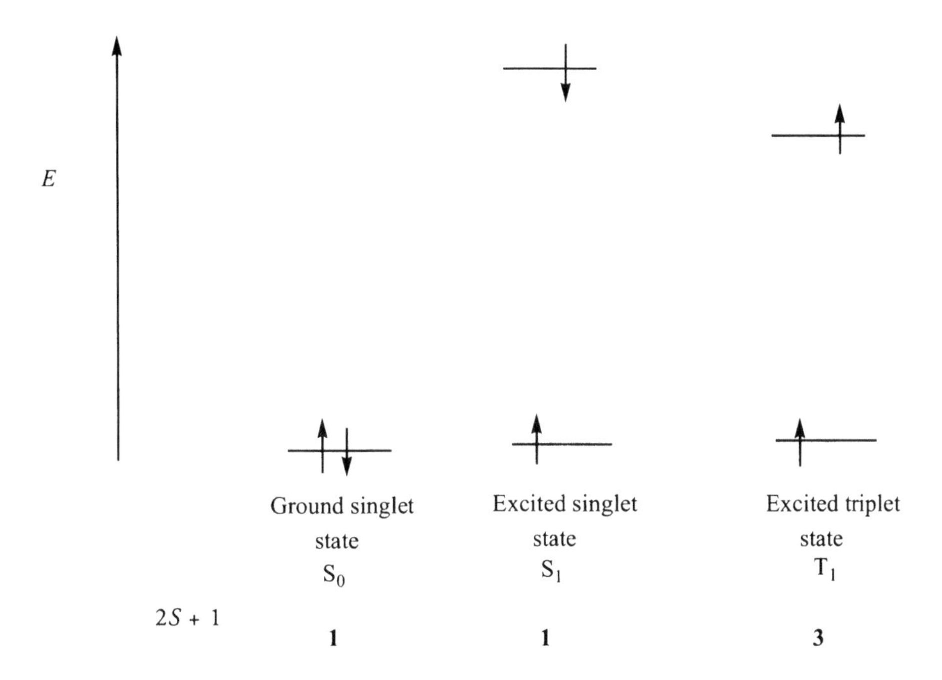

FIGURE 2.5 MULTIPLICITY OF GROUND AND EXCITED STATES: CONFIGURATION OF OUTER ELECTRONS SHOWN.

2.3.2.2 Probability

The energy of a given transition, such as those shown for haematoporphyrin in Figure 2.4, is determined by its wavelength (section 2.1): the molar absorption coefficient gives the probability or allowedness of the transition. This is related to the area under the absorption band, but is commonly spoken of in terms of the value of ε_{max}. There are various features which cause a transition to be 'forbidden' ie less probable, so lowering ε:

(i) *Spin*: transitions involving a change in multiplicity (S $\rightarrow$ T, or T $\rightarrow$ S) are forbidden. This restriction can be relaxed in the presence of heavy atoms, paramagnetic species, and by spin-orbit coupling. Nevertheless the transition $S_o \rightarrow T_1$ is strongly forbidden.

(ii) *Space*: orbitals involved in the transition need to overlap: they do so for the $\pi \rightarrow \pi^*$ transition of the carbonyl group, but not for the $n \rightarrow \pi^*$ transition, which is forbidden (ε 15 at λ_{max} 279 nm for acetone in hexane). (See Panel 2.3).

(iii) *Symmetry*: if the sign of the wavefunction changes on reflection through a centre of symmetry, then the symmetry is referred to as *ungerade* denoted *u*: if it is unchanged, it is termed *gerade* (German: even, direct) denoted *g*. Then $u \rightarrow g$ and $g \rightarrow u$ transitions are allowed, while $u \rightarrow u$ and $g \rightarrow g$ transitions are forbidden. Thus the $\pi \rightarrow \pi^*$ transition of the C = C double bond:

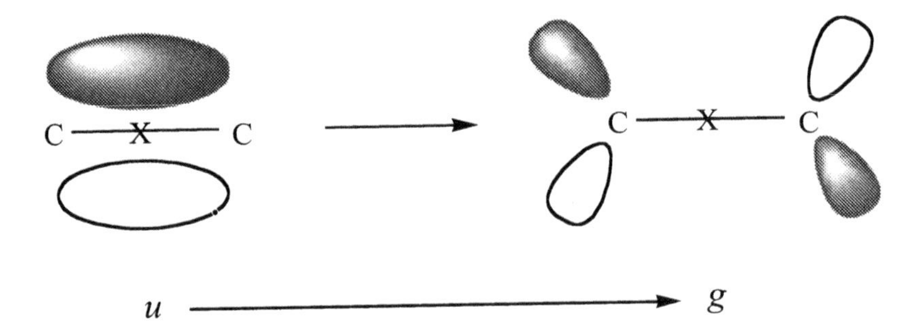

is an allowed transition on the symmetry criterion. It occurs as an intense band at about 165 nm in ethylene.

(iv) *Momentum*: transitions which cause a large change in linear or angular momentum of the molecule are forbidden.

The overlap and symmetry of orbitals in relation to the allowedness of electronic transitions in the carbonyl group of formaldehyde are illustrated schematically in Panel 2.3.

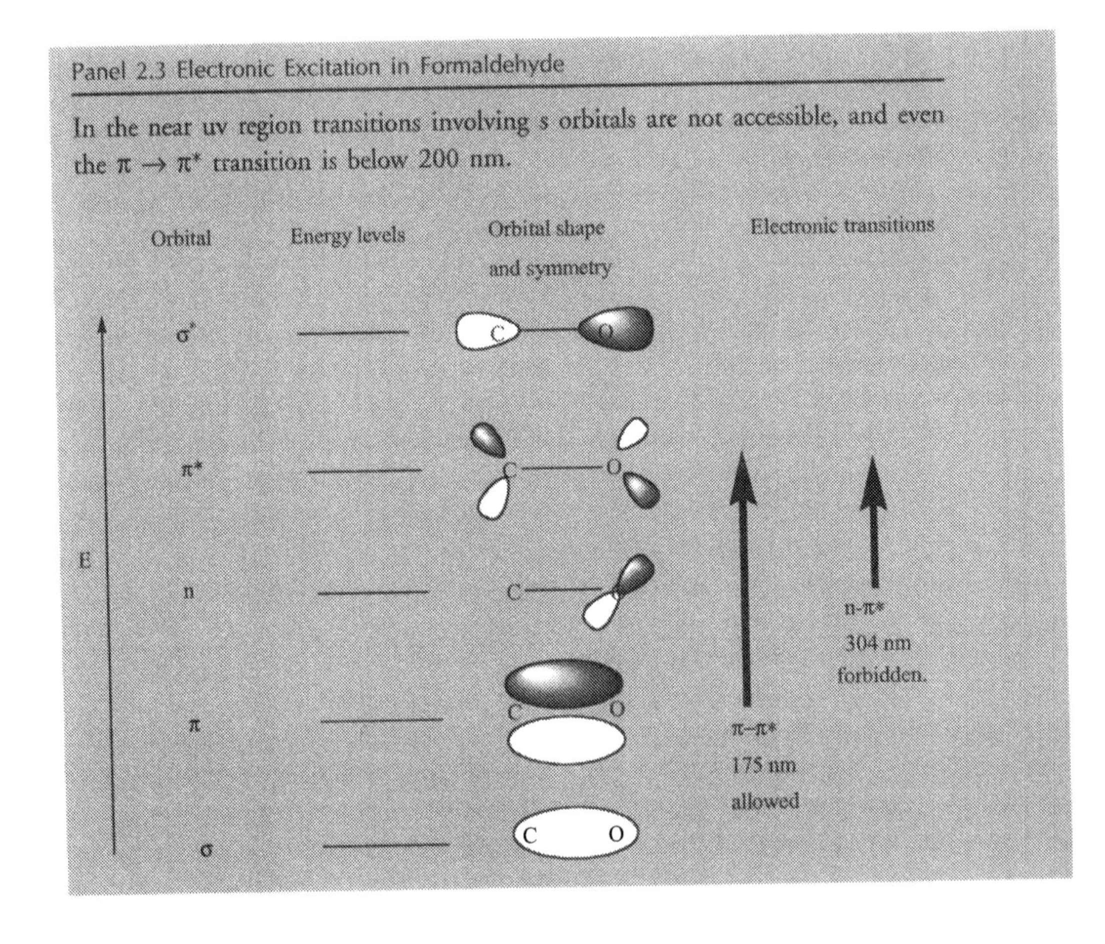

Panel 2.3 Electronic Excitation in Formaldehyde

In the near uv region transitions involving s orbitals are not accessible, and even the $\pi \rightarrow \pi^*$ transition is below 200 nm.

Panel 2.4 Photochemical Reaction Types

Reaction	Type
$\underset{S_0}{XYZ} \xrightarrow{h\nu} \underset{S_1 \text{ or } T_1}{XYZ^*} \longrightarrow XY\cdot + Z\cdot$	Photodissociation into radicals
$\longrightarrow XYZ^+ + e$	Photoionisation
$\longrightarrow XY^+ + \dot{Z}^-$	Internal electron transfer
$\longrightarrow XZY$	Intramolecular rearrangement
$\longrightarrow$ X—Y / Z (cyclic)	Photocyclisation
$\longrightarrow W + U$	Intramolecular cleavage
$\xrightarrow{RH} XYZ\dot{H} + \dot{R}$	H atom abstraction
$\xrightarrow{V} VXYZ$	Photoaddition
$\xrightarrow{V} XYZ^+ + V^-$	External electron transfer
$\xrightarrow{V} XYZ + V^*$	Photosensitisation

2.4 EMISSION

Subsequent to excitation, the excited state, be it singlet or triplet, can do one or more of four things:

(i) it can emit a photon;
(ii) it can convert to a different state. Changes between states of the same multiplicity are called *internal conversion* (eg $S_1 \rightarrow S_o^v$, vibrationally excited ground state) while changes between states of different multiplicity (which are formally spin forbidden) are called *intersystem crossing* (eg $S_1 \rightarrow T_1$);
(iii) it can undergo a chemical reaction. Some examples of known reaction types are collected in Panel 2.4; and
(iv) it can pass on its excitation to another (ground state) molecule (electronic excitation transfer).

Emission of a photon from the excited singlet state is called fluorescence, while emission from the triplet state is called phosphorescence. In both cases the wavelength of the light emitted is longer than the exciting wavelength, and in both cases the ground state, S_o, is reached. Since the change from the triplet, say $T_1 \rightarrow S_o$, is spin forbidden, this emission, and the triplet itself, are long-lived (typically ms – s): the $S_1 \rightarrow S_o$ emission is not spin forbidden, and fluorescence lifetimes, and lifetimes of the excited singlet state, are short (typically $< \mu s$). Because the triplet excited state is longer lived, it tends to be important in many photochemical reactions. Although fluorescence is easily observed in solution, phosphorescence is not: T_1 is sufficiently long lived to be deactivated by collision. Therefore phosphorescence has usually to be observed at low temperature in glasses such as EPA (ether: isopentane: ethanol = 5:5:2) which forms a transparent glass at 77 K.

The measurement of fluorescence is particularly important, since it is used both for detection and analysis. Fluorescence is measured with a spectrofluorimeter, which consists of a light source with a monochromator to provide the *excitation* wavelength. The cuvette is polished on all four sides since (unlike the cuvette in absorption spectroscopy where the beam is transmitted through the sample) in this case the emission is measured at right angles to the incident beam. The emitted beam is passed through a second monochromator and on to a detection device to generate the fluorescence spectrum.

Two types of spectrum can be obtained: the *excitation spectrum* is obtained by scanning the incident wavelengths for a set emission wavelength, while for the *emission spectrum* the dispersion of the emitted light is measured for a set exciting wavelength. Although fluorimetry is a very sensitive technique (metal-free porphyrins can be analysed at 10^{-8} M) it has some drawbacks which need to be recognised. Chief among these is the quenching of the emission, which effect increases with concentration (*concentration quenching*). Variable emission also occurs on aggregation (which always reduces fluorescence: Panel 12.1) and on interaction with other chromophores, so that direct measurements *in vivo* are particularly subject to error. The best way to analyse for photosensitiser in tissue is to extract it into solvent, thus getting rid of the bulk of tissue components, before making the fluorescence measurements. This is time consuming , and destroys the tissue: because of its sensitivity, convenience and non-invasive nature, direct fluorimetry *in vivo* is likely to remain popular in photobiology, in spite of its inherent ambiguities.

A last point on fluorescence. Whereas compounds of the porphyrin photosensitiser class generally show strong red fluorescence, this is considerably reduced if a substituent with a high atomic number (such as iodine) is present (*heavy atom effect*). In general, halogen substitution is found to increase ϕ_{isc} leading to enhanced yields of the triplet (and therefore high ϕ_{Δ}, as in eosin Y and rose bengal: Table 2.1). For metalloporphyrins, fluorescence is quenched where the metal has an incomplete *d* shell, since the empty orbitals here offer a pathway for the rapid deactivation of the excited states. Thus, whereas zinc(II) porphyrins (d_{10}) show strong orange-red fluorescence , complexes of the first transition series (including haems, of course) show no fluorescence to the naked eye.

2.5 JABLONSKI DIAGRAM

The ideas expressed above can be consolidated and developed using a modified Jablonski diagram, which is a schematic representation of relative state energies (Figure 2.6). This reveals that, besides electronic excitation levels mentioned up to now, there are additionally vibrational sub-levels (and, within those levels, rotational sub-levels, not shown here) to be considered.

The following points emerge from a consideration of Figure 2.6.

(i) In condensed phases, upper excited states (ie S_2, vibrationally excited S_1 and T_1) rapidly undergo collisional deactivation to give the lowest excited state (singlet or triplet) with v = 0 ('vibrational cascade'). *Emission occurs from this*

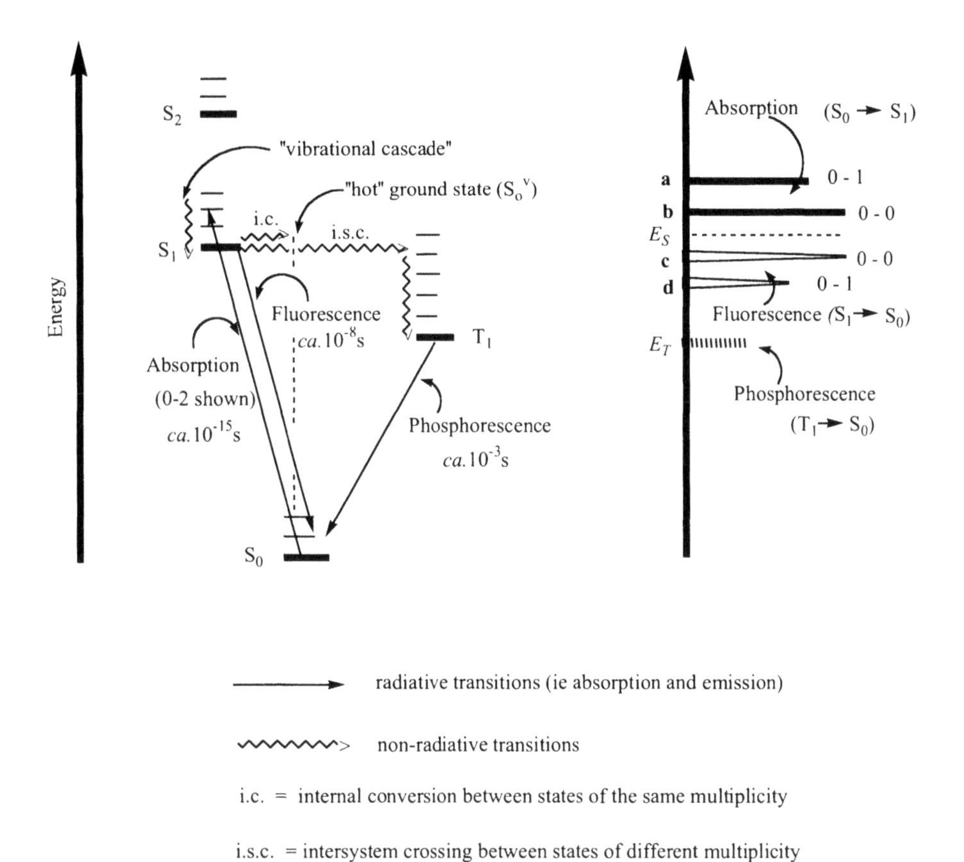

FIGURE 2.6 MODIFIED JABLONSKI DIAGRAM WHICH SHOWS IN A SCHEMATIC WAY THE ENERGIES OF THE VARIOUS STATES OF THE MOLECULE. JABLONSKI WORKED IN WARSAW, AND HIS ORIGINAL DIAGRAMS WERE MUCH SIMPLER (EG *NATURE*, 1933, **131**, 839–840). HERE, VIBRATIONAL SUB-LEVELS ARE SHOWN BUT ROTATIONAL SUB-LEVELS WITHIN THESE ARE OMITTED FOR CLARITY. ON THE RIGHT HAND SIDE SCHEMATIC ABSORPTION, FLUORESCENCE AND PHOSPHORESCENCE SPECTRA ARE SHOWN ON THE SAME (ARBITRARY) ENERGY SCALE. THE ENERGY OF THE SINGLET (E_S) IS TAKEN AS THE MEAN OF THE ENERGIES OF THE 0-0 BANDS IN ABSORPTION AND EMISSION: THE SEPARATION **b-c** IS CALLED THE STOKES SHIFT. THE ENERGY GAP **a-b** IS THE VIBRATIONAL SEPARATION IN THE EXCITED STATE, WHILE **c-d** IS THAT IN THE GROUND STATE. E_T IS DETERMINED FROM THE POSITION OF PHOSPHORESCENT EMISSION (SEE TEXT).

lowest excited state. This is an example of Kasha's rule — polyatomic molecules luminesce with appreciable yield only from the lowest excited state of a given multiplicity.

(ii) Fluorescence therefore occurs from S_1 (v =0) and is of lower energy than the absorption band. The emission is structured because the $v^* = 0 \rightarrow v = 1$ transition has a lower energy than the $v^* = 0 \rightarrow v = 0$ transition. In cases where both the absorption and fluorescence spectra show vibrational structure, there is often a rough mirror-image relationship between the two spectra. The 0–0 bands in absorption and emission are not coincident because the excited singlet state, short-lived as it is, relaxes to a lower energy conformation before it emits.

The separation of the two 0–0 bands is known as the Stokes shift: the energy of S_1 can be reasonably estimated from the mean of the two 0–0 bands (energy units).

(iii) T_1 cannot be efficiently populated by direct excitation. However intersystem crossing $S_1 \rightarrow T_1^v$ occurs isoenergetically followed by deactivation to T_1 (v = 0) (Kasha rule). The $S_1 \rightarrow T_1$ change is formally spin-forbidden, but it is a downhill process and in some molecules (eg benzophenone) intersystem crossing occurs so efficiently that fluorescence is quenched. Such compounds are useful for generating excited triplets by excitation transfer (sensitisation) in molecules where intersystem crossing is minimal (section 2.7).

The upshot is that $E_S > E_T$. The two electrons in the triplet excited state have parallel spins and by an extension of Hund's rule cannot occupy the same orbital (ie space): hence electron repulsion is lower than it is in the corresponding singlet state. As a consequence, for a given molecule, phosphorescence occurs at longer wavelengths than does fluorescence (Figure 2.6).

2.6 QUANTUM EFFICIENCY

Following the excitation process, the reaction process follows. The reaction process is actually a dark process which happens to the electronically excited species ie another photon is not required. Thus for substrate XYZ we have:

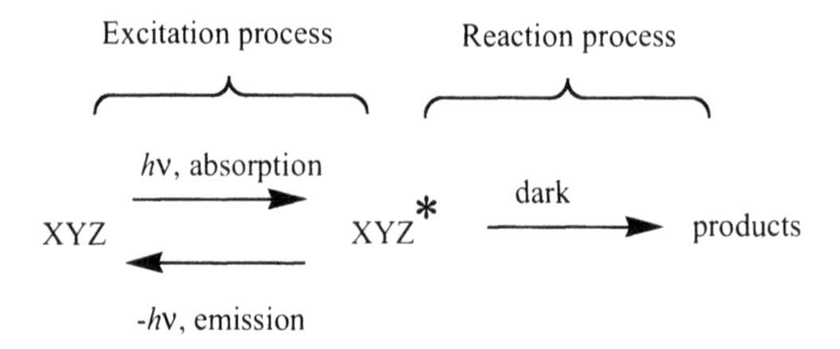

where the types of reaction are formulated in Panel 2.4.

For practical purposes it is usually necessary to know the percentage yields of products in the conventional way ie based on the mass of XYZ. However, from a photochemical point of view it is also important to know how efficiently the product is generated in terms of the photons absorbed. This is given by the *quantum yield* which may be defined as

$$\phi_{\text{product}} = \frac{\text{number of molecules of product formed per ml per second}}{\text{number of quanta absorbed per ml per second}}$$

Since quantum yield varies with wavelength, this quantity is most meaningfully measured with monochromatic light. ϕ_{product} cannot be greater than unity (this is another expression of the Stark-Einstein Law: see section 1.1 which relates an exception to this statement), and is sometimes very low. This results in long irradiation times, but these may still be acceptable if the overall chemical yield is good.

In determining ϕ, the numerator is obtained by chemical analysis, and the denominator from a measurement of the total incident energy (actinometer) and the fraction of that energy absorbed at the specific wavelength (from the molar absorption coefficient). *Actinometry* can be carried out in an absolute fashion, using a thermopile to measure the heat produced when the incident beam is totally absorbed by a blackened receiver. More conveniently, the measurement is made by chemical actinometry, in which the progress of a suitable photochemical reaction, which has already been calibrated against a thermopile, is followed. A good chemical actinometer is the potassium ferrioxalate system devised by Parker and Hatchard, in which photoreduction occurs to Fe(II), which is then estimated by forming the deep red complex with 1,10-phenanthroline. In practice, commercial light meters are commonly employed, although they need to be calibrated against a standard photoreaction.

The idea of quantum yield can also be usefully applied to the photophysical 'products' of the excitation process. Thus we have quantum yield of fluorescence ϕ_f, quantum yield of phosphorescence ϕ_p, quantum yield of intersystem crossing ϕ_{isc}, quantum yield of singlet ϕ_S and quantum yield of triplet ϕ_T. Again $\Sigma\phi_x \leq 1$. The ratio ϕ_p/ϕ_f covers an extensive range: >1000 (ie efficient intersystem crossing) shown by aromatic ketones, for example; and <1, shown by polycyclic aromatic hydrocarbons.This is significant for the problem now to be considered.

2.7 INTERMOLECULAR ELECTRONIC EXCITATION TRANSFER

This phenomenon was first demonstated by the Russians, and since it is important in the context of the topic of this book, it is proposed to look at one of the original examples in some detail. The work is due to Terenin and Ermolaev.

When 0.5 M acetophenone in an ethanol-ether glass at 90 K is irradiated at 366 nm, phosphorescent emission of acetophenone occurs at 388 nm, thus:-

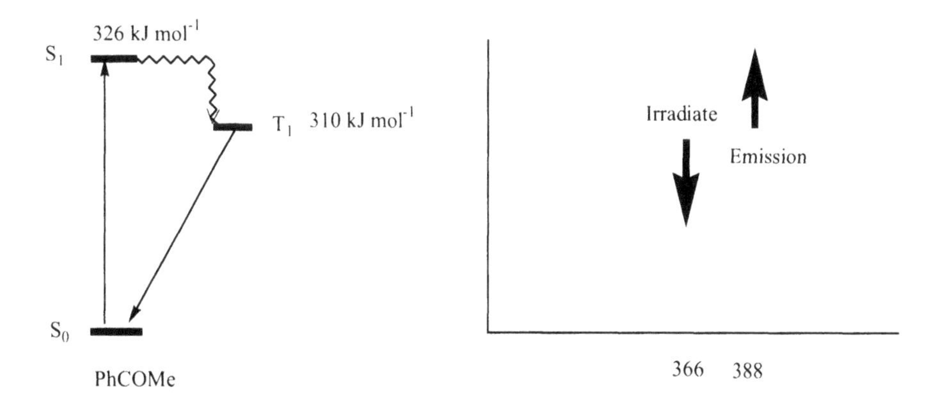

Now, if naphthalene (0.5 M) is added to the system, emission at 388 nm is quenched, and instead emission occurs at 470 nm, which corresponds to naphthalene phosphorescence.

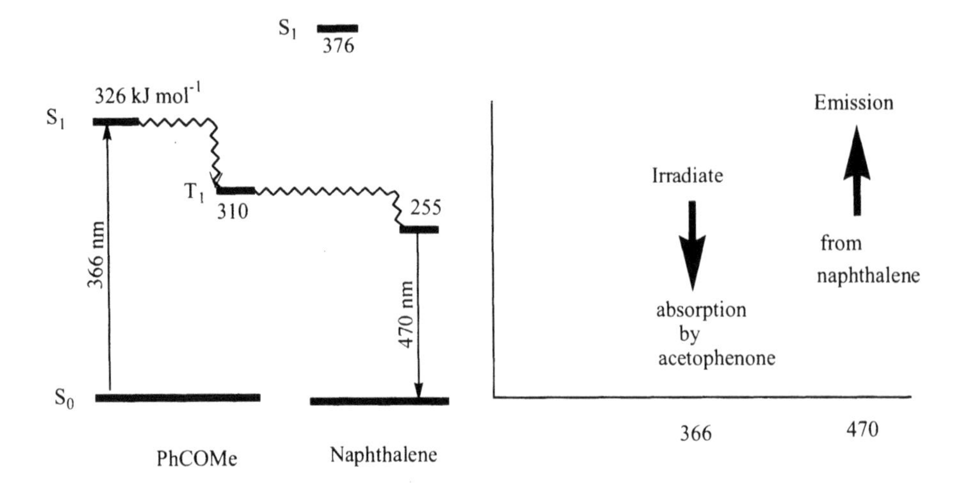

Irradiation at 366 nm does not cause naphthalene alone to emit (because it does not absorb — Grotthus-Draper Law — section 1.1). Hence the excitation of acetophenone, which results in the generation of the acetophenone triplet state (T_1, 310 kJ mol^{-1}) has, in the second experiment, been handed on to naphthalene. Naphthalene cannot absorb the energy itself, but can accept it from the acetophenone. The naphthalene triplet has been generated without naphthalene itself having been directly photochemically excited (since the energy available cannot reach the naphthalene S_1 at 376 kJmol^{-1}).

Thus the acetophenone, which we can call the donor, *D*, is the sensitiser of naphthalene phosphorescence, while the naphthalene, which we can call the acceptor, *A*, is the quencher of acetophenone phosphorescence. Thus the process in summary is:

$$D \xrightarrow{h\nu} D^*$$

$$D^* + A \rightarrow A^* + D$$

For this to work the donor triplet energy (E_S) must be larger than the acceptor triplet energy, and spin conservation must obtain, thus:-

$$D\,(T_1) \;+\; A\,(S_0) \longrightarrow D\,(S_0) \;+\; A\,(T_1)$$

$$\uparrow\uparrow \qquad \uparrow\downarrow \qquad\qquad \uparrow\downarrow \qquad \uparrow\uparrow$$

The total spin on each side of the equation is the same (unity). Consequently we have here a way of generating triplet species by photosensitisation in cases where intersystem crossing quantum yields are small.

2.8 SINGLET OXYGEN QUANTUM YIELDS

The next chapter is about singlet oxygen. One of the ways in which singlet oxygen is generated is by electronic excitation transfer from suitable sensitisers, and it is this sort of photosensitiser that is used in photodynamic therapy. The process can be summarised as follows (Sens = sensitiser):

$$\text{Sens}\ (S_0) \xrightarrow{h\nu} \text{Sens}\ (S_1) \xrightarrow{\text{isc}} \text{Sens}\ (T_1)$$
$$\text{Sens}\ (T_1) + {}^3O_2 \longrightarrow \text{Sens}\ (S_0) + {}^1O_2$$

This last equation is spin allowed, since the total spin is the same before and after reaction (Wigner rule), thus:

$$\begin{array}{lccccccc} & \text{Sens}\ (T_1) & + & {}^3O_2 & = & \text{Sens}\ (S_0) & + & {}^1O_2 \\ & \uparrow\uparrow & & \downarrow\downarrow & & \uparrow\downarrow & & \uparrow\downarrow \\ \text{Total spin} & & 0 & & & & 0 & \end{array}$$

but to proceed effectively the energy of the T_1 state must be above 94 kJ mol^{-1}, the energy of singlet oxygen above its ground state (see section 4.3.1 and Figure 4.3).

The photochemical efficiency with which various photosensitisers generate singlet oxygen is the singlet oxygen quantum yield, denoted by ϕ_Δ. If singlet oxygen is derived only from the triplet of the sensitiser (as is usually the case, but see below), it follows that ϕ_Δ must be less than ϕ_T, but in fact dioxygen is a very efficient quencher of many of these triplet states, and ϕ_Δ values are often not far short of ϕ_T. Some illustrative values are given in Table 2.1 and some additional values referring particularly to macrocyclic tetrapyrroles are provided in Table 4.3. It will be seen from Table 2.1 that the ϕ_Δ values are substantial (>0.3) for many of the compounds quoted. However, for 8-methoxypsoralen the quantum yield of singlet oxygen is low (0.009), and this substance is thought to act mainly by cycloaddition, not via production of singlet oxygen (section 5.5). And illustrating a point made in section 2.4, whereas the aluminium and zinc complexes (d^0 and d^{10}, respectively) of the phthalocyanines listed have appreciable ϕ_Δ values (*ca* 0.4), the corresponding copper(II) complex, which is d^9, has a ϕ_Δ value of zero because excited states tend to be rapidly deactivated in transition metal complexes where the metal has an incompletely filled *d* shell.

In principle, it is possible for the singlet oxygen quantum yield to reach a higher value than unity. This becomes feasible, for example, if both the $S_1 - T_1$ and the $T_1 - S_0$ energy gaps are about, or somewhat larger than, 94 k J mol^{-1}, thus permitting the spin allowed processes:

$$\text{Sens}\ (S_0) \xrightarrow{h\nu} \text{Sens}\ (S_1)$$

$$\text{Sens}\ (S_1) + {}^3O_2 \longrightarrow \text{Sens}\ (T_1) + {}^1O_2$$

$$\text{Sens}\ (T_1) + {}^3O_2 \longrightarrow \text{Sens}\ (S_0) + {}^1O_2$$

TABLE 2.1 SOME SINGLET OXYGEN QUANTUM YIELDS ϕ_Δ FOR A RANGE OF PHOTOSENSITISERS.

	Substance	Solvent	ϕ_Δ	Reference
A. General				
	Acridine	C_6H_6	0.83	1*
	Benzophenone	C_6H_6	0.35	1*
	Eosin Y	H_2O	0.61	1*
	8-Methoxypsoralen	D_2O	0.009	1
	Methylene blue	MeOH	0.51	1*
	Rose bengal	MeOH	0.80	1*
		H_2O	0.76	1*
B. Porphyrins and Related Compounds[2]				
	Porphyrin	CCl_4	0.75	1
	Haematoporphyrin	MeOH	0.52	1
	Tetraphenyl-porphyrin (TPP)	C_6H_6	0.66	1*
	–, Zn complex	C_6H_6	0.72	1*
C. Phthalocyanines				
	Phthalocyanine –, Mg complex	C_6H_5N	0.40	1
	Phthalocyanine, sulphonated –, Cl A1 complex	H_2O	0.34	1
	–, Cu complex	H_2O	0.0	1
	–, Zn complex	H_2O	0.45	1

[1]F. Wilkinson, W.P. Helman and A.B. Ross, *J. Phys. Chem. Ref. Data*, 1993, **22**, 113–262. The asterisks refer to averages of a range of values.
[2]Additional values for this series in Table 4.3.

For example, rubrene (5,6,11,12-tetraphenyltetracene) is reported to have a ϕ_Δ value of 1.8 in toluene at an oxygen pressure of 10 atmospheres. However, the photosensitisers which have so far been found suitable for PDT do not have a large enough $S_1 - T_1$ energy gap (see Figure 4.3), and in any case the likely requirement for a high oxygen pressure makes the application of this effect rather remote.

Panel 2.5 Some Useful Definitions

(See also earlier sections, particularly 2.1 and 2.3, and *Pure and Applied Chemistry*, 1988, **60**, 1055–1106; 1996, **68**, 2223–2286).

Spectral Band Properties

Bathochromic: shift to lower energies; red shift.
Hypsochromic: shift to higher energies; blue shift.
Hyperchromic: increase in intensity of a spectral band.
Hypochromic: decrease in intensity of a spectral band.
Chromophore: that part of a molecule or other system responsible for the absorption of light, and in which the electronic transition predominantly occurs. (analogously, *fluorophore* and *luminophore* for the emission of light).
Auxochrome: a substituent on a chromophore, which, while having little chromophoric character of its own, is able to cause intensification of the colour, usually by both bathochromic and hyperchromic effects.

General, Alphabetical

Action Spectrum: a plot of the photoresponse (of whatever sort, but usually biological) per number of incident photons against wavelength.
Biphotonic Excitation: the simultaneous absorption of two photons (either the same or different wavelength), the energy of excitation being the sum of the two energies. This usually requires high light intensities, as obtain with laser sources.
El-Sayed Rules: in intersystem crossing (see Figure 2.6) from the lowest singlet state to the lowest triplet state, a change in the orbital type increases the rate of the process. So $^1n,\pi^* \rightarrow {}^3\pi,\pi^*$ is faster than $^1n,\pi^* \rightarrow {}^3n,\pi^*$.
Emission Spectrum: plot of emitted light intensity against its wavelength or wave number. When it is adjusted for wavelength-dependent variations in the response of the detector it is called a corrected emission spectrum. The *excitation spectrum* is the plot of the emitted light intensity against the wavelength of the light used for *excitation*.
Excimer: a complex formed by the interaction of an electronically excited molecule with a molecule of the same species but in the ground state.
Exciplex: a complex formed by the interaction of an electronically excited molecule with a ground state molecule of a different substance.
Half-width: the width of a spectral band at a height half of that at the band's maximum. Best given in wavenumber or frequency units.

Hund's Rule of Maximum Multiplicity: Of the different multiplets resulting from different configurations of electrons in degenerate orbitals of an atom, those with the greatest multiplicity have the lowest energy.

Kasha Rule: in polyatomic molecules, luminesence occurs essentially only from the lowest excited state of a given multiplicity *ie* higher excited states usually rapidly lose energy to give S_1 or T_1 as the case may be. "Vibrational cascade".

Kasha-Vavilov Rule: the quantum yield of luminescence is independent of the wavelength of the exciting light. Exceptions occur.

Lifetime (τ): for a species decaying by a first order process, the time to reach $1/e$ of its initial concentration. It is equal to reciprocal of the sum of all the (pseudo)unimolecular rate constants responsible for the decay.

Half-life ($\tau_{1/2}$), the time for concentration to halve, is also used.

Radiative lifetime is the lifetime of an excited species in the absence of radiationless transitions; it is the reciprocal of the first order rate constant for the radiative step.

Luminescence: spontaneous emission of radiation from an electronically or vibrationally excited species not in thermal equilibrium with its surroundings. Includes fluorescence (from S_1), phosphorescence (from T_1), but many other phenomena including chemiluminescence and bioluminescence.

Multiplicity: the number of possible orientations, given by $2S + 1$, of the spin angular momentum corresponding to a given total spin quantum number (S) for the same spatial electronic wave function. Thus, in this nomenclature an organic radical, with one unpaired electron, has $S = 1/2$, and is a doublet species.

Photobleaching: a light-induced loss of absorption or emission intensity.

Stokes Shift: the difference (in cm^{-1}) between the maxima of the absorption band and the luminescence band arising from the same electronic transition.

Wigner Spin Conservation Rule: when electronic excitation energy transfer occurs between species the total spin angular momentum of the *system* does not change.

Light Intensity Parameters

Fluence (H_0): the total radiant energy traversing a small transparent imaginary spherical target containing the point under consideration, divided by the cross sectional area of the target. The product of fluence rate and time of irradiation. The SI unit is $J\ m^{-2}$, but this is an impracticably large unit, and $J\ cm^{-2}$ is commonly used.

Fluence Rate (E_0): the rate of fluence. Four times the ratio of the radiant power, P, incident on a small transparent imaginary spherical volume element containing the point under consideration, divided by the surface area of that sphere. Operational unit is $mW\ cm^{-2}$. This parameter reduces to irradiance, E, for a parallel and perpendicularly incident beam which is not reflected or scattered by the target or its surroundings.

Irradiance (E): the radiant power, P, of all wavelengths incident on an infinitesimal element of surface containing the point under consideration divided by the area of that element. Operational unit $mW\ cm^{-2}$.

Radiance (L): for a *parallel* beam, the radiant power leaving or traversing an infinitesimal element area of surface in a given direction, divided by the orthogonally projected area of that element in a plane normal to the given direction. Unit mW cm^{-2}.

Radiant Power (P) = Radiant Flux; power emitted, transferred or received as radiation. Unit is J s^{-1} = W.

Spectral Irradiance (E_λ); irradiance at wavelength λ per unit wavelength interval. Useful unit mW cm^{-2} nm^{-1}.

BIBLIOGRAPHY

J.G. Calvert and J.N. Pitts, *Photochemistry*, Wiley, New York, 1966. Old, but coverage and approach thorough and excellent.

P.W. Atkins, *Physical Chemistry*, 6th edition, 1998, Oxford University Press. Chapter 17, 'Electronic Transitions'. Basic treatment, clear and concise.

R.W. Ditchburn and others, 'Light', in The New Encyclopaedia Britannica, Vol. 23, pp. 1–28 (1995). Useful background summary with bibliography.

P. Suppan, *Chemistry and Light*, Royal Society of Chemistry, London, 1994. Recent treatment of basics with good coverage.

S.E. Braslavsky and K.N. Houk, 'Glossary of terms used in photochemistry', *Pure Appl. Chem.*, 1988, **60**, 1055. See also S.E. Braslavsky, 'Selected terms and symbols in photochemistry', *J. Photochem. Photobiol., B. Biol.*, 1987, **1**, 261. Updated by J.W. Verhoeven, *Pure Appl. Chem.*, 1996, **68**, 2223–2286. Approved definitions as seen by an international committee.

F. Wilkinson, W.P. Helman and A.B. Ross, 'Quantum yields for the photosensitised formation of the lowest electronically excited singlet state of molecular oxygen in solution', *J. Phys. Chem. Ref. Data*, 1993, **22**, 113–262. A labour of love. A very detailed and excellently referenced and indexed source of ϕ_Δ values.

A. Gilbert and J. Baggott, *Essentials of Molecular Photochemistry*, Blackwell Scientific Publications, Oxford, 1991. Introductory chapters provide clearly written material for further reading. A physical approach: for an organic approach see W. Horspool and D. Armesto, *Organic Photochemistry*, Ellis Horwood — PTR Prentice Hall, New York and London, 1992.

J.V.E. Roche, C. Whitehurst, P. Watt, J.V. Moore and N. Krasner, 'Photodynamic therapy (PDT) of gastrointestinal tumours: a new light delivery system', *Lasers Med. Sci.*, 1998, **13**, 137–142. Progress with an inexpensive light source for PDT based on a small high pressure xenon arc.

CHAPTER 3

SINGLET OXYGEN

"13 Novr. 1847
9235. *Oxygen* in coal gas was very magnetic and the experiment was very beautiful. The ammonia, in uniting with the muriatic acid of the issuing oxygen, made a cloud which fell well and quickly; but on making the magnet active, the cloud instantly ran back, and accumulated about the poles, occupying the magnetic field, so that the pole ends were hid in a cloud of smoke an inch or more in diameter. When the magnet was thrown out of action, this instantly fell and cleared away, or if allowed to fall only a little and then the magnetic force renewed, it rushed back again into the magnetic field."

Michael Faraday, Diary
(describing the discovery of the paramagnetism of ground state oxygen)

"Over a period of time all of us change our minds, but I think Kautsky's work stands unique in terms of its cleverness. It was just too early for people to use."

Michael Kasha, 1970

3.1 GENERAL

Molecular oxygen (dioxygen is the correct name, but it is seldom used) is unusual amongst common molecules in having a triplet ground state (Panel 3.1).

The two electronically excited states immediately above the ground state are both singlet states: one, designated the $^1\Sigma_g^+$ state, is very short lived, and rapidly decays to the lower singlet state, designated $^1\Delta_g$, which has a lifetime of a few microseconds in the condensed phase. The properties of these three states are summarised in Table 3.1.

The lifetime of the $^1\Sigma_g^+$ state in solution (less than a picosecond) is so short that for practical purposes we can regard this state, if formed in solution, as being deactivated immediately to the $^1\Delta_g$ state, and from now on the latter species will be referred to as "singlet oxygen". The lifetime of singlet oxygen in solution varies with the solvent (Table 3.2) and is longer in deuteriated solvents than in the corresponding protiated solvents. This behaviour is ascribed to the stronger infrared absorption of the protiated solvents at 1270 nm and 1070 nm where the 0–0 and 1–0 (emission) bands of singlet oxygen occur. Interaction between the vibrational levels of the solvent molecules and the electronic/vibrational levels of the singlet oxygen occurs to deactivate the singlet oxygen more effectively in the protiated solvent.

Panel 3.1 Ground State Dioxygen

In a simplified description of the molecular orbital (MO) approach to chemical bonding (Linear combination of atomic orbitals — LCAO — approximation) we consider the overlap of atomic orbitals to give molecular orbitals, for example as follows:

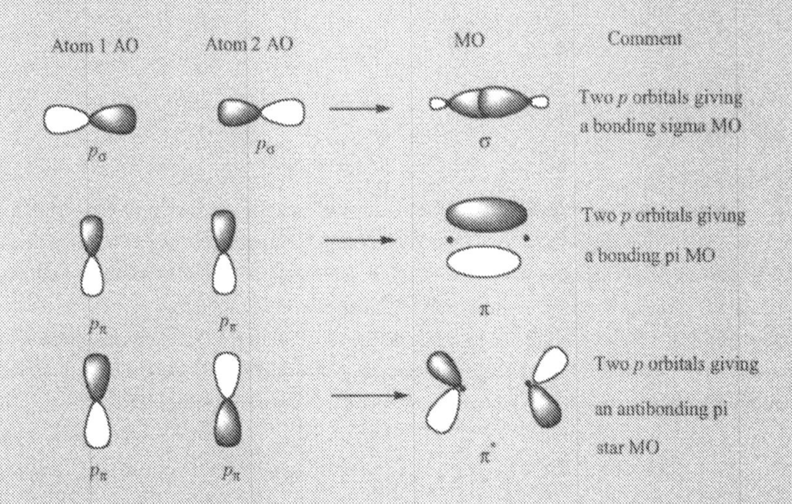

The shading refers to the sign of the wavefunction, and the asterisk denotes an antibonding orbital.

For homonuclear diatomic molecules composed of elements towards the end of the second period, the orbital energies are approximately as shown in the following scheme.

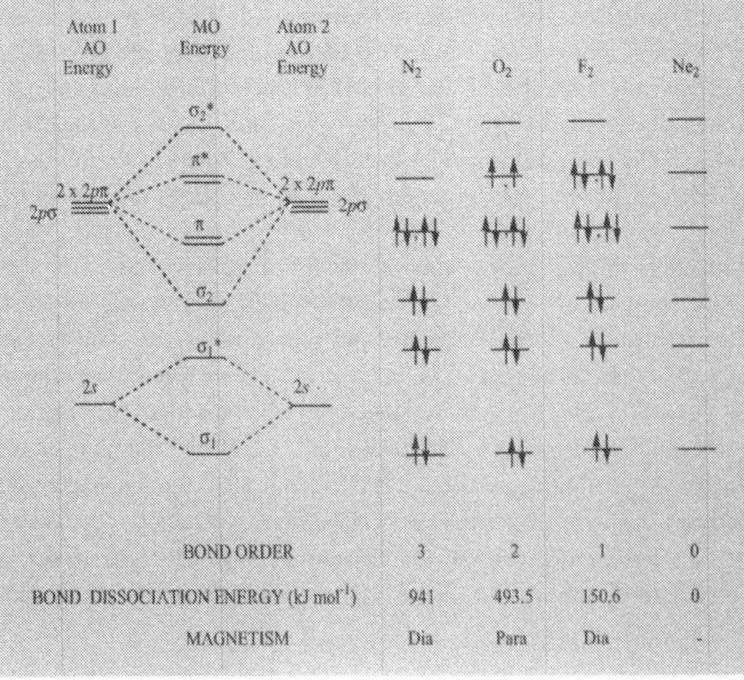

Thus, if the electrons of the outer shell ($2s$, $2p$) are fed into these molecular orbitals starting at the lowest energy, we have the following situations.

Dinitrogen: $2 \times 2s^2p^3 = 10e$. Electronic configuration is $(\sigma_1)^2(\sigma_1^*)^2(\sigma_2)^2(\pi)^4$ – all the electrons are paired, and the molecule is diamagnetic. The bond order (bonding electron pairs minus antibonding electron pairs) is 3.

Difluorine: $2 \times 2s^2p^5 = 14e$. The electronic configuration is $(\sigma_1)^2(\sigma_1^*)^2(\sigma_2)^2(\pi)^4(\pi^*)^4$. Again all the electrons are paired, the molecule is diamagnetic and the bond order is one.

Dioxygen is the exceptional case. Now there are 12 electrons to fit in ($2 \times 2s^2p^4$). Building goes as expected until the 2π level is filled, and then there are two electrons over. These have to go into the next energy level, which happens to be two π^* orbitals of equal energy (orbitals of equal energy and the same symmetry are termed degenerate). The lowest energy arrangement is that with the two electrons being placed one in each π^* orbital, and with the same spin (an extension of the application of Hund's Rule of Maximum Multiplicity). Hence the ground state of dioxygen has two unpaired electrons, and is a triplet state (multiplicity = $2S + 1$ where S is the total spin = $2 \times (1/2 + 1/2) + 1 = 3$). Because of the unpaired electrons the molecule is paramagnetic, but the bond order is 2.

Thus ground state dioxygen cannot properly be represented as O=O: this is a reasonable representation of $^1\Delta_g$ singlet oxygen, however.

For a more rigorous treatment of the subject see M. Kasha and D.E. Brabham in *Singlet Oxygen* (Eds. H.H. Wasserman and R.W. Murray), Academic Press, New York, 1979, pp. 1–33.

TABLE 3.1 THE THREE LOWEST ELECTRONIC STATES OF DIOXYGEN.

State Designation	Common Name	Energy		Lifetime in Condensed Phase	Electronic Configuration of HOMO
		kJ mol^{-1}	ν cm^{-1}		
$^1\Sigma_g^+$	—	155	13120	$<10^{-9}$ s	↑ ↓
$^1\Delta_g$	Singlet oxygen, $^1O_2$	94	7882	~ 10 μs	↑↓ –
$^1\Sigma_g^-$	Oxygen, $^3O_2$	0	0	∞	↑ ↑

TABLE 3.2 LIFETIMES OF SINGLET OXYGEN ($^1\Delta_G$) IN VARIOUS SOLVENTS.

Solvent	H_2O	MeOH	C_6H_6	CS_2	CCl_4	C_6F_6	D_2O	[Air, 1 atm.]
τ (μs)	2	7	24	200	700	3900	20	~76000

3.2 GENERATION OF SINGLET OXYGEN

3.2.1 Chemical Methods

a. Hydrogen peroxide and sodium hypohalite:

$$NaOCl + H_2O_2 \rightarrow H_2O + NaCl + {}^1O_2$$

$$HO{-}OH \quad {}^{-}OCl \longrightarrow O{-}O \longrightarrow O{=}O + Cl^{-}$$

b. N-Chlorosuccinimide and alkaline hydrogen peroxide

$${}^{-}OOH \longrightarrow \quad + {}^1O_2 + Cl^{-}$$

These processes are related, and the formation of singlet oxygen occurs efficiently.

c. Thermolysis of endoperoxides

$$\xrightarrow[\text{e.g. } C_6H_6\text{, reflux}]{\Delta} \quad + {}^1O_2$$

Singlet oxygen , generated in this way from a water-soluble naphthalene endoperoxide at 37^0C, has been shown to inactivate a number of enveloped viruses (eg HIV Type 1, HSV Type 1) but is reported to have no effect on non-enveloped viruses (eg, poliovirus 1) (see also section 14.5.3.5).

d. Thermolysis of phosphite-ozone complexes

To prepare the ozone complex, ozone gas is passed through the dichloromethane solution of the phosphite at –70°C until the blue colour of excess dissolved ozone is seen, and the latter is then removed at –70°C with a stream of nitrogen.

$$(PhO)_3P + O_3 \xrightarrow[-70^o]{CH_2Cl_2} (PhO)_3P(O_3) \xrightarrow[-30^o]{\text{in } CH_2Cl_2} (PhO)_3PO +$$

3.2.2 Physical Methods

a. By direct excitation

Although not commonly done, it is possible to form singlet oxygen by direct excitation. This is a spin forbidden process, and therefore inefficient. It requires a pressure cell in which oxygen is dissolved in a good solvent for it (such as hexafluorobenzene) under high pressure (140 atmospheres), and irradiation with an intense light source (YAG laser) in the region of the 0–1 transition (1070 nm). For observing reactions with organic substrates this is a hazardous procedure because of the risk of explosion, and it has had rather little application in preparative work.

b. By microwave discharge

Microwave discharge in a stream of oxygen at 1–10 mm generates a mixture of singlet oxygen and atomic oxygen, the latter being scrubbed out by passing the gas stream over mercuric oxide. Reaction is then carried out either in the gas phase or by passing the emergent gas (which contains about 6% $^1O_2$) into a solution of the reactant.

$$^3O_2 \xrightarrow[< 10\text{mm}]{\text{microwave discharge}} {}^1O_2 + O \xrightarrow{\text{HgO}} {}^1O_2$$

(g)

, MeOH, (l)

O

O–O

MeOH, (l)

HOO   O   OMe

c. By photosensitisation

This is by far the most common method of producing singlet oxygen in the laboratory, and it is the one that is directly relevant to the subject of this book.

The study of the reaction of photosensitisers, oxygen and organic substrates has had two phases, one in the 1930s, and one starting in the 1960s and continuing to the present. Two mechanisms were advanced, thus:

Schönberg (1935) / Schenck (1948, 1964) — Sensitiser–oxygen complex theory

$$\text{Sens}(S_0) \xrightarrow[\text{2. intersystem crossing}]{1.\ h\nu} \text{Sens}(T_1)$$

$$\text{Sens}(T_1) + {}^3O_2 \longrightarrow \text{Sens.}O_2 \xrightarrow{\text{A}} AO_2 + \text{Sens}(S_0)$$

Kautsky (1931) / Foote (1968) — Singlet oxygen theory

$$\text{Sens}(S_0) \xrightarrow[\text{2. isc}]{1.h\nu} \text{Sens}(T_1)$$

$$\text{Sens}(T_1) + {}^3O_2 \longrightarrow \text{Sens}(S_0) + {}^1O_2$$

$$\text{A} + {}^1O_2 \longrightarrow AO_2$$

The first step is the same in each. The mechanisms differ in that in the Schönberg mechanism the oxidant is a complex between the excited sensitiser and dioxygen, whereas in the mechanism as first advanced by Kautsky the oxidant is singlet oxygen. While the Schönberg mechanism may still operate in certain circumstances, the bulk of the evidence favours the singlet oxygen interpretation. In fact, Kautsky described a very elegant and simple experiment which depends on the supposition that singlet oxygen can exist in the gas phase, whereas the excited sensitiser–dioxygen complex is expected to have a very low vapour pressure. Two samples of silica gel particles were taken, and the photosensitiser acriflavine (an acridinium quaternary salt, **3**) was adsorbed on to one sample, and (colourless) leuco-malachite green (**4**) was adsorbed on to the other. The samples were dried, intimately mixed, and irradiated. At appropriate oxygen tensions (about 10^{-3} mm worked well) the colourless leuco-dye particles were oxidised and became green. Kautsky proposed that an activated form of oxygen was generated on the photosensitised particles, and diffused to oxidise the leuco-dye on a neighbouring particle (Figure 3.1).

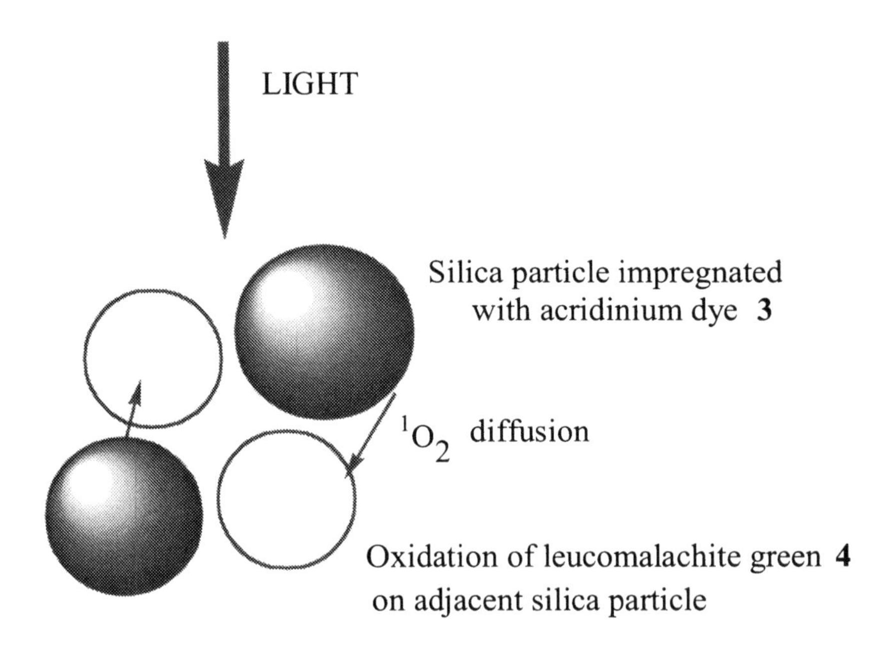

FIGURE 3.1 PHOTOOXIDATION OF LEUCOMALACHITE GREEN ADSORBED ON SILICA GEL. KAUTSKY'S DEMONSTATION OF THE GENERATION AND DIFFUSIBILITY OF SINGLET OXYGEN. (*BER. DEUTSCH. CHEM. GESELL.*,1933, **66**, 1588).

In contemporary terms the process can be written:

3 Acriflavine (S_0) $\xrightarrow{h\nu}$ Sens (S_1) $\xrightarrow{isc}$ Sens (T_1)

$$\text{Sens }(T_1) + {}^3O_2 \longrightarrow \text{Sens }(S_0) + {}^1O_2$$

$$[Me_2N\text{-}C_6H_4]_2CHPh + {}^1O_2 \longrightarrow \text{Malachite green} + OOH^-$$

4 Leucomalachite green

Similar experiments have been carried out more recently. In one of these, shown in Figure 3.2, the gas phase addition of singlet oxygen to 9,10-diphenylanthracene (section 3.3.3) was observed when air was passed through an irradiated fluidised bed of silica gel particles, some of which were impregnated with rose bengal, the others with the substrate.

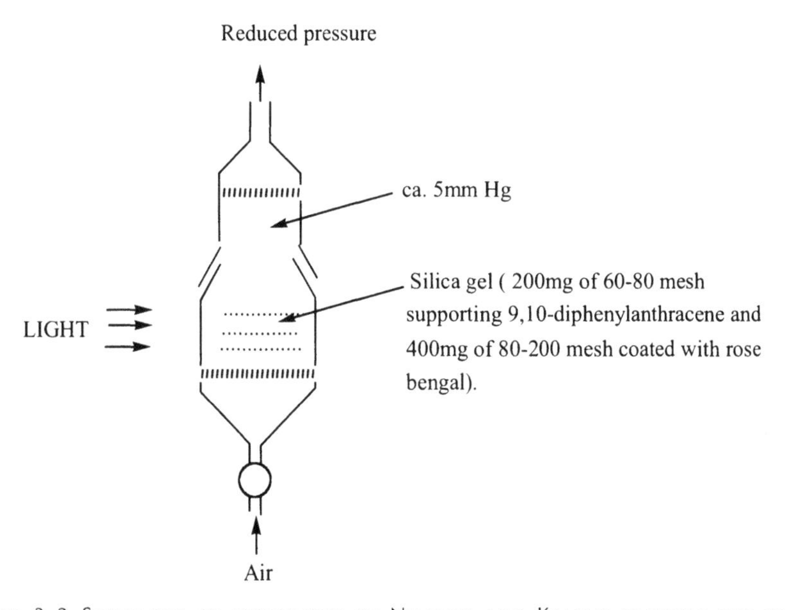

FIGURE 3.2 SCHEMATIC OF EXPERIMENT BY NILSSON AND KEARNS SHOWING THE SET-UP FOR THE REACTION OF GAS PHASE SINGLET OXYGEN PRODUCED BY PHOTOSENSITISATION WITH ROSE BENGAL. THE SUBSTRATE WAS 9,10-DIPHENYLANTHRACENE WHICH GAVE THE 9,10-ENDOPEROXIDE. (*PHOTOCHEM. PHOTOBIOL.*, 1974, **19**, 181–184).

Panel 3.2 Comparisons of Singlet Oxygen Sources

1. Product distributions

(a)

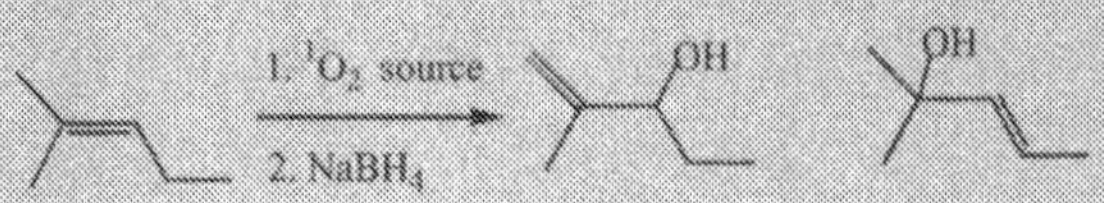

1. $^1O_2$ source
2. $NaBH_4$

$^1O_2$ SOURCE	PRODUCT DISTRIBUTION, % relative abundance	
Rose bengal, *hv*, O_2, MeOH	51	49
$Ca(OCl)_2$, H_2O_2	48	52

(b)

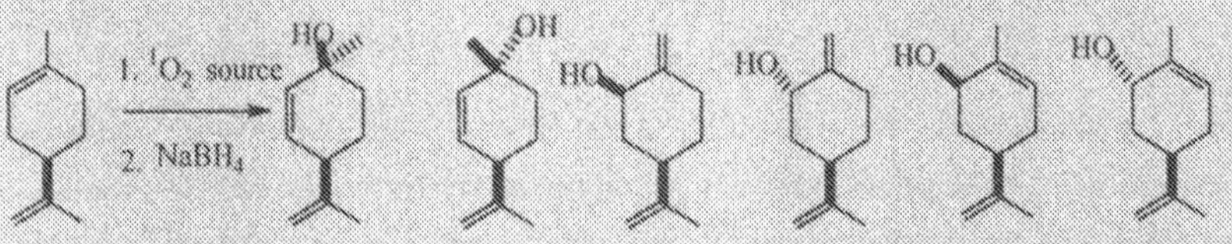

1. $^1O_2$ source
2. $NaBH_4$

(+) - Limonene — *trans*-Carveol

MB, O_2, *hv*, MeOH	36	10	21	19	4	9
RB, O_2, *hv*, MeOH	34	10	21	20	5	10
NaOCl, H_2O_2	34	9	24	18	7	9
(Addition of 2,6-di-t-butylphenol, a radical inhibitor, to this reagent gives similar results)						
Microwave discharge, MeOH, -50°, O_2	41	9	22	18	3	7

2. Relative reactivities

$$A + {}^1O_2 \rightarrow AO_2$$

k_{rel} (1.00 = 2,3-dimethylbut-2-ene)

	Photooxygenation	OCl^-/H_2O_2
2,5-Dimethylfuran	2.4	3.8
Cyclopentadiene	1.2	1.1
1-Methylcyclopentene	0.05	0.03
1-Methylcyclohexene	0.004	0.006

(G.O. Schenck, K. Gollnick, G. Buchwald, S. Schroeter and G. Ohloff, *Liebig's Ann. Chem.*, 1964, **674**, 93–117. C.S. Foote, S. Wexler and W. Ando, *Tetrahedron Lett.*, 1965, 4111–4118. C.S. Foote, *Acc. Chem. Res.*, 1968, **1**, 104–110).

The generation of singlet oxygen by microwave discharge, or by chemical procedures such as the sodium hypochlorite-hydrogen peroxide reaction, and *the similarity in product distribution to that of the photosensitised oxidation with a variety of substrates* confirms the singlet oxygen interpretation. Examples are given in Panel 3.2, which also provides evidence against a free radical mechanism. In the reaction of (+)-limonene (example 1.b), the addition of a radical inhibitor does not change the ratio of products, and optical activity is retained. Thus, in the last product shown (*trans*-carveol) the observed optical rotations are $[\alpha]_D^{20} = -178°$ (rose bengal and methylene blue reactions) and $[\alpha]_D^{20} = -131°$ (hypochlorite, hydrogen peroxide reaction). However, if benzophenone is used as the photosensitiser not only is the product distribution markedly different, but the corresponding $[\alpha]_D^{20}$ value for *trans*-carveol is $-18°$, indicating extensive racemisation. This arises because benzophenone not only has a much higher E_T value (286 kJ mol^{-1}) than methylene blue (142 kJ mol^{-1}) or rose bengal (167 kJ mol^{-1}), but the carbonyl $n \rightarrow \pi^*$ excited state of benzophenone is an avid hydrogen abstractor. This process competes with energy transfer to triplet oxygen, and results in racemisation, thus:-

Ph_2CO^*, $-H^{\bullet}$

$+ Ph_2\dot{C}OH$

1. $^3O_2$
2. Reduction

OH OH HO HO

enantiomers

enantiomers

3.3 CHEMICAL REACTIONS

There are three main types of chemical reaction of singlet oxygen: because these processes are simple to carry out, and show regio and stereospecificity, they are often useful in organic synthesis.

3.3.1 Alkenes – the "ene" Reaction

Examples of this have already been referred to in Panel 3.2. The basic process involves the formation of an allylic hydroperoxide, with a concerted shift of the double bond, thus:

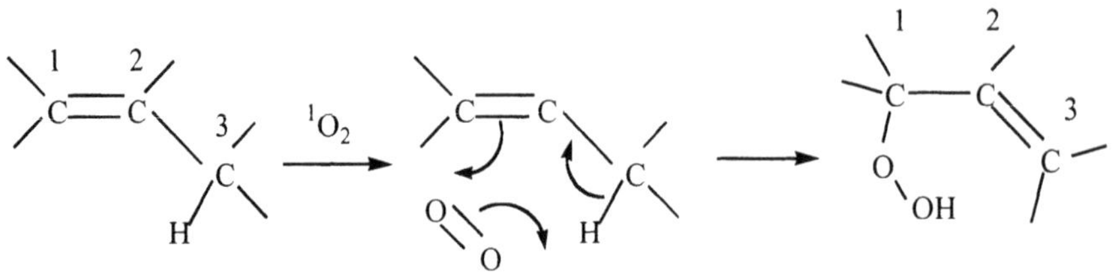

The reaction has alternatively been regarded as proceeding via the perepoxide:

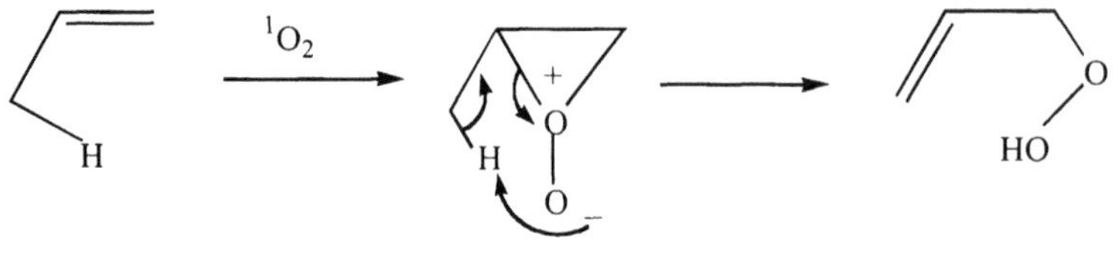

perepoxide

If the hydroperoxide is not desired as the product, it is usual to reduce it (eg with sodium borohydride) to generate the allylic alcohol which is easier to handle.

Singlet oxygen is an electrophile, and attacks more highly alkylated double bonds in preference to less highly alkylated ones. Thus in (+)-limonene (Panel 3.2 (1b)) it is the trisubstituted double bond that is preferentially attacked. The reaction is also subject to steric hindrance. For example, α-pinene (**5**) is attacked almost exclusively on the open α-face and regioselectively at C-3 that is, away from the dimethylcyclobutano feature.

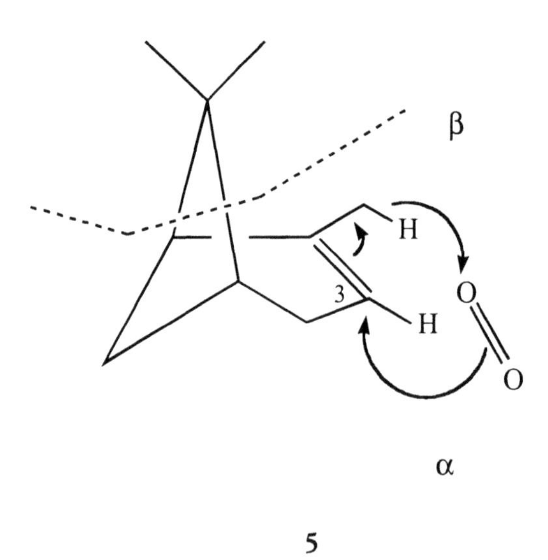

5

This substrate can be used to distinguish between singlet oxygen and radical pathways since the distribution of products is very distinctive.

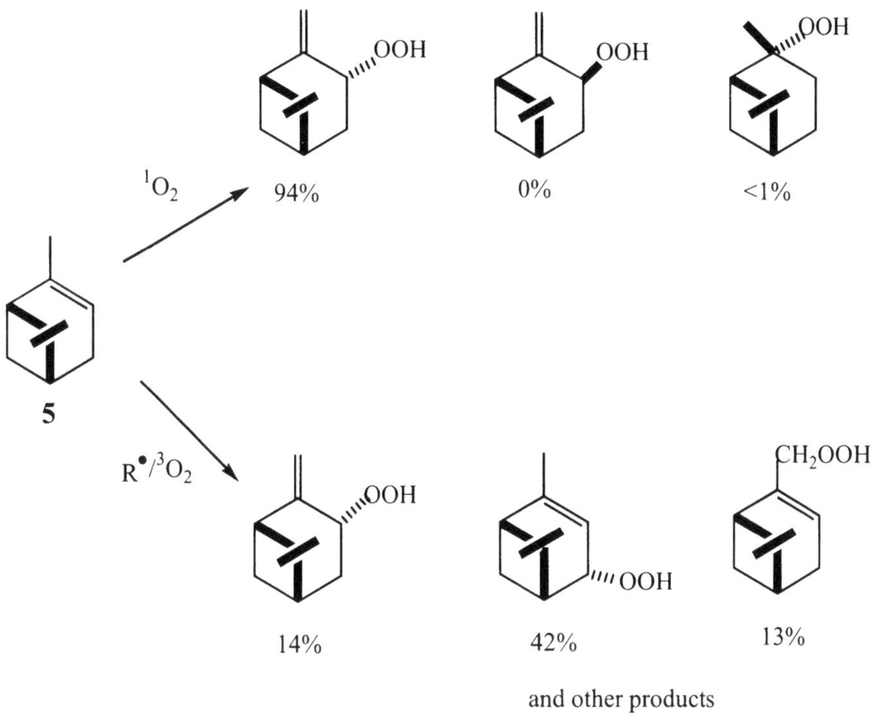

OOH
OOH
OOH
$^1O_2$
94%
0%
<1%
5
R•/$^3O_2$
OOH
CH2OOH
OOH
14%
42%
13%
and other products

3.3.2 Electron-rich Alkenes

Electron-rich (low oxidation potential) alkenes are generally observed to react by cleavage to give two carbonyl fragments, for example:

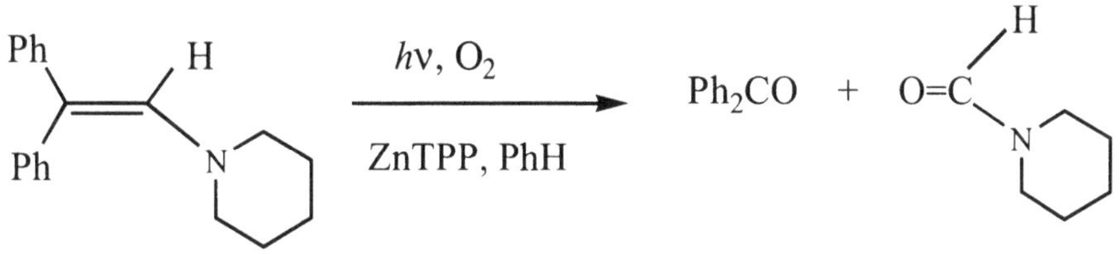

Ph
Ph
H
N
hν, O2
ZnTPP, PhH
Ph2CO + O=C
H
N

O
O
hν, O2
rose bengal, acetone
O
CHO
CHO
O

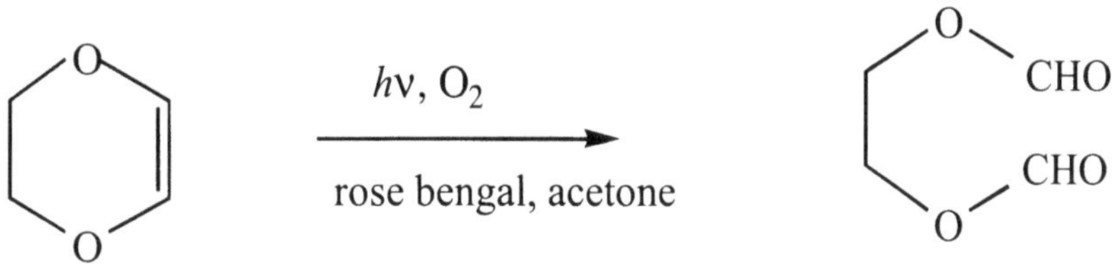

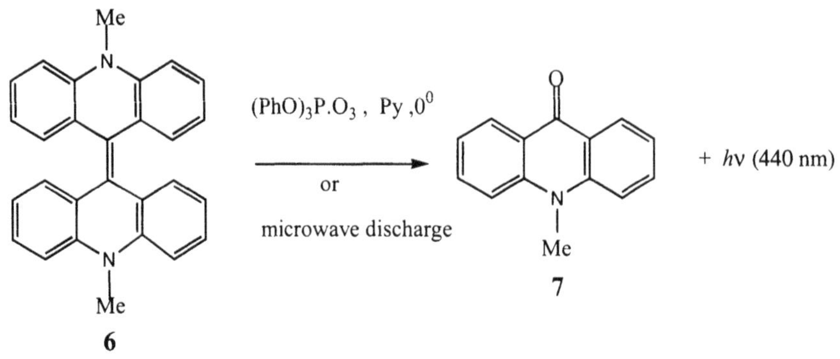

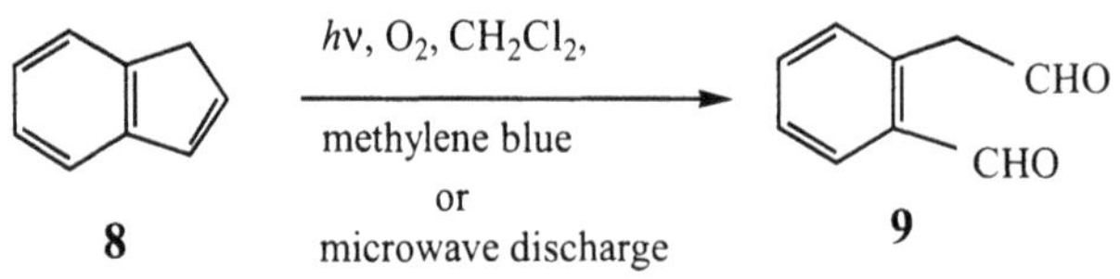

Me
N
(PhO)$_3$P.O$_3$, Py ,0^0
or
microwave discharge
N
Me
6
O
N
Me
7
+ $h\nu$ (440 nm)

$h\nu$, O_2, CH_2Cl_2,
methylene blue
or
microwave discharge
8
CHO
CHO
9

In mechanistic terms, this type of reaction does not necessarily involve an allylic hydroperoxide intermediate (foregoing section), and generally does not do so. Thus in the reaction of indene (**8**) to give homophthaldehyde (**9**), the authentic hydroperoxide has been shown not to react in this way under the reaction conditions. In some cases the formation of allylic hydroperoxides occurs concomitantly with the cleavage progress, for example with 1,2-dimethylindene.

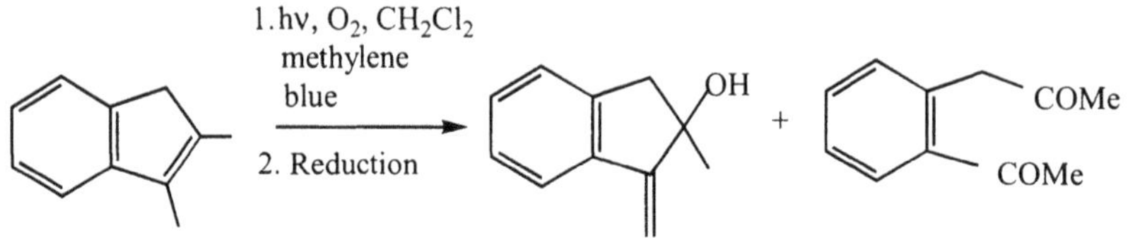

1. hv, O_2, CH_2Cl_2
methylene
blue
2. Reduction
OH
+
COMe
COMe
65%
via ene addition
30%
via 1,2-cycloaddition

The intermediate in the cleavage reaction is the 1,2-dioxetane formed by 1,2-cycloaddition. The 1,2-dioxetanes are thermally unstable, but they can be isolated in suitable cases:

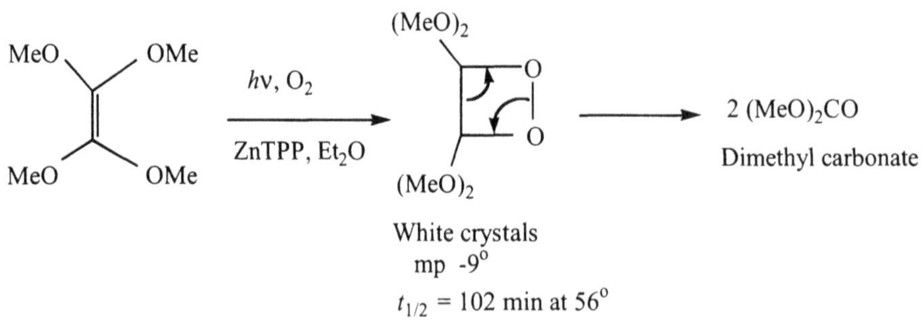

MeO
OMe
MeO
OMe
$h\nu$, O_2
ZnTPP, Et_2O
$(MeO)_2$
O
O
$(MeO)_2$
White crystals
mp -9°
$t_{1/2}$ = 102 min at 56°
2 $(MeO)_2CO$
Dimethyl carbonate

The cycloaddition is stereospecific: thus *cis*-1,2-diethoxyethene (**10**) gives the *cis*-1,2-dioxetane (**11**), while *trans*-1,2-diethoxyethene (**12**) gives the *trans*-product under the conditions indicated.

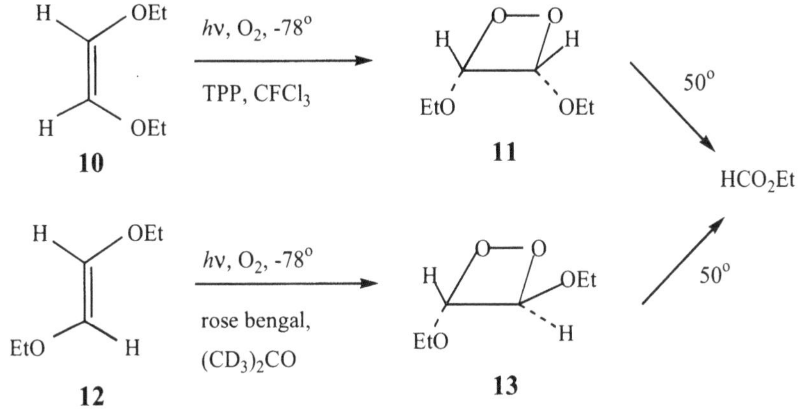

(When the reaction of the *trans* compound (**12**) is photosensitised by *meso*-tetraphenylporphyrin in $CFCl_3$, a mixture of (**11**) and (**13**) is produced. This may indicate that the reaction is not entirely concerted; or that the *trans*→*cis* isomerisation (**12**→**10**) is photocatalysed by TPP).

Although perepoxide and biradical intermediates have been considered, the bulk of the evidence favours a concerted $2\pi + 2\pi$ cycloaddition. A symmetry allowed [2s + 2a] route is possible in the antarafacial mode:

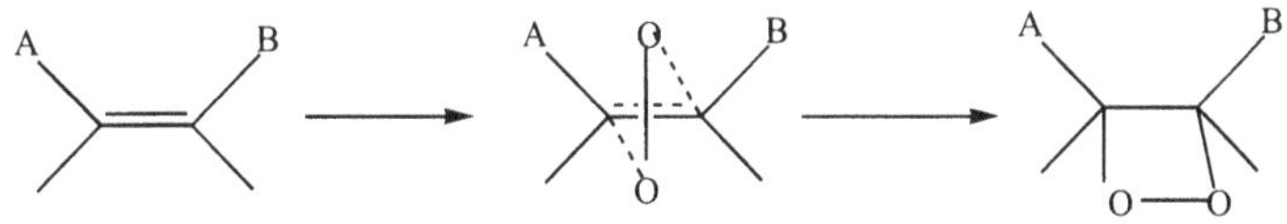

Because the transition state here has a difficult geometry, alternatives have been considered. The most attractive involves an initial charge transfer, which is conceivable where (as here) the HOMO of the electron-rich alkene is higher in energy than the LUMO of singlet oxygen. In these circumstances a [2s + 2s] addition is allowed suprafacially, thus:

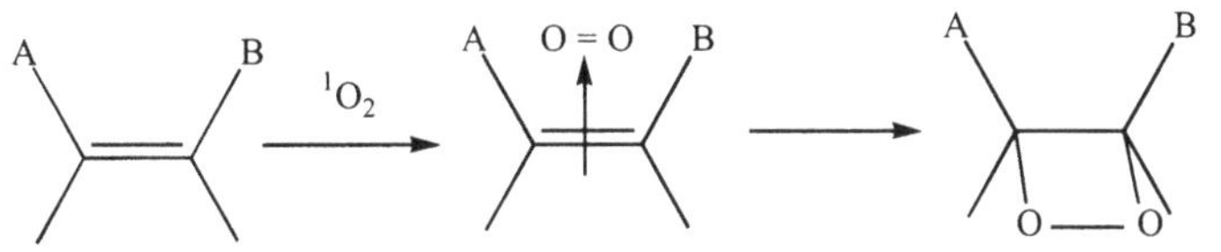

Orbital conservation concepts are also important in the thermal decomposition of the 1,2-dioxetane system, which is an allowed concerted process, provided that one of the carbonyl fragments is in the excited state. Thus chemiluminescence is often observed in the decomposition of 1,2-dioxetanes. Sometimes it is necessary to add a fluorescent molecule to visualise the emission, but in other cases the carbonyl cleavage products are themselves luminescent, as in the formation of N-methylacridone (7) from (**6**) shown above.

3.3.3 Conjugated dienes — Diels-Alder Addition

Unlike triplet oxygen, singlet oxygen readily adds to 1,3-dienes to give a ring system, called an *endoperoxide*.

O
O
O
O

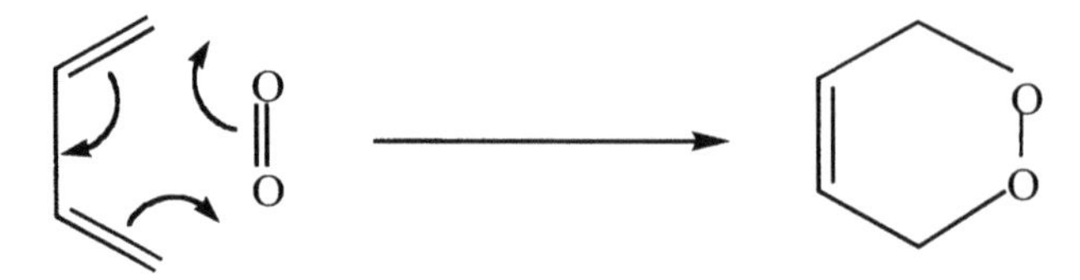

Like the 1,2-dioxetanes (above) these cyclic peroxides are often rather unstable, and some of them decompose with explosive violence. The reaction by which they are formed is a symmetry allowed concerted $4\pi + 2\pi$ cycloaddition.

LUMO of
dienophile
(singlet oxygen)
HOMO of
diene
HOMO of
dienophile
LUMO of
diene

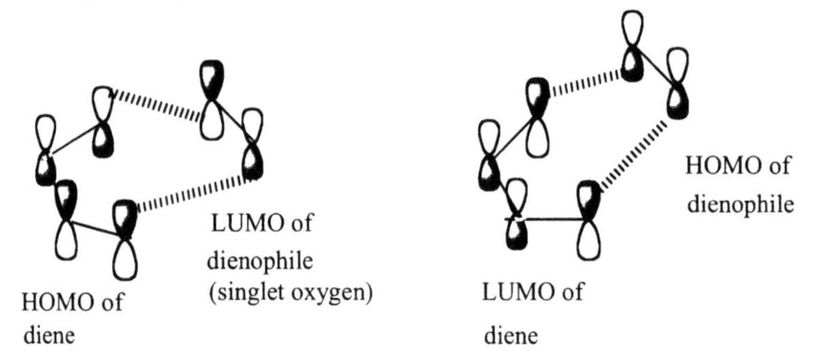

Indeed, some of the dienes react so rapidly that they can be used as chemical traps to divert the course of a reaction under study and provide evidence for a singlet oxygen pathway (section 4.4.3.2).

In some cases the endoperoxides can be detected as intermediates, and occasionally they can be isolated. Thus cyclopentadiene gives a rather noxious explosive endoperoxide: as is often the case with fragile molecules, this can be stabilised by appropriate phenyl substitution.

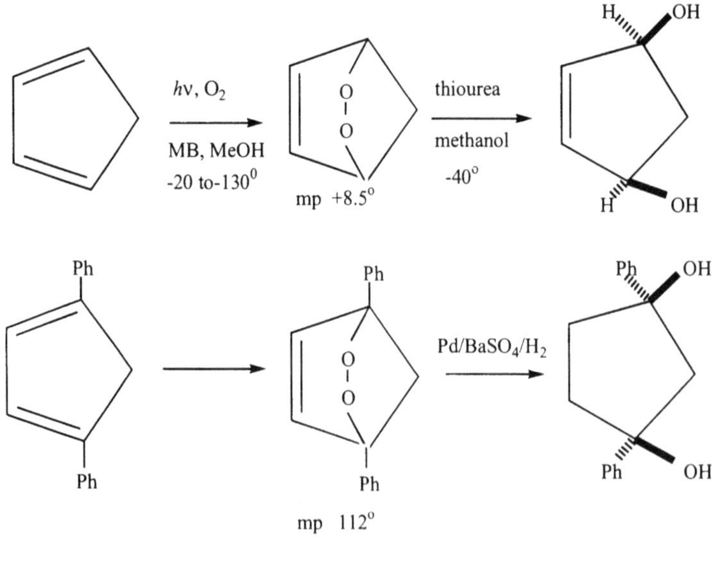

hν, O2
MB, MeOH
-20 to-130°
O
O
mp +8.5°
thiourea
methanol
-40°
H
OH
H
OH
Ph
Ph
Ph
O
O
Ph
mp 112°
Pd/BaSO4/H2
Ph
OH
Ph
OH

The synthetic value of this reaction in generating *cis* hydroxy functions is evident. Another synthetic aspect is illustrated in the first synthesis of ascaridol (**15**) from α-terpinene (**14**).

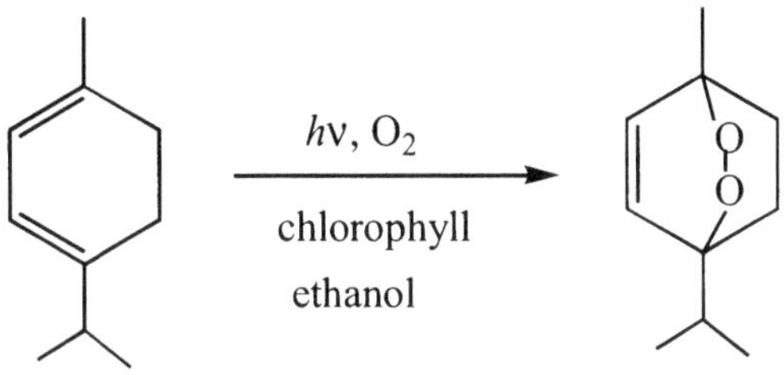

$h\nu$, O_2
chlorophyll
ethanol

14 **15** Ascaridol

Ascaridol occurs naturally as a component of chenopodium oil. Schenck and Ziegler, who carried out this synthesis in 1941, expressed the view that ascaridol may be generated in the plant by a photosensitised reaction involving chlorophyll, rather than by an enzymatically controlled process. That natural ascaridol is optically inactive supports this view.

Extended aromatic systems give endoperoxides which can be manipulated at room temperature (again the effect of aryl substitution).

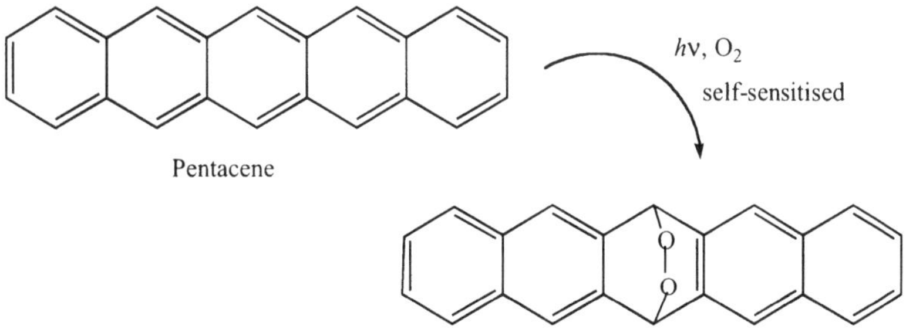

Pentacene

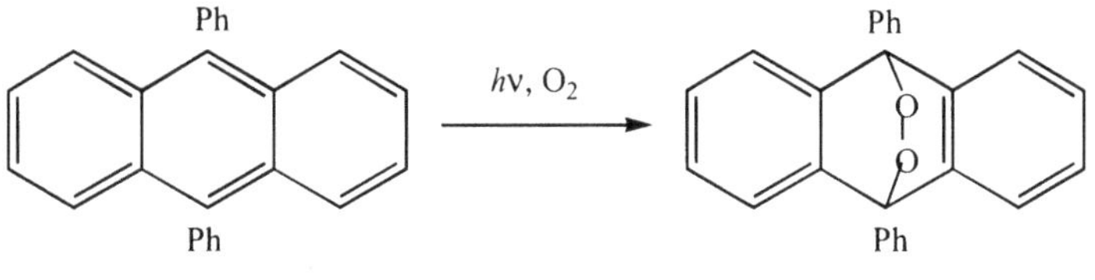

$h\nu$, O_2
self-sensitised

Ph
Ph
$h\nu$, O_2
Ph
Ph

9,10-Diphenylanthracene

Thermolysis in solution reverses the cycloaddition, and singlet oxygen is generated, in some cases (eg 9,10-diphenylanthracene endoperoxide) in a preparatively useful way. Tetraphenylcyclopentadienone reacts rapidly and is decolourised: the intermediate is not isolated but loses carbon monoxide spontaneously to give *cis*-1,2-dibenzoyl-1,2-diphenylethene.

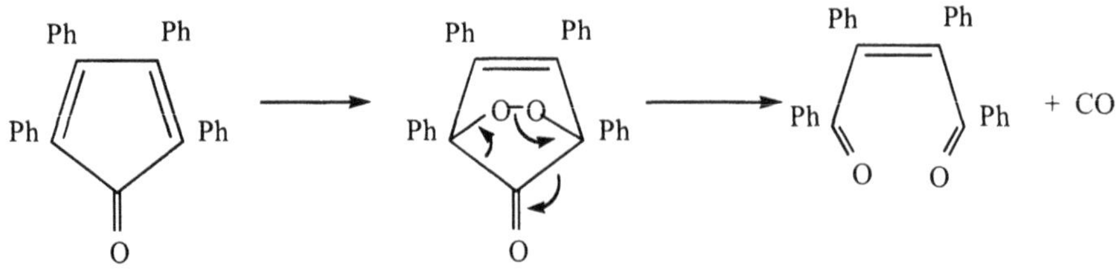

Amongst heteroaromatic compounds furans and isobenzofurans merit special mention. These substrates are very reactive, and are sometimes used (with characterisation of products) to test for singlet oxygen generators. In both cases the endoperoxide is an ozonide; it can be detected but is not isolated. 2,5-Dimethylfuran and 1,3-diphenylisobenzofuran are the key compounds used.

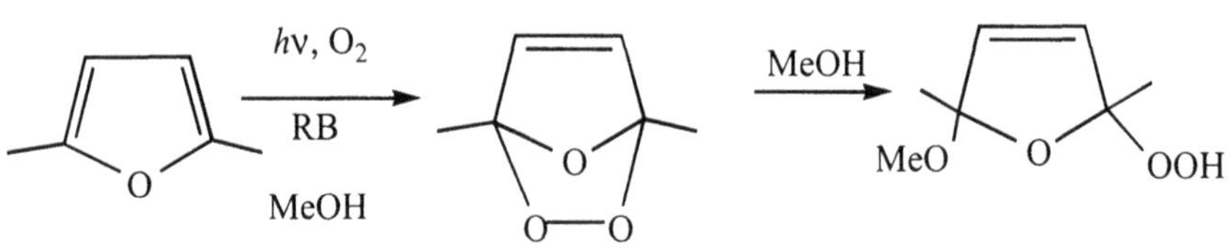

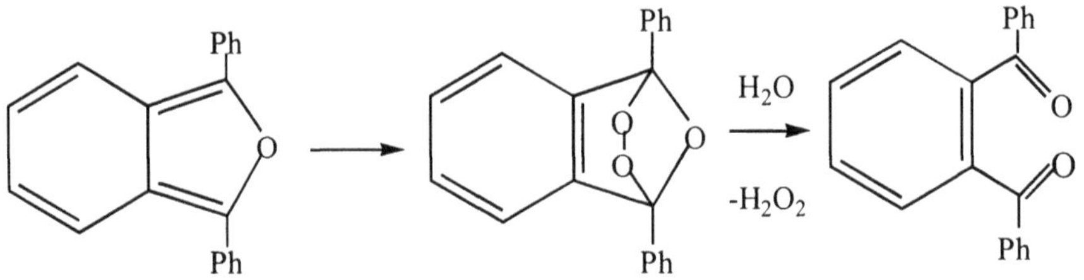

The synthetic value of these reactions has already been alluded to, and is important in the present context, since it underlines the effectiveness of the photochemical generation of singlet oxygen by sensitisation. Two further examples are given in Panel 3.3.

3.3.4 Miscellaneous Reactions

Singlet oxygen is an electrophile and may be expected to react with a variety of other electron-rich systems and centres. In general these reactions are less well understood than those considered above.

Panel 3.3 Singlet Oxygen in Synthesis

1. First synthesis of (±) abscisic acid (leaf senescence signal)

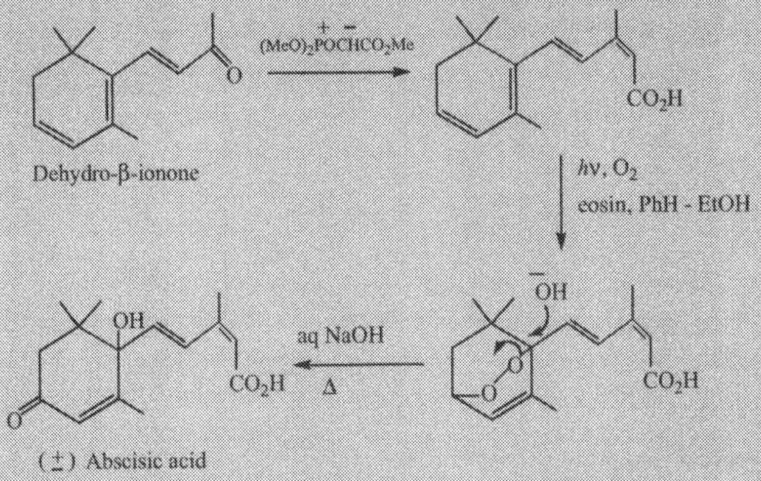

$(MeO)_2\overset{+}{P}O\overset{-}{C}HCO_2Me$

CO_2H

Dehydro-β-ionone

$h\nu$, O_2
eosin, PhH - EtOH

$\bar{O}H$

aq NaOH
Δ

CO_2H

(±) Abscisic acid

(J.W. Cornforth, B.V. Milborrow and G. Ryback, *Nature*, 1965, **206**, 715).

2. Synthesis of methanol propentdyopent adducts (models for the photooxygenation products of bilirubin).

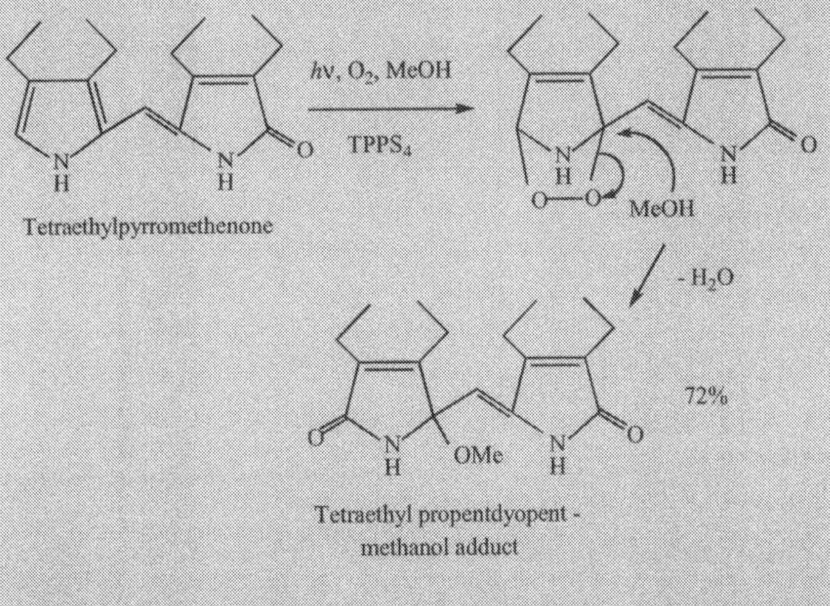

$h\nu$, O_2, MeOH
$TPPS_4$

Tetraethylpyrromethenone

MeOH

$- H_2O$

72%

OMe

Tetraethyl propentdyopent - methanol adduct

(R. Bonnett, S. Ioannou and F.S. Swanson, *J. Chem. Soc., Perkin Trans.* **1**, 1989, 711–714)

(i) Electron-rich aromatics eg phenols

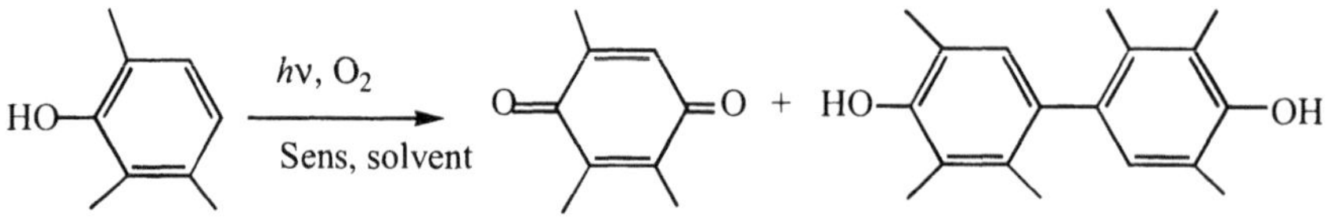

$h\nu$, O_2 / Sens, solvent

Singlet oxygen product

Phenol radical coupling product

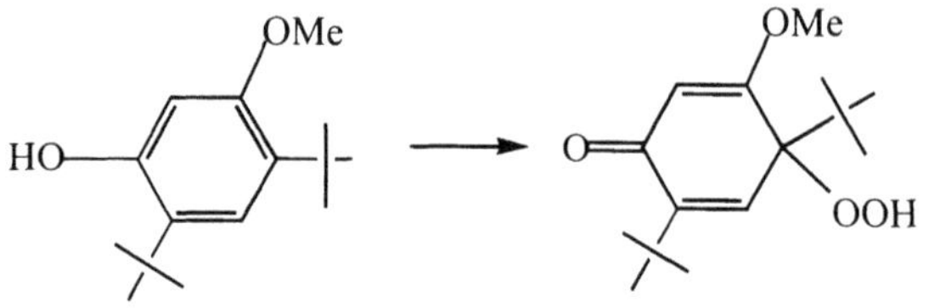

OMe HO OMe O OOH

The second reaction proceeds with chemical and photochemical sources of singlet oxygen at the same rate to give the same product, and possibly involves a 1,4-endoperoxide as an intermediate.

(ii) Amines

Amines both react with, and physically quench, singlet oxygen. The products of chemical reaction are aldehydes, possibly formed as follows:

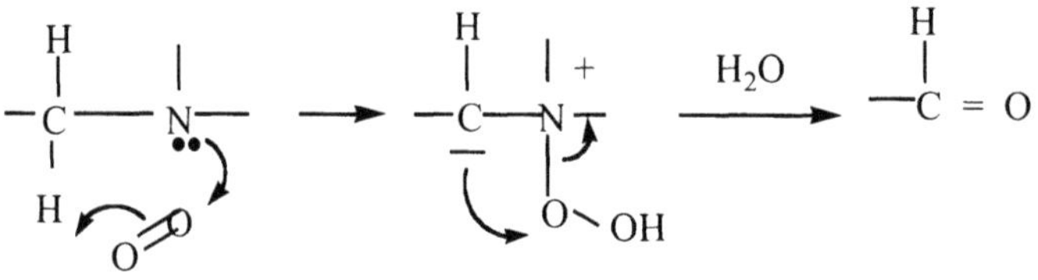

H_2O

(iii) Sulphur compounds

Sulphides generate sulphoxides, which in turn give sulphones:

$$RSR \xrightarrow[\text{sens, solvent}]{h\nu,\ O_2} RSOR \xrightarrow[\text{sens, solvent}]{h\nu,\ O_2} RSO_2R$$

BIBLIOGRAPHY

H.H. Wasserman and R.W. Murray (editors) *Singlet Oxygen*, Academic Press, New York, 1979.

C.S. Foote, J.S. Valentine, A. Greenberg and J.F. Liebman (editors) *Active Oxygen in Chemistry*, Blackie Academic and Professional, London,1995. Chapter 4 concerns singlet oxygen (C.S. Foote and E.L. Clennan, pp.105–140), and the other chapters deal with hydroperoxides, and with superoxide and hydroxyl radical chemistry. This book is the second volume of a series: volume 3 is called *Active Oxygen in Biochemistry*, and is also useful, especially as a source of references.

Chapter 4

PHOTODYNAMIC ACTION

"........ a refined definition in the spirit of Schenck would be that Type I photosensitization is characterized by electron transfer to/from the excited sensitizer (including simultaneous transfer of a proton corresponding to the transfer of a hydrogen atom), resulting in a free radical pathway, while Type II photosensitization is characterized by energy transfer from the excited sensitizer. This is fully compatible with the original definition, can be applied for any photosensitization process and does not require the reclassification of the less common case, when electron transfer occurs to oxygen......"

Tamàs Vidòczy, 1992

4.1 DEFINITION

Photodynamic action, which is the expression of the photodynamic effect, refers to the damage or destruction of living tissue by visible light in the presence of a photosensitiser and oxygen. In principal, light in the very near infrared region could be employed, but sensitisers absorbing there are rather uncommon, and in most cases irradiation is with visible light. The light source may be incoherent (eg a tungsten lamp) or coherent (eg a laser): the sensitiser must absorb the light and the excited sensitiser then activates oxygen in some way (section 4.4). Molecular oxygen is essential for the process to occur, and in most examples is derived from the atmosphere.

4.2 DISCOVERY AND DEVELOPMENT

The effect was discovered in the winter semester of 1897-98 at the Ludwig-Maximillian University in Munich where Oscar Raab was a medical student. It appears that as part of his career-building programme he spent time in the pharmacology laboratory of Professor H. von Tappeiner examining the effect of dyestuffs on paramecia, which are unicellular organisms of the Protista class. He observed, using low concentrations of acridine (**16**) as the photosensitiser, that the paramecia were killed in the presence

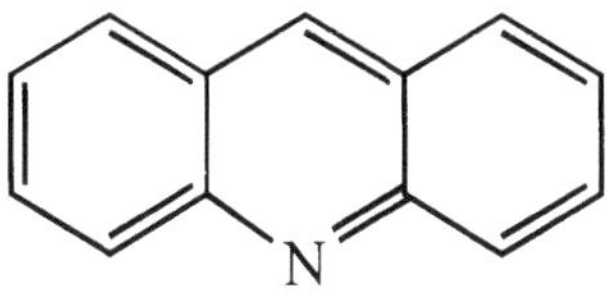

N

16

of daylight, but that in darkness they survived. The result was published in 1900, and stimulated much further activity, so that by 1903 von Tappeiner and Jesionek were proposing various dermatological applications for photosensitisers (such as eosin **24**) some of which they demonstrated. In 1904, von Tappeiner and Jodlbauer used the term "photodynamische wirkung" for the first time, emphasising the requirement for oxygen in the process. This German term has been generally translated as "photodynamic action", but this seems an awkward rendering and the equivalent and more natural-sounding "photodynamic effect" will generally be used here (not exclusively though: where the older rendering is felt to be more appropriate it will be retained).

Experiments followed in rapid succession extending the variety of living systems and photosensitisers studied. Clearly the subject had caught the scientific imagination. Using the photohaemolysis of erythrocytes as a model, Hasselbach showed in 1909 that the photosensitiser increased the rate of the photoreaction only in the presence of oxygen, confirming the earlier conclusions of von Tappeiner and Jodlbauer. In Hans Fischer's organic research laboratory in Munich (1912) it was shown that haematoporphyrin had a strong photodynamic effect in the mouse whereas, under the conditions used (subcutaneous injection) mesoporphyrin did not. On the other hand, against the paramecium the activities were reversed.

Several preliminary comparisons of this sort were reported, culminating in the demonstration of the photodynamic effect in Man. Friedrich Meyer-Betz was an Austrian physician. He had been a coworker of Hans Fischer, and had become familiar with the phototoxicity of haematoporphyrin in the mouse. (He had collaborated with Fischer in the comparisons with mesoporphyrin referred to in the last paragraph). Working in the medical clinic at Königsberg (now Kaliningrad in Russia) he carried out a remarkable experiment on himself.

Between 5.45 am and 6.15 am on October 14th 1912 Meyer-Betz injected himself intravenously with 200 mg of haematoporphyrin. (The porphyrin was dissolved in 10 ml of 0.1M NaOH, and diluted with 300 ml of physiological saline, the whole being heat and filter sterilised). The next day was overcast, and nothing spectacular occurred. However October 16th 1912 was a sunny day in Königsberg and a photosensitised reaction set in — a prickling and burning sensation, with those regions of his body which had been exposed to sunlight developing erythema (reddening) and oedema (swelling), which intensified. Meyer-Betz had exposed the right side of his face more than the left side, and this shows clearly on the photograph in Figure 4.1a, taken on October 17th. By October 19th the swelling had receded somewhat, (Figure 4.1b, which begins to give us an impression of his normal appearance), but the photosensitivity remained for several weeks until the haematoporphyrin (which was probably not a pure substance) had worked its way out of his body. In the gloom of the Königsberg winter the effects were not observed, and by the Spring of 1913 the sensitisation had disappeared. In modern health and safety terms this has to be regarded as a very foolhardy experiment (especially with such a massive dose), but nonetheless, at a time when such heroic experiments were not regarded as unexceptionable, it did demonstrate very clearly the photodynamic effect of porphyrin sensitisers in human beings.

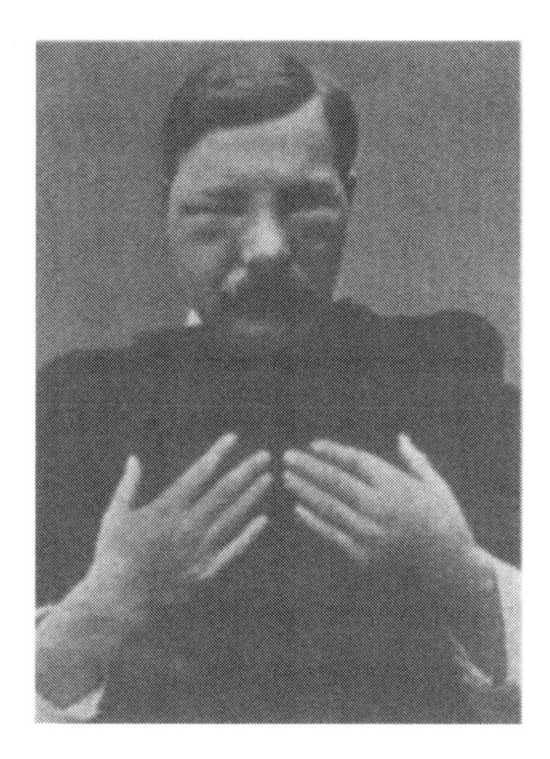

FIGURE 4.1 THE MEYER-BETZ SELF-PHOTOSENSITISATION EXPERIMENT (SEE TEXT). (TOP) THREE DAYS AFTER INJECTION; THE RIGHT SIDE OF THE FACE HAS BEEN MOST EXPOSED TO SUNLIGHT. (BOTTOM) FIVE DAYS AFTER INJECTION: THE OEDEMA IS REDUCING, BUT LESIONS ON HAND AND FACE ARE PRESENT. PHOTOSENSITIVITY REMAINED FOR SEVERAL WEEKS.

Many years later it was realised that evidence for this effect was already available in certain of the porphyrias, where photodynamic damage had been recorded in photographs of patients taken in the late nineteenth century. The *porphyrias* are a group of diseases associated with enzymatic deficiencies in the biosynthetic pathway to haem (also called protohaem, because it is the iron(II) complex of protoporphyrin). This is a multistep pathway, and since it will be important to refer to it again later, it is summarised in Panel 4.1. Some of the malfunctions of the pathway lead to the production of metal-free porphyrins in the body. The porphyrins are circulated in the blood stream, penetrate to the dermal and epidermal layers, and photosensitisation results. Thus, in the condition known as *erythropoietic protoporphyria,* the ferrochelatase enzyme, which is responsible for inserting iron into protoporphyrin, is not functioning properly. Photosensitised damage results from the enhanced levels of protoporphyrin in the plasma (see Panel 4.1). In severe cases of photosensitisation considerable damage is caused to exposed parts of the body (hands, face).

Two other examples will serve to illustrate the range of the photodynamic effect. In 1933 Rimington and Quin reported from South Africa on a condition which affected sheep, a photosensitivity disease called geel-dikkop (*Afrikaans,* "yellow thickhead"). The substance causing the photosensitivity was isolated: it proved not to be one of the plant-derived photodynamic agents such as hypericin (**20**, section 4.3.2) but a porphyrin (phytoporphyrin **18,** otherwise known as phylloerythrin: Figure 4.2). The skeleton of this porphyrin closely resembles that of chlorophyll *a* (**17**) and it is derived from it through the action of the bacterial flora of the gut. The photosensitivity disease presumably arises by pathological absorption of this photosensitiser through the gut wall and into the blood stream.

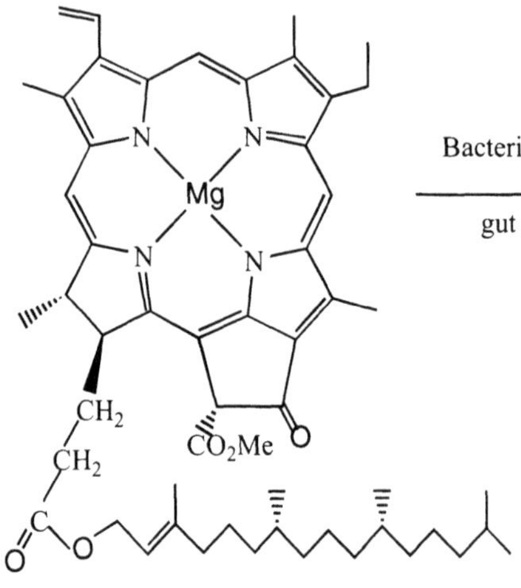

Bacteria in gut

17 Chlorophyll *a* (in diet)

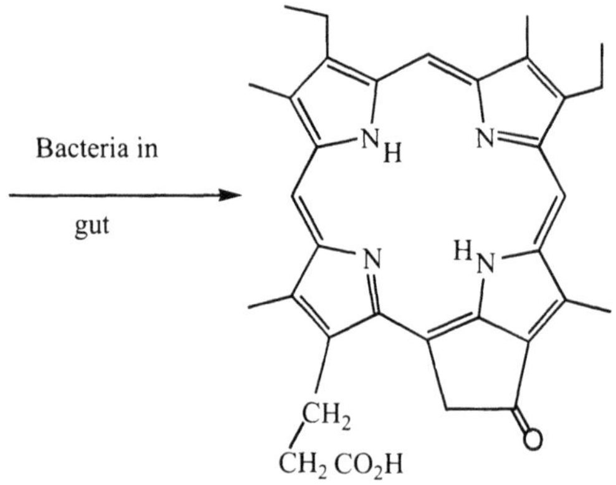

18 Phytoporphyrin (also called phylloerythrin); pathological transport through gut wall into bloodstream leads to photosensitisation of exposed tissues.

FIGURE 4.2 INTERPRETATION OF THE PHOTOSENSITISATION DISEASE (GEEL-DIKKOP) IN SOUTH AFRICAN SHEEP.

Panel 4.1 Porphyrin and Haem Biosynthesis in Relation to Porphyrias

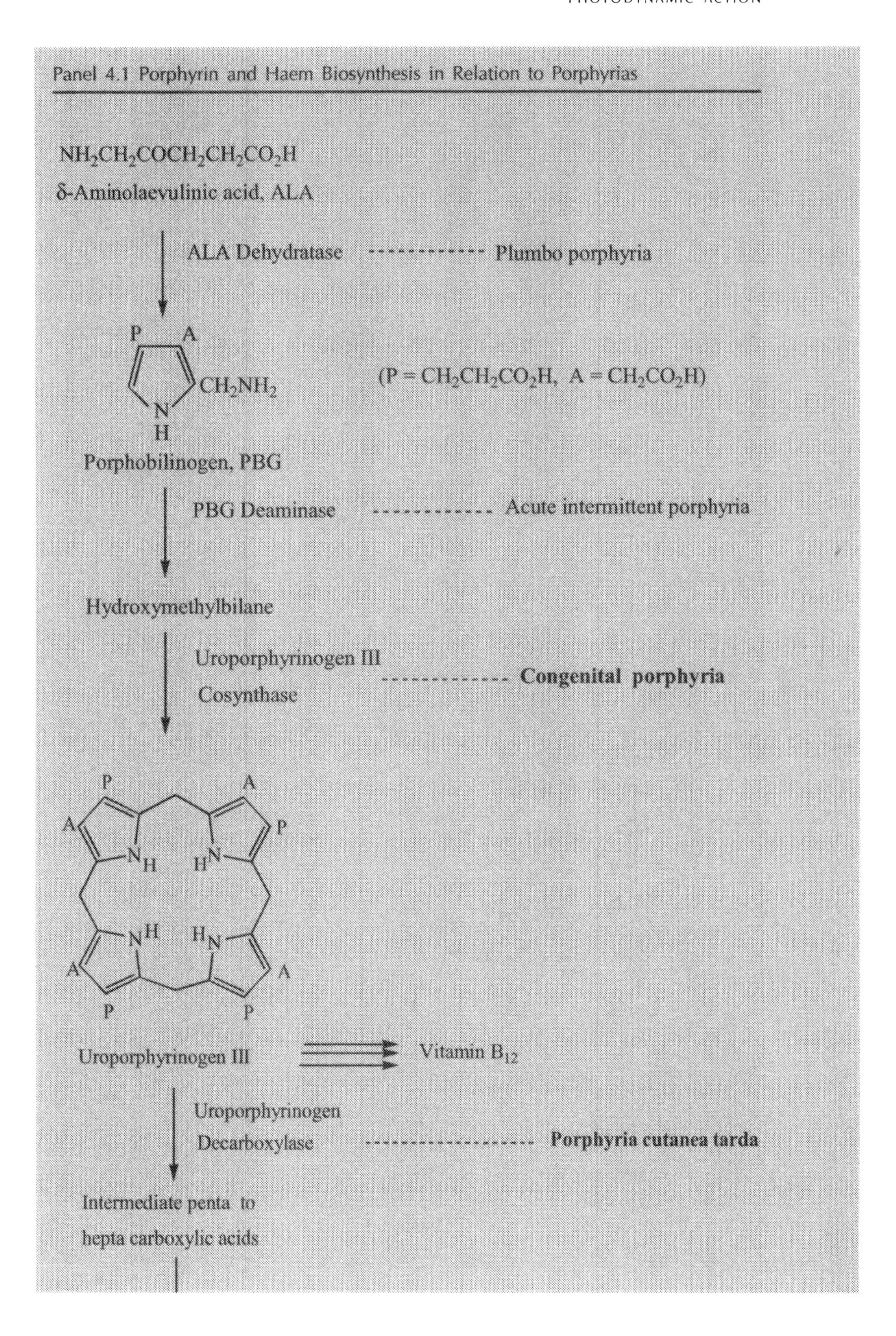

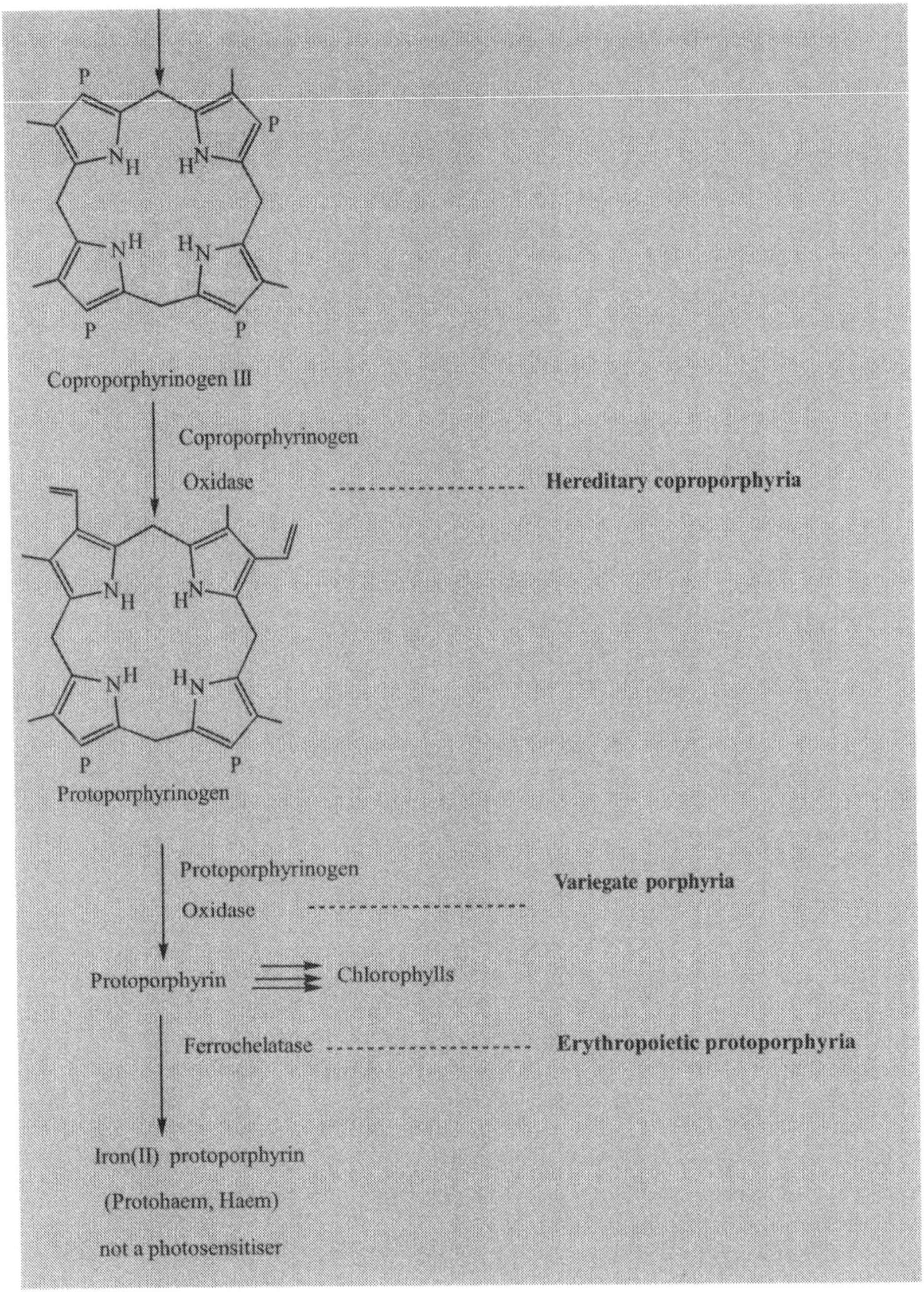

P P NH HN P NH HN P P

Coproporphyrinogen III

Coproporphyrinogen Oxidase ------------------- **Hereditary coproporphyria**

NH HN NH HN P P

Protoporphyrinogen

Protoporphyrinogen Oxidase ------------------- **Variegate porphyria**

Protoporphyrin ⇛ Chlorophylls

Ferrochelatase ------------------- **Erythropoietic protoporphyria**

Iron(II) protoporphyrin

(Protohaem, Haem)

not a photosensitiser

1. The biosynthetic pathway to protohaem (left hand side) operates at the hexahydroporphyrin (porphyrinogen) level, with side extensions leading to the vitamin B_{12} and chlorophyll series. The porphyrinogens are colourless, and are not expected to be photosensitisers: but reduced activity of an enzymatic step leads to the accumulation of the intermediate before the blockage. If this intermediate is a porphyrinogen, then adventitious chemical oxidation to the porphyrin occurs, and this causes a photodynamic effect.
2. The various recognised porphyrias are indicated on the right-hand side of the panel, opposite the site of the relevant enzymatic malfunction. Photosensitising conditions are in bold. Note that if one of these enzymes is completely deactivated then the organism will die, since internally generated haems are essential for living things.
3. Besides porphyrinogens of the III series, porphyrinogens of the I series also occur in small amounts, but do not appear to have a vital function. The corresponding porphyrins are detected, however, in porphyria cases. Uroporphyrinogen I has the structure:

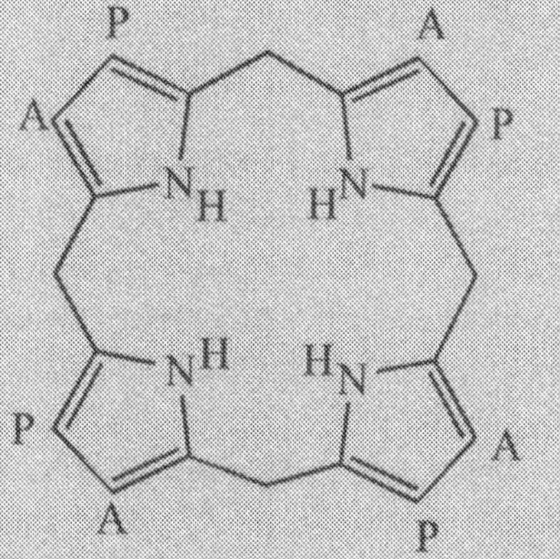

P A A P NH HN NH HN P A A P

4. Note that iron(II) protoporphyrin (protohaem, usually simply called haem because it is by far the most important of the iron complexes in biological terms) is not a photodynamic sensitiser. Had it been so life would have had not only to have evolved initially at modest depths below sea/water level (which it probably did anyway) but to have stayed there, or gone underground.
5. The synthesis of porphobilinogen, uroporphyrinogen and coproporphyrinogen occur in the cytoplasm of the cell: the other steps occur in the mitochondria. The entire process is controlled by the level of protohaem produced, which is able to modulate ALA synthase activity and the production of δ-aminolaevulinic acid. Thus when haem concentration is excessive, the first step is switched off (negative feedback control). This is an important factor in photosensitation of cancer using endogeneous protoporphyrin (section 8.5).

The second example comes from the plant kingdom, and concerns the mode of action of diphenylether herbicides, such as acifluorfen methyl (**19**). These herbicides are effective in the light, but not in the dark. They work by disrupting haem and chlorophyll biosynthesis in the plant by interfering with the activity of protoporphyrinogen oxidase (Panel 4.1). The excess protoporphyrinogen which begins to accumulate diffuses into the leaf structure and is oxidised non-enzymatically (and possibly also enzymatically) to protoporphyrin. The protoporphyrin in the presence of light and oxygen exerts a photodynamic effect, thus destroying the exposed growing parts of the plant.

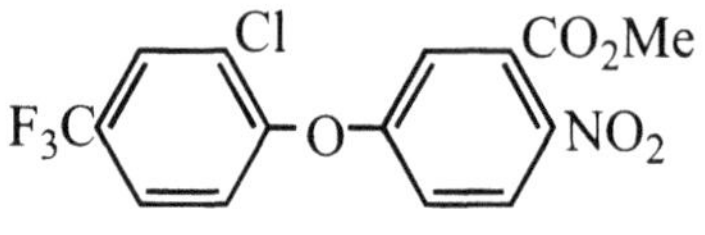

19 Acifluorfen methyl

4.3 PHOTODYNAMIC AGENTS — STRUCTURAL TYPES

4.3.1 General Considerations

Photodynamic activity is not limited to any one structural type. The conditions for activity are as follows:-

(i) A chromophore must be present with absorption in the visible region or thereabouts.

(ii) If singlet oxygen formation is involved, as is commonly the case, the excited triplet state of the molecule must be efficiently formed by intersystem crossing.

(iii) The triplet energy E_T should be greater (but not too much greater) than 94 kJ mol^{-1} (the energy of singlet oxygen above the ground state), and energy transfer (Sens $T_1 \rightarrow {}^3O_2$) should occur efficiently. Chromophores with absorption only in the ultraviolet may not be effective because E_T is too large and energy transfer is not efficient. Chromophores with their lowest energy absorption band in the infrared (*ca.* > 900 nm, that is the *singlet* energy is less than ~ 130 kJ mol^{-1}) may be ineffective because the *triplet* energy is likely to be less than 94 kJ mol^{-1}.

(iv) If the substance is subject to aggregation under the conditions employed then the non-associated ("monomeric") form is found to have the sharper visible absorption spectrum and is the more effective photodynamic species (Panel 12.1).

4.3.2 Structural Types

Examples of photodynamic sensitisers include the following:

(i) *Extended Quinones.* The typical substance here is hypericin (**20**) a natural product present in the foliage of *Hypericum* spp. (eg St. John's Wort) which is phototoxic to grazing animals. It forms solvated blue black needles from pyridine, in which it dissolves to give a cherry red solution with a red fluorescence.

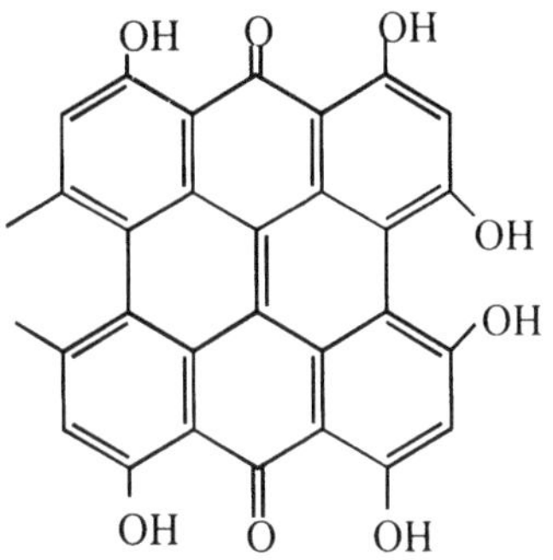

OH O OH

OH

OH

OH O OH

20 Hypericin

λ_{max} (EtOH)
590 nm (ε 41 600), 548(23 500),
508 (8 700), 473 (13 000).

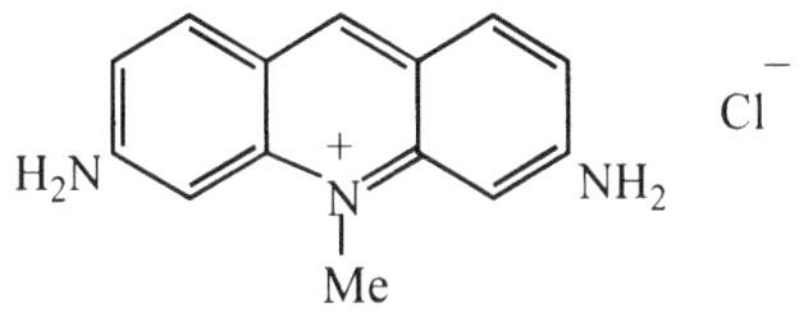

H_2N $\overset{+}{N}$ NH_2 Cl^-

Me

3 Acriflavine

λ_{max} (MeOH) 460 nm (ε 54 800)

(ii) *Acridine Dyes.* Acriflavine (also known as trypaflavine) is an example (**3**). This is a synthetic quaternary salt which is marketed as a deep orange powder, normally also containing the unmethylated compound as its hydrochloride. It is freely soluble in water, less so in ethanol, and insoluble in ether and chloroform. Acridine orange (**21**) is a related compound. It is not a quaternary ammonium salt but is available either as the hydrochloride hydrate (λ_{max} 492 nm) or as the zinc chloride double salt (λ_{max} 489 nm).

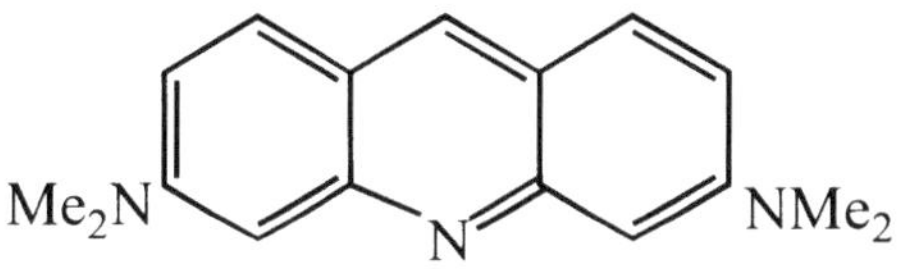

Me_2N N NMe_2

21 Acridine orange (free base)

(iii) *Phenothiazine Dyes.* Methylene blue (**22**) is the type material, again a synthetic substance and sometimes employed to generate singlet oxygen in chemical synthesis. It forms a dark green crystalline trihydrate, the crystals having a bronze lustre, and is soluble in water and chloroform, but not in ether. Toluidine blue (drawn at **23** with an alternative canonical structure for the phenothiazinium nucleus) is a related compound.

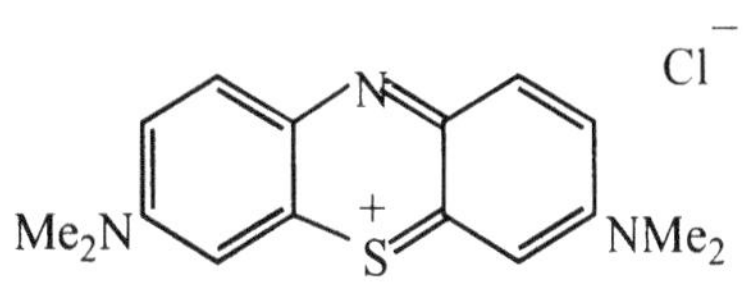

N Cl^-

Me_2N $\overset{+}{S}$ NMe_2

22 Methylene blue

λ_{max} (EtOH) 660 nm (ε 88 000)

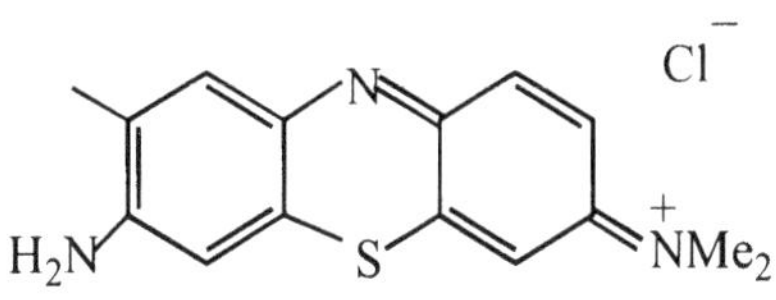

N Cl^-

H_2N S $\overset{+}{N}Me_2$

23 Toluidine blue

λ_{max}(EtOH) 626 nm (ε 63 000)

Thionine (as the hydrochloride) has the structure shown for methylene blue with NH_2 replacing each NMe_2 group (λ_{max} 598 nm in aqueous solution at pH 7.0).

(iv) *Xanthene Dyes.* The compounds that are amongst the most active photosensitisers are the halogenated derivatives of fluorescein, again synthetic compounds. The substance generally used to colour red ink is eosin Y (**24**), which is a tetrabromofluorescein: rose bengal (**25**) is a tetraiodotetrachloro derivative of fluorescein. These are deep red crystalline compounds which are soluble in water and alcohol, but not in ether. Rose bengal has been frequently used as a singlet oxygen generator in synthetic and mechanistic chemistry. The triethylammonium salt (rather than the sodium or potassium salt) is available for use in non-polar solvents.

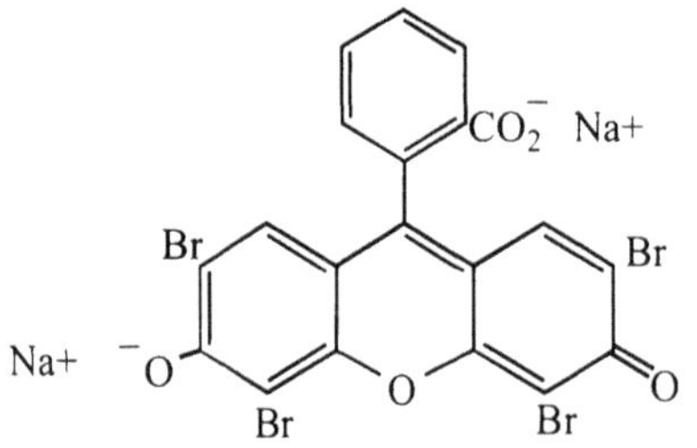

24 Eosin Y
λ_{max}(aqueous, pH 9.1)
520 nm (ε 100 000)

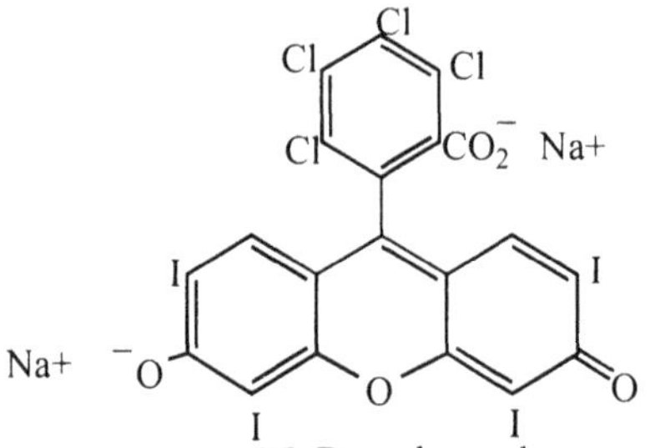

25 Rose bengal
λ_{max} 549 nm

The halogen atoms in these molecules increase intersystem crossing (section 2.4) and so promote formation of the triplet species required for singlet oxygen generation (section 3.2.2).

(v) *Cyanine Dyes.* The cyanine dyes were originally developed as sensitisers for incorporation into photographic emulsions. An example here is merocyanine 540 (**26**) which has found various applications in biology, for example as a fluorescent axonal stain. In water it aggregates, but can be kept in the monomeric form in aqueous buffered detergent solution.

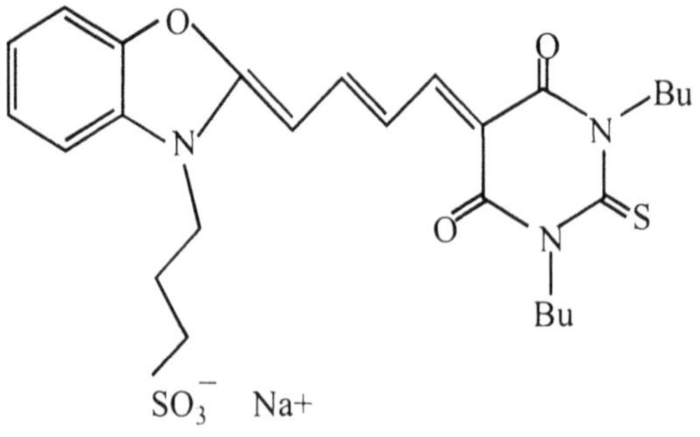

26 Merocyanine 540
λ_{max}(EtOH) 559 nm (ε 138 000)

(vi) *Porphyrins.* The porphyrins (and the related chlorins and phthalocyanines below) have three general advantages as photodynamic agents: (a) intense absorption in the visible, so that little material is required; (b) aromatic stability, so that the photosensitiser is fairly rugged with respect to self-destruction; and (c) generally, a low toxicity in the dark. Typically porphyrins are deep purple crystalline solids which in chloroform give reddish purple solutions which show a strong red fluorescence in ultraviolet light (using a medium pressure mercury source,preferably filtered through Wood's glass to give the λ 366 nm emission. Section 2.2.3). However the basic skeleta are hydrophobic so that if water solubility is required, appropriate solubilising groups must be introduced. Sulphonic acid, carboxylic acid and hydroxy functions have been used for this purpose (section7.2.4). Some of the substances are derived from natural products, for example haematoporphyrin (**2**) which is prepared from blood by treatment with sulphuric acid or (in the laboratory) from protohaemin (chloroiron (III) protoporphyrin) using hydrogen bromide with aqueous work up. Commercial samples are of poor purity (section 6.3).

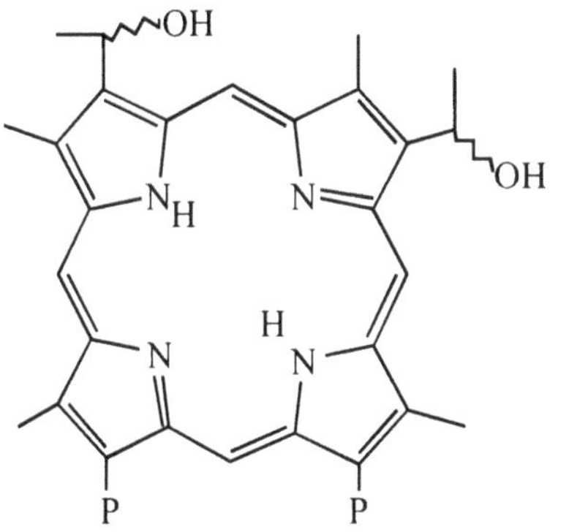

2 Haematoporphyrin
λ_{max} (acetone)
see Figure 2.4

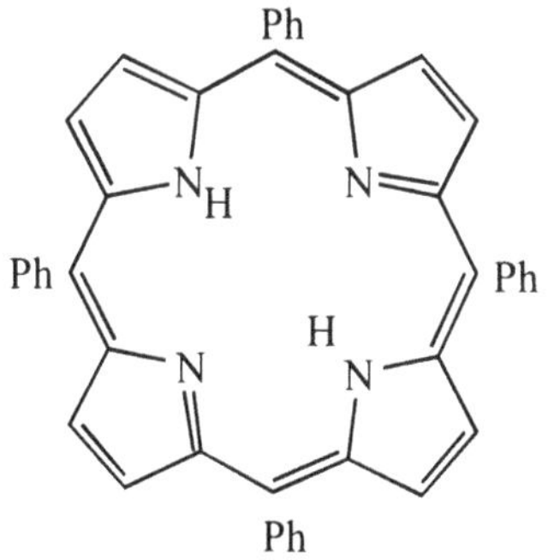

27 5,10,15,20-Tetraphenylporphyrin
$\lambda_{max}(CHCl_3)$ 418 nm (ε 411 000), 515 (17 300), 551 (8 060), 590 (6 400), 645 (6 220).

The simplest synthetic porphyrins are the *meso*-tetraphenylporphyrins (eg the parent, **27**) which can be prepared from an α,α′-free pyrrole and a benzaldehyde under conditions of acid catalysis (Rothemund-Adler or Lindsey conditions, see section 8.4). Yields are modest (~20%), but this is endured with a smile since this is a one-pot reaction from (generally) readily available precursors.

Since we shall meet them again, Panel 4.2 provides a note on the nomenclature of porphyrins.

(vii) *Chlorins and Bacteriochlorins.* Chlorin has one ββ′ double bond of porphyrin saturated: bacteriochlorin has two such bonds reduced, and they are disposed opposite to one another. These are dark solids which dissolve (eg in chloroform) to give green solutions which again show a red fluorescence in UV light (366 nm). Natural examples come from chlorophyll *a* (**17**) and from bacteriochlo-

Panel 4.2 A Note on Porphyrin Nomenclature

Since porphyrin compounds will be encountered frequently, a note on nomenclature is appropriate.

The parent substance is called porphyrin: the β,β′-positions of the pyrrole rings are called β-positions; the carbon in between the pyrrole rings is called the *meso*-position (or methine bridge). Numbering is as shown (IUPAC system):

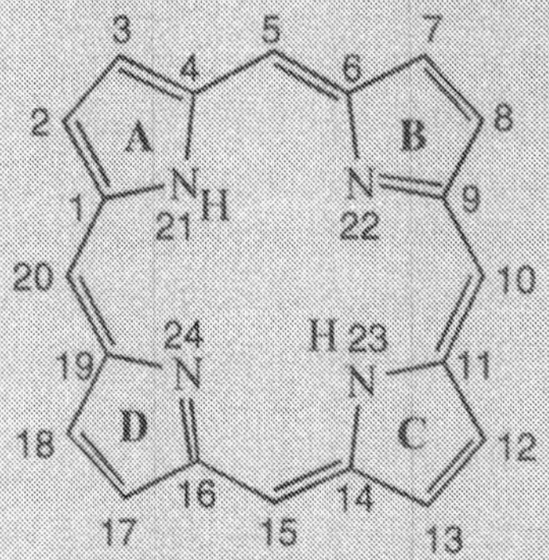

IUPAC Numbering System

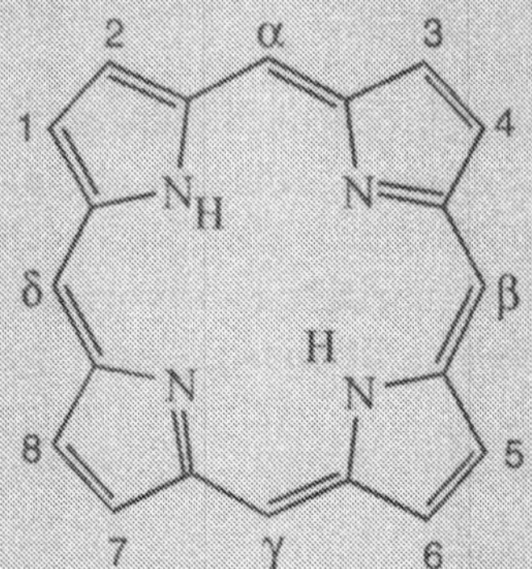

Older (Fischer) Numbering System

Conventionally the imino hydrogens are placed on N-21 and N-23: the four rings are labelled A-D, the "extra" ring in chlorophyll *a*, called the isocyclic ring, being labelled E.

There is an older numbering system due to Hans Fischer. But, as shown above on the right, this numbers only half the positions, and is not convenient for ^{13}C nmr spectroscopy or X-ray analysis, for example. It is gradually falling into disuse and will not be used here.

Another point of nomenclature has been already met (Panel 4.1) in discussing the isomers of uroporphyrin and coproporphyrin. This is referred to as *type isomerism*. The natural porphyrins are derived from porphobilinogen (see Panel 4.1), a pyrrole which has acetic acid (A) and propionic acid (P) residues at the β-positions. There are four ways of arranging four rings of this sort in a macrocyclic array, and, following a practice introduced by Fischer, they are denoted by Roman numerals in the following way:

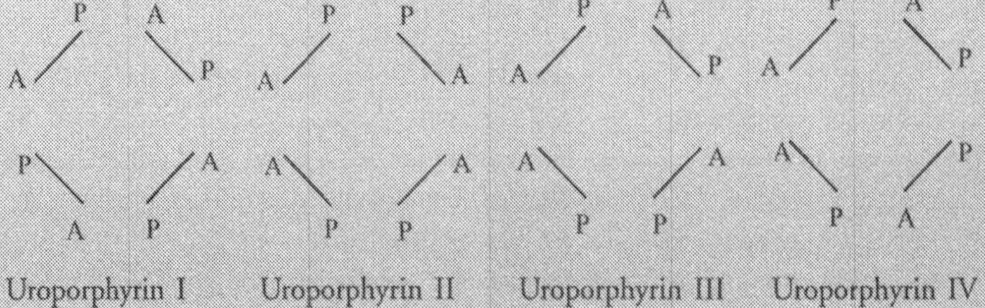

Uroporphyrin I

Uroporphyrin II

Uroporphyrin III

Uroporphyrin IV

(In this shorthand representation the lines represent the ββ′-bonds of the rings A-D). Thus the uroporphyrins and coproporphyrins found in porphyrias are of the I and III types.

Nomenclature in this series is described in detail in G.P. Moss, *Eur. J. Biochem.*, 1988, **178**, 277–328.

rophyll *a*: for example chlorin e_6 (**28**) is derived from chlorophyll *a*. Alternatively the substances may be made synthetically, for example by reduction of the corresponding porphyrin with diimide, N_2H_2. *meso*-Tetraphenylbacteriochlorin (**29**) is an example here.

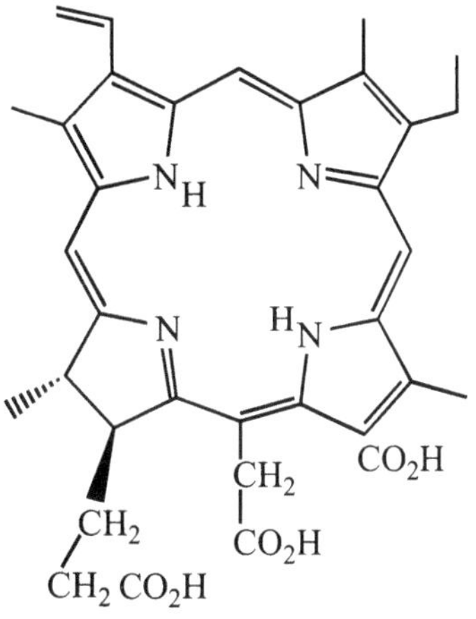

28 Chlorin e_6
λ_{max}(acetone) 400 nm (ε 127 000), 490(9 000), 528(4 600), 608(4 500), 664(41 000).

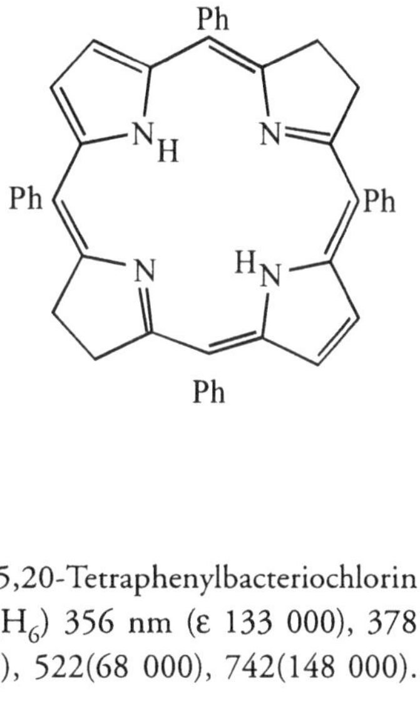

29 5,10,15,20-Tetraphenylbacteriochlorin
λ_{max}(C_6H_6) 356 nm (ε 133 000), 378 (172 000), 522(68 000), 742(148 000).

(viii) *Phthalocyanines*. This is a group of synthetic substances which has a well established commercial importance because of their pigmentary and electroactive properties (section 10.5.1).

Phthalocyanine is a tetraazatetrabenzoporphyrin (**30**), which has a chromophore with an intense band at about 700 nm. The parent substance as its copper(II) complex is used widely as a blue pigment ("Monastral Blue").

The parent compounds and the corresponding zinc complexes are virtually insoluble in water, and for use as photodynamic sensitisers they must be solubilised either by aromatic sulphonation, or by using a liposomal or emulsifying delivery system.

In contrast to the phthalocyanines the naphthalocyanines are not intensely coloured to the eye (yellowish green-grey) because they have their intense lowest energy electronic transition in the near infra red region.

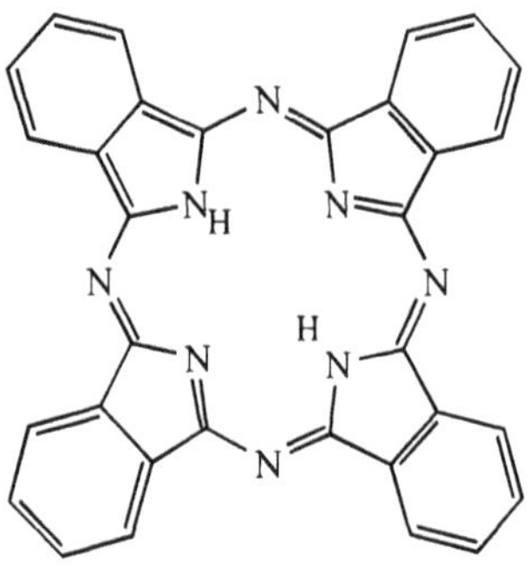

30 Phthalocyanine (Pc)
λ_{max}($C_{10}H_7Cl$) 665nm (ε152 000), 698(162 000)

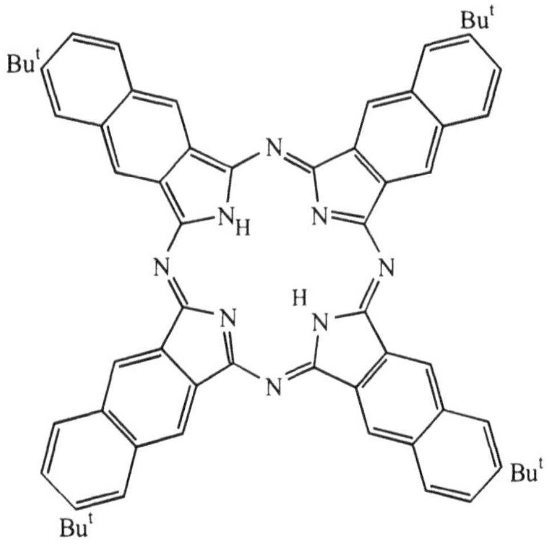

31 Tetra-t-butylnaphthalocyanine (Bu_4NPc)
λ_{max}(PhCl) 365nm (ε7 080),398i(33 100), 700(45 700), 751i(50 100),784(269 000)

4.4 MECHANISM OF PHOTODYNAMIC ACTION

There are two broad mechanistic categories by which light, in the presence of a photosensitiser and dioxygen, can promote chemical reaction in a substrate, including damage to living tissue. They are referred to as Type I and Type II processes. There has been some discussion about the definitions of these two processes; here we shall use the generally accepted one, as outlined in the quotation from Vidòczy at the chapter title.

4.4.1 The Type I Mechanism — Electron Transfer

In this mechanism the excited state generates a radical species, for example by electron transfer from (or to) a substrate, or by hydrogen atom abstraction from a substrate. The radical species then reacts with ground state oxygen so that overall the reaction is a photochemically initiated autoxidation, thus:

$$\mathrm{Sens} \xrightarrow{h\nu} \mathrm{Sens}^{*}$$

$$\mathrm{Sens}^{*} + \mathrm{A} \xrightarrow{e \text{ transfer}} \mathrm{Sens}^{\bullet +} + \mathrm{A}^{\bullet -}$$

$$\mathrm{Sens}^{*} + \mathrm{AH_2} \xrightarrow{\text{H atom transfer}} \mathrm{SensH}^{\bullet} + (\mathrm{AH})^{\bullet}$$

$$(\mathrm{AH})^{\bullet} + {}^{3}\mathrm{O_2} \longrightarrow \mathrm{AH\text{-}OO}^{\bullet} \longrightarrow \text{products}$$

A simple example is provided by the photoautoxidation of a secondary alcohol in the presence of benzophenone as the sensitiser. The overall reaction is:

$$\mathrm{Me_2CHOH} \xrightarrow[\mathrm{Ph_2CO}]{h\nu,\ \mathrm{O_2}} \mathrm{Me_2CO} + \mathrm{H_2O_2}$$

and the mechanism is written:

$$\mathrm{Ph_2CO\,(S_0)} \xrightarrow{h\nu} \mathrm{Ph_2CO\,(S_1)} \xrightarrow{\text{i.s.c}} \mathrm{Ph_2CO\,(T_1)}$$

$$\underset{n \rightarrow \pi^*}{\mathrm{Ph_2CO\,(T_1)}} + \mathrm{Me_2CHOH} \xrightarrow[\overset{\bullet}{\mathrm{H}}\text{ abstraction}]{} \mathrm{Ph_2\overset{\bullet}{C}OH} + \mathrm{Me_2\overset{\bullet}{C}OH}$$

With this free radical initiation step the photochemistry is finished. Subsequent dark steps include a propagation sequence, thus:

$$\mathrm{Me_2\overset{\bullet}{C}OH} + {}^{3}\mathrm{O_2} \longrightarrow \mathrm{Me_2C(OH)\text{-}O\text{-}O^{\bullet}}$$

$$\mathrm{Me_2C(OH)\text{-}O\text{-}O^{\bullet}} + \mathrm{Me_2CHOH} \longrightarrow \mathrm{Me_2C(OH)\text{-}O\text{-}OH} + \mathrm{Me_2\overset{\bullet}{C}OH}$$

which progresses until terminated by a quenching of radicals:

$$Ph_2\dot{C}OH + Me_2C(OH)(O\text{-}O^{\bullet}) \longrightarrow Ph_2CO + Me_2C(OH)(O\text{-}OH)$$

The elimination of hydrogen peroxide from the hydroxyhydroperoxide is a slow thermal process:

$$Me_2C(OH)(O\text{-}OH) \xrightarrow{\Delta} Me_2CO + H_2O_2$$

A biological process which can proceed by a radical pathway is lipid peroxidation. In this case a photochemically generated species abstracts a hydrogen atom from an allylic position of an unsaturated lipid, thus:

$$R \xrightarrow{h\nu} R^{\bullet}$$

$$R^{\bullet} + \text{-CH}_2\text{-CH=CH-} \xrightarrow{-RH} \text{-}\dot{C}H\text{-CH=CH-} \longleftrightarrow \text{-CH=CH-}\dot{C}H\text{-}$$

The allylic radical then reacts with triplet oxygen to give a mixture of hydroperoxides (and their decomposition products):

$$\text{-}\dot{C}H\text{-CH=CH-} \longrightarrow \text{-CH(OOH)-CH=CH-} + \text{-CH=CH-CH(OOH)-}$$

Hence the product mixture is more complex than it is for a singlet oxygen reaction (section 3.3.1).

A particular example of electron transfer is electron transfer to dioxygen, giving the superoxide radical anion, thus:

$$\text{(a)}\quad \text{Sens} \longrightarrow \text{Sens}^*$$

$$\text{Sens}^* + {}^3O_2 \longrightarrow \text{Sens}^{\bullet +} + O_2^{\bullet -}$$

superoxide

$$\text{(b)}\quad \text{Sens}(T_1) + \text{Substrate} \longrightarrow \text{Sens}^{\bullet -} + \text{Substrate}^{\bullet +}$$

$$\text{Sens}^{\bullet -} + {}^3O_2 \longrightarrow \text{Sens}(S_0) + O_2^{\bullet -}$$

Superoxide itself is not particularly reactive, and is readily oxidised back to dioxygen. However the protonated form, the hydroperoxyl radical, undergoes a dismutation which occurs spontaneously but which is also catalysed by widely distributed enzyme systems (superoxide dismutases) to give oxygen and hydrogen peroxide.

$$O_2^{\bullet -} \underset{pK_a\ 4.16}{\overset{H^+}{\rightleftharpoons}} HOO^{\bullet} \xrightarrow{2\,x} H_2O_2 + O_2$$

This process occurs adventitiously as a minor pathway in oxygen transport in erythrocytes. (This is not a photochemical process, of course). Methaemoglobin is formed, and a specific enzyme (methaemoglobin reductase) is present to reduce it back to the iron(II) form. When generated, here or elsewhere in the living organism, the hydrogen peroxide, which is a toxic substance, is enzymatically destroyed. The enzymes most commonly involved are catalases:

$$2H_2O_2 \xrightarrow{\text{catalase}} 2H_2O + O_2$$

or glutathione peroxidases:

$$2GSH + H_2O_2 \xrightarrow{\text{glutathione peroxidase}} GSSG + 2H_2O$$

(where GSH represents the reduced form of the tripeptide glutathione)

However, when hydrogen peroxide is generated in a photochemical process *in vivo* in a situation where catalases (or other peroxidases) are ineffective or absent, then it is conceivable, in the presence of traces of transition metal ions, that hydroxyl radicals will be generated. Such species are very reactive and are expected to promote radical reactions, including autoxidation processes, in biomolecules, thus:

$$H_2O_2 + Fe(II) \xrightarrow{\text{Fenton reaction}} HO^{\bullet} + OH^{-} + Fe(III)$$

$$\text{BiomoleculeH} + HO^{\bullet} \longrightarrow \text{Biomolecule}^{\bullet} + H_2O$$

$$\text{Biomolecule}^{\bullet} + {}^3O_2 \longrightarrow \text{Biomolecule-OO}^{\bullet} \longrightarrow \text{products}$$

4.4.2 The Type II Mechanism — Energy Transfer

In this mechanism electronic excitation energy is transferred from the excited triplet of the sensitiser to triplet dioxygen, to give the sensitiser in its ground state, and singlet oxygen, thus:

$$\text{Sens }(S_0) \xrightarrow{h\nu} \text{Sens }(S_1) \xrightarrow{\text{i.s.c}} \text{Sens }(T_1)$$

$$\text{Sens }(T_1) + {}^3O_2 \longrightarrow \text{Sens }(S_0) + {}^1O_2$$

$$\text{Biomolecule} + {}^1O_2 \longrightarrow \text{products}$$

This mechanism has been discussed in section 3.2.2, and the reactivity of singlet oxygen has been categorised (section 3.3). Amongst the biomolecules which react readily with singlet oxygen are:

(i) unsaturated lipids (cf section 3.3.1)

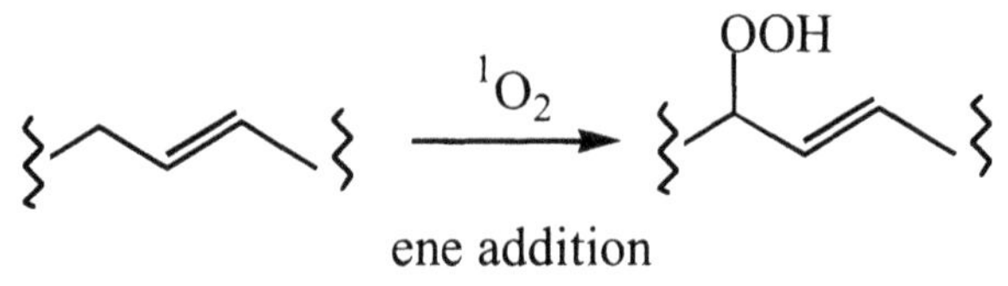

$^1O_2$

OOH

ene addition

(ii) cholesterol, characteristically giving the 5α-hydroperoxide (cf section 3.3.1)

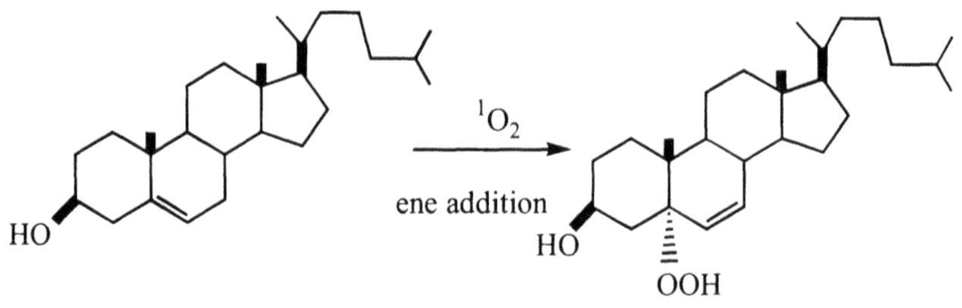

$^1O_2$

ene addition

HO

HO

OOH

(autooxidation gives predominantly the 7α and 7β hydroperoxides)

(iii) α-amino acid derivatives

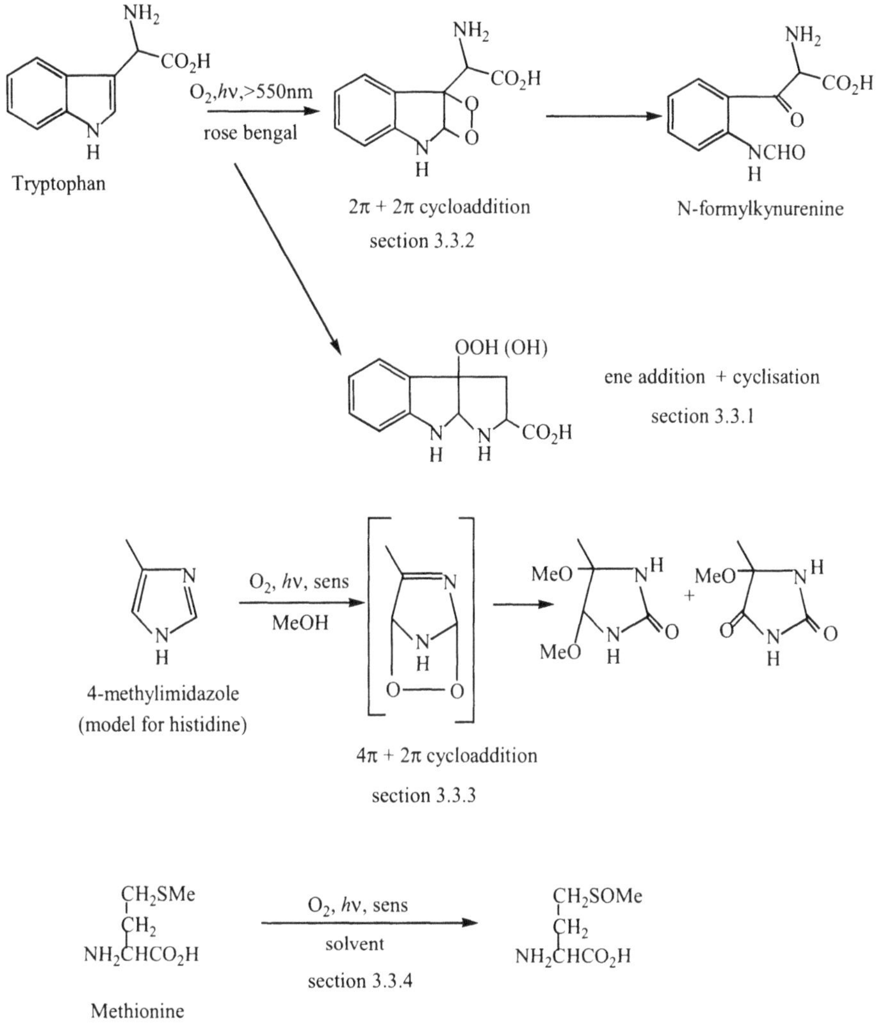

Tryptophan

O_2, $h\nu$, >550nm

rose bengal

$2\pi + 2\pi$ cycloaddition

section 3.3.2

N-formylkynurenine

OOH (OH)

ene addition + cyclisation

section 3.3.1

4-methylimidazole

(model for histidine)

O_2, $h\nu$, sens

MeOH

$4\pi + 2\pi$ cycloaddition

section 3.3.3

CH_2SMe
CH_2
NH_2CHCO_2H

Methionine

O_2, $h\nu$, sens

solvent

section 3.3.4

CH_2SOMe
CH_2
NH_2CHCO_2H

Histidine reacts most rapidly, while, in aqueous methanol, tryptophan actually quenches singlet oxygen faster than it reacts with it. The rates of reaction of some α-amino acids with singlet oxygen generated photophysically (Nd-YAG laser) in D_2O, and by photosensitisation with methylene blue in aqueous methanol, are shown in Table 4.1.

Unsaturated lipids, including cholesterol, and proteins are major components of the membranes of cells, and hence membrane damage is expected to be important in PDT (section 13.2.2).

TABLE 4.1 SECOND ORDER RATE CONSTANTS FOR REACTIONS WITH SINGLET OXYGEN. k_r IS THE RATE CONSTANT FOR CHEMICAL REACTION, k_q IS THE RATE CONSTANT FOR PHYSICAL QUENCHING. ALL VALUES IN UNITS OF $10^7 M^{-1}s^{-1} \pm 30\%$.

$^1O_2$ Source	Nd-YAG laser D_2O		Methylene blue, $h\nu$, H_2O-MeOH	
α-Amino acid	k_r	k_r+k_q	k_r	$k_r + k_q$
Histidine	10	10	0.7	5
Tryptophan	3	5.1	0.4	4
Methionine	1.6	1.7	0.5	3
Tyrosine	0.8	nd	—	—

(I.B.C. Matheson and J. Lee, *Photochem. Photobiol.*, 1979, **29**, 879–881).

(iv) purine bases (eg of DNA,RNA)

Guanine $\xrightarrow[\text{sens, solvent}]{O_2,\ h\nu,}$ Guanidine + CO_2 + Parabanic acid

This reaction is thought to involve both Type I and Type II pathways. The mechanism is not obvious (one possibility is shown in Panel 4.3), but is important because guanine residues appear to be selectively destroyed in DNA (both in solution and in cells) on irradiation with sensitisers. On a similar theme, the photodynamic loss of activity of certain enzymes is associated with the destruction of tryptophan (eg in chymotrypsin), histidine (eg in phosphoglucomutase) and methionine (eg in phosphoglucomutase) residues.

As a result of this reactivity of common biomolecules, the lifetime of singlet oxygen in the cell is much reduced from the 2 μs in water (Table 3.2). Moan and Berg estimate a value of <0.04 μs, with a diffusion pathlength of <0.02 μm (section 12.5.1).

Both Type I and Type II processes may occur together. It is often helpful to describe the situation in terms of a modified Jablonski diagram (see section 2.5), as shown in Figure 4.3, where the left hand side of the diagram refers to the sensitiser (heavy horizontal lines) and the right hand side to dioxygen molecular species.

Tryptophan provides an example where Type I and Type II reactions can give the same products, but in different proportions (Table 4.2). This example is instructive because it shows the close relationship between the two processes.

4.4.3 Distinguishing between Type I and Type II Photooxygenation Processes

The distinction between the two mechanistic types is not easily made, and some of the so-called distinguishing tests (such as lifetime in deuteriated solvents, quench-

Panel 4.3 Photooxygenation of Guanine — Mechanistic Rationalisation

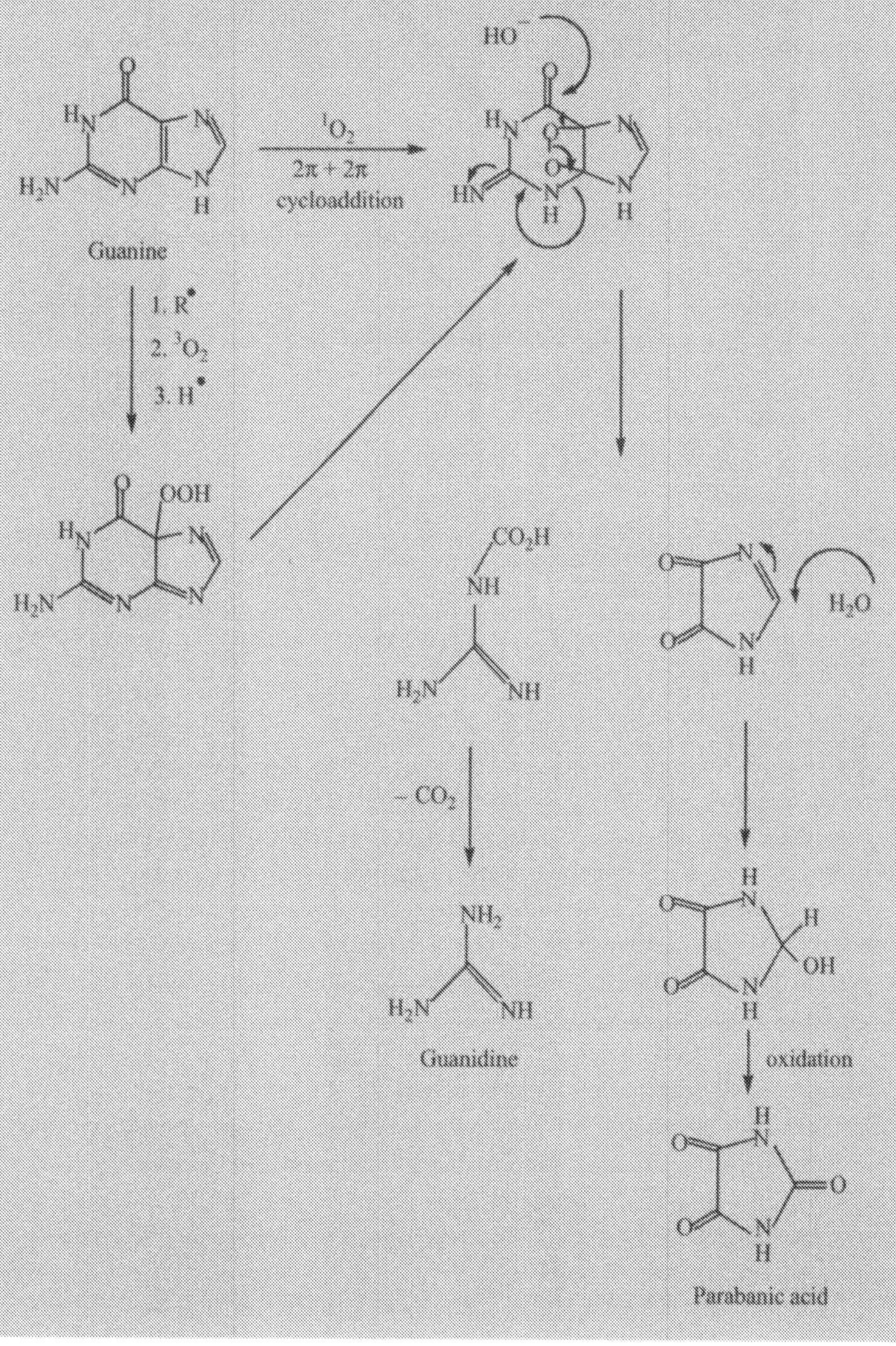

Guanine

$^1O_2$
$2\pi + 2\pi$
cycloaddition

1. R$^{\bullet}$
2. $^3O_2$
3. H$^{\bullet}$

HO^-

H_2O

$- CO_2$

Guanidine

oxidation

Parabanic acid

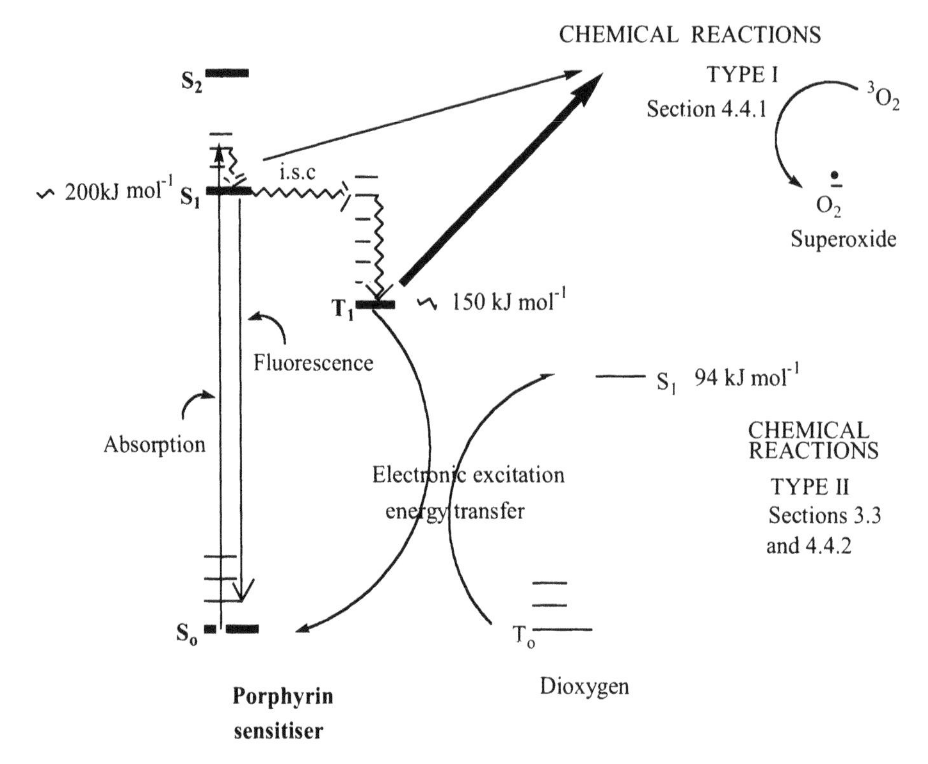

FIGURE 4.3 MODIFIED JABLONSKI DIAGRAM TO SHOW ORIGIN OF TYPE I AND TYPE II PHOTOOXYGENATION PROCESSES.

ing by azide, detection of spin-trapped hydroxyl radical) do not reliably distinguish between singlet oxygen and superoxide intermediates. Moreover, proving a species is present (and minute amounts of radical species can be detected by ESR spectroscopy) does not prove that it is involved in the main pathway of the reaction.

Considerable care is needed both in experiment and interpretation. It seems to me that there are at present simply not enough reliable methods available. To reach a reasonably secure conclusion several lines of experiment need to be explored. The most trustworthy methods appear to be:

(i) luminescence detection for singlet oxygen
(ii) the reaction of cholesterol: it gives as the *main* product(s) the 5α-hydroperoxide (Type II) or a mixture of the 7α- and 7β-hydroperoxides (Type I)
(iii) quenching by BHT, mannitol (Type I)
(iv) quenching by 9,10-diphenylanthracene with formation of the endoperoxide (Type II)
(v) same products and *product distribution* as with an alternative source of singlet oxygen (eg $NaOCl/H_2O_2$; microwave discharge).

In considering the next two sections describing the "tests" that have been employed, these caveats need to be borne in mind.

TABLE 4.2 PHOTOOXYGENATION OF TRYPTOPHAN UNDER VARIOUS CONDITIONS.

Source of $^1O_2$	Conditions	% Product yield A	% Product yield B	Reaction type deduced
Endoperoxide	35°C	23	42	Type II
Rose bengal/O_2	>300 nm	30	1	Type II, + photodecomposition of **B**
Rose bengal/O_2	>500 nm	23	44	Type II
Thionine/O_2	>550 nm	22	23	Type I + Type II

The endoperoxide used was

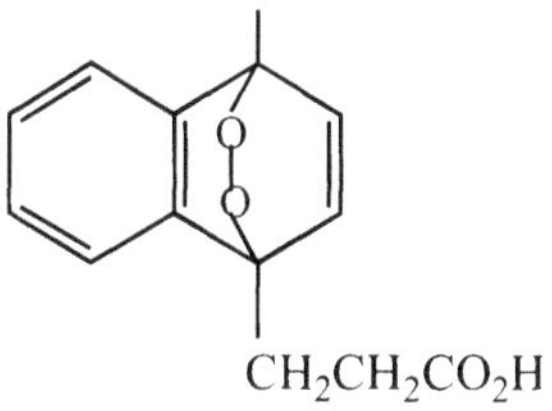

$CH_2CH_2CO_2H$

A is N-formylkynurenine

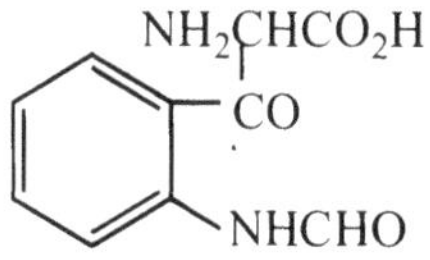

NH_2CHCO_2H CO NHCHO

B is the hydroxy tricyclic product

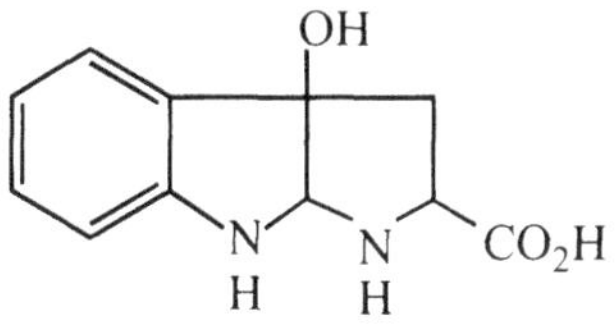

OH N H N H CO_2H

(K. Inoue, T. Matsuura and I. Saito, *Bull. Chem. Soc. Japan*, 1982, 55, 2959–2964)

4.4.3.1 Tests for radical intermediates (Type I mechanism)

Electron transfer processes are in principle detectable by the methods used to demonstrate the existence of free radicals. These include:

(i) quenching of the reaction by the addition of antioxidants such as α-tocopherol or butylated hydroxytoluene (**32**, BHT, 2,6-di-tert-butyl-4-methylphenol). These hindered phenols contain readily abstracted hydrogen atoms giving relatively unreactive resonance stabilised phenoxy radicals, so breaking the autooxidation chain.

R• + [BHT] ——→ RH + [phenoxy radical]

32 BHT

(ii) Inhibition of the reaction on addition of mannitol

R• (•OH) + mannitol ——→ RH (H_2O) + mannitol radical —R•→ RH + keto sugar

tert-Butanol and benzoic acid are also used as chain-breaking agents.

(iii) Electron spin resonance spectroscopy, either directly, or, more usefully, indirectly using the spin-trapping technique.

However, singlet oxygen may interfere with this test, since it is reported to form an adduct with DMPO which has a complex chemistry (Figure 4.4), including the generation of the same spin adduct (of •OH) as formed with a Type I process (see also Figure 4.5).

R• + DMPO ——→ nitroxide radical

5,5-Dimethyl-1-pyrroline-1-oxide (DMPO, spin trapping reagent)

Nitroxide radical characteristic ESR spectrum

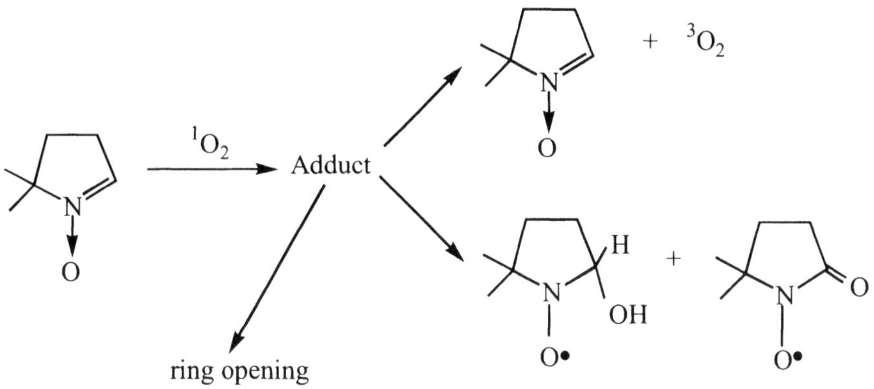

FIGURE 4.4 REACTION OF SINGLET OXYGEN WITH 5,5-DIMETHYL-1-PYRROLINE 1-OXIDE.

(iv) Reactions involving superoxide as an intermediate are quenched by the enzyme superoxide dismutase. Dismutation (background or enzymatic) leads to hydrogen peroxide, which in the presence of iron(II) can generate the hydroxyl radical (section 4.4.1). If this is responsible for the observed effect then catalase (to remove H_2O_2) or sequestering agents (to remove traces of transition metal ions) will reduce the rate.

4.4.3.2 Tests for singlet oxygen (Type II mechanism)

Evidence for singlet oxygen formation, and possibly for a singlet oxygen pathway, may be provided by a combination of factors, as follows:

(i) The sensitiser has a high quantum yield of singlet oxygen production (ϕ_Δ). Table 4.3 shows values for ϕ_T and ϕ_Δ for a range of tetrapyrrolic photosensitisers.

TABLE 4.3 TRIPLET QUANTUM YIELDS (ϕ_T) AND QUANTUM YIELDS OF SINGLET OXYGEN PRODUCTION, (ϕ_Δ), FOR VARIOUS PHOTOSENSITISERS.

	Solvent	ϕ_T	ϕ_Δ
Protoporphyrin dme	PhH	0.80	0.57 (O_2)
Haematoporphyrin	MeOH/H_2O	0.83	0.65
Uroporphyrin III	H_2O	0.93	0.52
TPP	PhH	0.67	0.63
m-THPP	MeOH	0.69	0.46
Photoprotoporphyrin dme, isomer A	PhH	0.65	0.67
m-THPC	MeOH	0.89	0.43
m-THPBC	MeOH	0.83	0.43
Bacteriochlorophyll a	PhH	0.32	0.32
Porphycene	PhMe	0.42	0.30
Zinc(II) phthalocyanine tetrasulphonic acid	MeOH	0.47	0.43
Si naphthalocyanine	PhH	0.39	0.35

Principally from data tabulated by D.J. McGarvey and T.G. Truscott, 'Photodynamic Therapy of Neoplastic Tissue', (ed. D. Kessel), CRC Press, Boca Raton, 1990. Vol. II, pp.179–189.

(ii) The reaction mixture displays a luminescence due to singlet oxygen. There are two sorts of emission, and both are very weak in solution. Luminescence from a single molecule occurs at 1270 nm.

$$^1O_2 \rightarrow {}^3O_2 + h\nu \ (1270 \text{ nm})$$

while two molecules can interact to produce emission at higher energy ("dimol luminescence").

$$2\ ^1O_2 \rightarrow 2\ ^3O_2 + h\nu \ (634,\ 704 \text{ nm})$$

The 1270 nm luminescence arises by a spin forbidden process ($S_1 \rightarrow T_0$), but occurs in a region generally free from complicating luminescence from substrate molecules. Although weak, this luminescence provides perhaps the most reliable direct evidence for singlet oxygen formation.

(iii) The reaction rate as diminished by singlet oxygen quenchers, such as β-carotene, sodium azide, and DABCO (**33**). β-Carotene (**34**) (and other carotenoids) quench by an energy transfer mechanism:

$^1O_2$ + β-Carotene (S_0) E_T <94 kJ mol^{-1} → $^3O_2$ + β-Carotene (T_1) → heat

whereas with DABCO (1,4-diazabicyclo[2.2.2] octane) quenching an electron transfer mechanism is involved. It will be noticed that for one-electron transfer, the intermediate necessarily has to be formulated as a superoxide-like species.

33 DABCO + $^1O_2$ ⇌ $^1[DABCO^{+\bullet} \ldots O_2^{-\bullet}]$ ⇌ $^3[DABCO^{+\bullet} \ldots O_2^{-\bullet}]$ → DABCO + $^3O_2$ + Δ

33 DABCO

Other electron-rich quenchers (amines, phenols) also quench by electron transfer, although chemical modification of the quencher usually occurs as a competing process.

(iv) Chemical traps for singlet oxygen lower the rate of the reaction under study if singlet oxygen is involved. Substances which react particularly rapidly with singlet oxygen, and which therefore constitute excellent traps, include 1,3-diphenylisobenzofuran, 2,5-dimethylfuran and 9,10-diphenylanthracene. The latter appears to be quite specific for singlet oxygen. As a confirmation, the products of singlet oxygen reaction should be identified (section 3.3.3). Bilirubin reacts very rapidly with singlet oxygen. In this case the visible chromophore (λ_{max} 450 nm) of the trap is destroyed, and the trapping process can be readily followed spectroscopically (see also section 5.6).

(v) The products of the photochemical process are also formed in the same proportions from the substrate molecule in a dark reaction with a known singlet oxygen source, such as NaOCl-H_2O_2 (section 3.3.1) or microwave discharge (section 3.3.2).

(vi) Singlet oxygen has a longer lifetime in D_2O than in H_2O (section 3.1). Hence for reactions involving singlet oxygen the rate will be faster in D_2O (and other deuteriated solvents) than in H_2O (and other protiated solvents). This effect will only be seen for substrates of modest reactivity, where deactivation by solvent molecules determines singlet oxygen concentration.

4.4.4 The Overall Mechanistic Picture

As foreshadowed in section 1.1, as an adventitious photobiological process, the photodynamic effect is expected to be mechanistically complex. This has proved to be the case: there is even evidence for the photosensitiser moving from one biological compartment to another on irradiation.The overall conclusion at present is that the predominant mechanism is a Type II process via singlet oxygen, with a contribution from Type I processes. The major targets of singlet oxygen are unsaturated lipids and certain α-amino acid side chains (section 4.4.2). These are important components of membranes, and damage to various cell membranes would be expected to lead to the death of the cell.

The relative importance of the two chemical mechanisms has not generally been settled in quantitative terms, and may in any case be expected to vary from one tissue to another. It has been very difficult to obtain convincing mechanistic evidence from biological experiments. Clear evidence for a Type I mechanism is particularly thin although the crosslinking of certain proteins (spectrin in erythrocytes) under photodynamic conditions has been taken as evidence for radical intermediates. The evidence for a singlet oxygen mechanism is rather better, but even here there is no clear report of singlet oxygen luminescence from cell culture experiments, except in cases where the photosensitiser is on the surface of the cells. Presumably singlet oxygen generated within the cell is rapidly removed by reactive biomolecules which are all around it. However, the most effective photodynamic agents have high ϕ_Δ values *in vitro* (Table 4.3); and there are several reports that singlet oxygen quenchers or traps ameliorate photodynamic action. Thus oral β-carotene (**34**) is used to help sufferers from the photosensitising porphyrias in the summer months; and 1,3-diphenylisobenzofuran protects cells grown in culture from photodynamic damage by m-THPC.

It is clear that this is a topic which will continue to provide a fertile field for mechanistic study. The reaction pathways are dependant on the substrate, the photosensitiser, the solvent (including pH), aggregation and wavelength. *In vivo* the localisation and compartmentalisation of the sensitiser, the drug-light interval, and the possible metabolism of the drug during this period, are all expected to be significant.

In conclusion three examples of increasing biological complexity are provided for purposes of illustration, though it has to be said that not one of them represents a truly *in vivo* situation.

(i) The first example is purely chemical although the substrate is a key biomolecule (J.M. Wessels, C.S. Foote, W.E. Ford and M.A.J. Rodgers, *Photochem. Photobiol.*, 1997, **65**, 96–102). Tris(2,2′-bipyridyl) ruthenium chloride ($Ru[bpy]_3Cl_2$) sensitises the formation of singlet oxygen when irradiated in aerated aqueous solution (ϕ_Δ = 0.18), and electron transfer is negligible.

$$Ru(bpy)_3^{2+} \xrightarrow{h\nu} Ru(bpy)_3^{2+*} \xrightarrow{^3O_2} {}^1O_2 + Ru(bpy)_3^{2+}$$

However when the same photosensitiser is adsorbed onto colloidal particles of antimony-doped tin dioxide, then irradiation leads to the ready transfer of an electron into the conduction band of the semiconductor, leading to the generation of $Ru(bpy)_3^{3+}$.

$$Ru(bpy)_3^{2+} \xrightarrow{h\nu} Ru(bpy)_3^{2+*} \xrightarrow{SnO_2} Ru(bpy)_3^{3+} + SnO_2(e)$$

The latter complex can oxidise tryptophan by electron transfer. When tryptophan concentration is determined by following its fluorescence at 368 nm, it is found that, under equivalent conditions of light absorption, the rate of photooxidation for the Type I process is about 70 times that of the Type II process.

(ii) The second example concerns the photodynamic rupture of red blood cells (*photohaemolysis*). Although the cells are metabolically active, they are not in their natural environment, and biologists quite properly regard these, and, indeed, cell and tissue culture experiments in general, as being *in vitro*. From a chemical point of view, however, the system is much more complex than that encountered in homogeneous solution.

Photohaemolysis leads initially to leakage of potassium ions from the cell, and then to cell rupture with leakage of haemoglobin. There is usually an induction period, especially noticeable for the latter process. When erythrocytes impregnated with bacteriochlorin *a* (**35**, called bacteriochlorin e_6 by Fischer) are suspended in PBS and irradiated with a laser diode source emitting at 757 nm, it is found that both potassium leakage and haemolysis depend on light dose and photosensitiser dose. The effects of additives is shown in Table 4.4. None of the quenching effects is strikingly large, but a modicum of evidence for both Type I and Type II reactions can be found.

This result prompted the photochemical study of a simpler system: this same photosensitiser in phosphate buffer (pH 7.2) in the presence of 5,5-dimethyl-1-pyrroline-1-oxide (DMPO, section 4.4.3.1.iii) to pick up radical intermediates.

TABLE 4.4 EFFECT OF ADDITIVES ON THE PHOTOHAEMOLYSIS OF HUMAN ERYTHROCYTES IN PBS BUFFER AT ROOM TEMPERATURE. PHOTOSENSITISER — BACTERIOCHLORIN *a*, CONTINUOUS ILLUMINATION AT 757 NM.

Additive	Test for	Relative time to 50% haemolysis	Inference
None	—	1	—
Azide	$^1O_2$	1.33	Type II
Histidine	$^1O_2$	1.22	Type II
Tryptophan	$^1O_2$ + R·	1.56	Type I/II
D_2O	$^1O_2$	1.02	[not Type II]?
Mannitol	R·	1.38	Type I
Glycerol	R·	1.25	Type I
Superoxide dismutase	O_2^{-}·	1.09	not O_2^{-}· (?)
BHT	R·/ROOH	1.10	not lipid peroxidation (?)

(M. Hoebeke, H.J. Schuitmaker, L.E. Jannink, T.M.A.R. Dubbelman, A. Jakobs and A. van der Vorst, *Photochem. Photobiol.*, 1997, **66**, 502-508)

ESR experiments showed that the DMPO-OH spin adduct (**36**) was formed. Because the formation of this adduct could be inhibited *but only to a maximum of 50%* by either a singlet oxygen trap (anthracene 9,10-dipropionic acid, reacting by cycloaddition) or by a superoxide quencher (superoxide dimutase), it has been proposed that under these conditions both Type I and Type II mechanisms occur in equal measure (Figure 4.5).

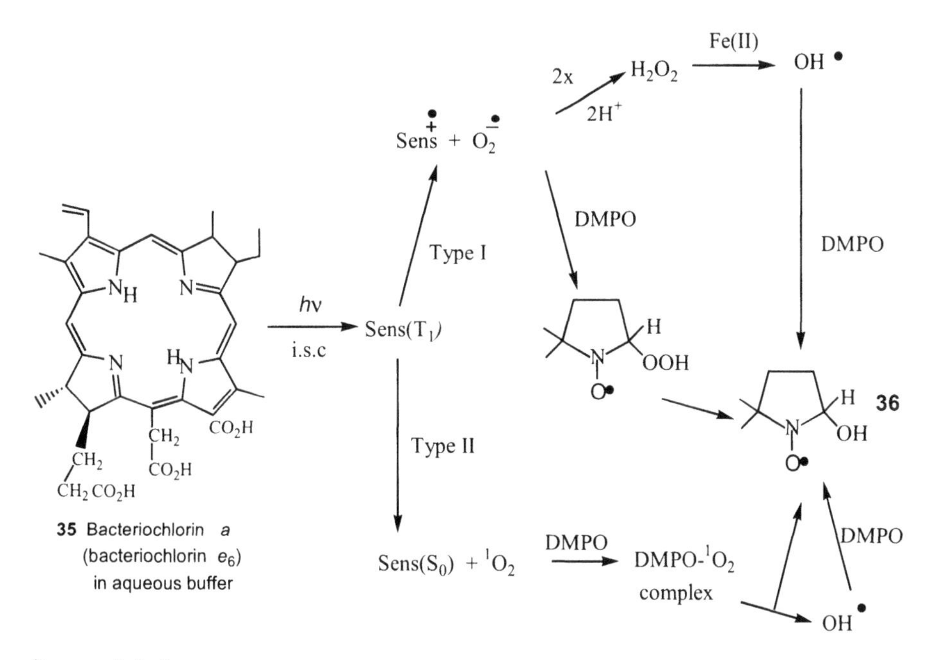

FIGURE 4.5 RATIONALISATION OF PHOTOSENSITISATION BY BACTERIOCHLORIN A IN WATER AS INVOLVING BOTH TYPE I AND TYPE II MECHANISMS.

TABLE 4.5 QUANTUM YIELDS OF SINGLET OXYGEN FORMATION ϕ_{Δ}, AND OF SUPEROXIDE FORMATION ϕ_{SUP}, IN RELATION TO CELL KILL, CELL UPTAKE, AND OCTANOL/WATER PARTITION COEFFICIENT FOR A SERIES OF PROTOPORPHYRIN DIAMIDES (ALL VALUES ARE MEANS ± SE).

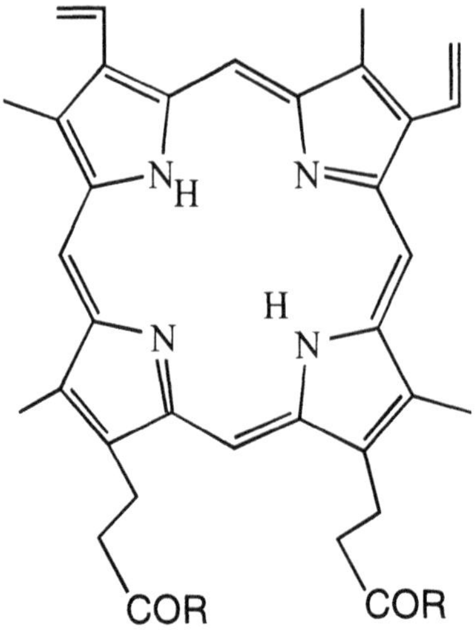

	Experiments *in vitro*			Experiments in melanoma cell culture	
R	ϕ_{Δ}	ϕ_{sup}	P[a] Octanol/H_2O	Do[b] J cm^{-2}	Uptake µg/ml
$NH(CH_2)_2NMe_3I$	0.64±0.03	$8.1\times10^{-6}\pm1.3\times10^{-5}$	4±0.1	0.02±0.01	2.5±0.7
$NH(CH_2)_3NMe_3I$	0.33±0.01	$2.7\times10^{-5}\pm1.9\times10^{-5}$	1.7±0.1	0.07±0.05	1.2±0.1
$NH(CH_2)_3NMe_2$	0.19±0.01	$9\times10^{-5}\pm2.5\times10^{-5}$	17.8±3.9	0.06±0.02	1.5±0.2
[HpD]	<0.01	$6.3\times10^{-5}\pm3\times10^{-5}$	0.7±0.5	0.15±0.05	0.7±0.1
$NH(CH_2)_3CH_3$	<0.01	0	45.2±21.7	0.56±0.15	272±165
NH_2	<0.01	$1.8\times10^{-5}\pm4.8\times10^{-6}$	5.5±0.9	22.7±29.3	2.7±0.5

[a]Partition coefficient between octanol and water: high values indicate high lipophilicity.
[b]Clonogenic cell survival analysed using a single-hit multi-target model with a quenched component: D_o describes the exponential unquenched portion of the curve. Lower D_o means a steeper curve and a greater cell kill.
(A.K. Haylett, F.I. McNair, D. McGarvey, N.J.F. Dodd, E. Forbes, T.G. Truscott and J.V. Moore, *Cancer Letters*, 1997, **112**, 233–238)

(iii) Protoporphyrin may be converted into carboxyamide derivatives which are generally more soluble in water than is protoporphyrin itself. Using solutions in D_2O-PBS (pH 7) the quantum yields of singlet oxygen formation (ϕ_{Δ}, measured from the time-resolved luminescence at 1270 nm, section 4.4.3.2.ii) and superoxide formation ($\phi_{superoxide}$, measured from the ESR signal of the spin adduct of superoxide with 5,5-dimethyl-1-pyrroline-1-oxide, section 4.4.3.1.iii) may be measured for irradiation of the photosensitizer solution at 355 nm. The results are shown in Table 4.5. In some cases ϕ_Δ cannot be measured under the conditions described: this is probably due to a combination of factors — the fact that $^1O_2$ luminescence is very weak, and the aggregation of some photosensitisers, such as HpD which is employed for comparative purposes. Nevertheless the reported $\phi_{superoxide}$ values are all very low, and when comparisons between ϕ_Δ and $\phi_{superoxide}$ can be made, the former is much the higher.

These sensitisers have also been compared with respect to their photodynamic activity against melanoma cells in tissue culture (the cells become depigmented under these conditions). The photosensitisers with appreciable uptake and high ϕ_Δ are the most effective in causing cell kill, and more effective than HpD. For protoporphyrin di(N-butylamide) (Table 4.5, fifth entry) the low ϕ_Δ appears to be balanced by a high uptake value, so that some activity in cell kill is observed. Protoporphyrin diamide, on the other hand, has low ϕ_Δ and uptake values and is virtually inactive. There is no correlation, across this series at any rate, between apparent $\phi_{superoxide}$ and cell kill.

BIBLIOGRAPHY

H.F. Blum, *Photodynamic Action and Diseases Caused By Light*, American Chemical Society Monograph Series No. 85. Reinhold Publishing Corp., New York, 1941. A detailed and perceptive account of photodynamic action as then known.

A.F. McDonagh and D.M. Bissel, 'Porphyria and porphyrinology-the past fifteen years', *Seminars in Liver Disease*, 1998, **18**, 3–15. A very readable summary, slightly popularised, of recent work on the porphyrias.

B.W. Henderson and T.J. Dougherty (eds.), *Photodynamic Therapy — Basic Principles and Clinical Applications*, Dekker, New York, 1992.

CHAPTER 5

SOME OTHER EXAMPLES OF PHOTOMEDICINE

5.1 INTRODUCTION

Before proceeding into the main topic of photodynamic therapy, it is appropriate to consider briefly some other areas where light is used (or has been used) in the treatment of disease. Historical and general descriptive aspects of some of these topics have already been introduced in section 1.2. Here we shall concentrate on molecular aspects, including chemical mechanisms where they are known or can be inferred.

5.2 SUNSCREENS

5.2.1 Sunlight and the Skin

The surface of the human body (with the mouth shut!) is composed entirely of protein-rich materials — skin, hair, nail: it is this surface, and essentially the skin, that is the entry port of photons from the sun. Those entering the eye initiate a visual function: this section is about what happens to the rest.

Figure 5.1 shows the structure of the skin in a schematic way, and defines the terms used to refer to the various layers. Whereas the dermis (~125 to 2000 μm from the surface) is supplied with blood vessels, the epidermis (0 to ~125 μm) is not. The epidermis consists of layers of cells (keratinocytes) which originate in the basal cell layer, and which over the course of about two months travel to the surface. They gradually become flattened, die and flake off (the flakes, incidentally, contributing appreciably to domestic dust, and the sustenance of the house mite). Thus the surface is constantly being renewed from below. Scattered among the basal cells are specialised cells called melanocytes, responsible for melanin production and tanning; while within the stratum malpighii other specialised cells (prickle cells) occur. In contrast to the epidermis, the dermis is much thicker, and is rich in connective tissue and fibres (largely collagen) which give the skin its mechanical toughness. The subcutaneous tissue below the dermis is lipid rich, and seems to act as a shock absorbing layer.

The radiation penetrating through the skin is attenuated by three factors — reflection, scattering and absorption. Scattering occurs especially from the outer dead stratum corneum layer.

The radiation typically comes from the sun (the emission from which resembles that of a black-body source at 5900 K), 99% of the total irradiance (1390

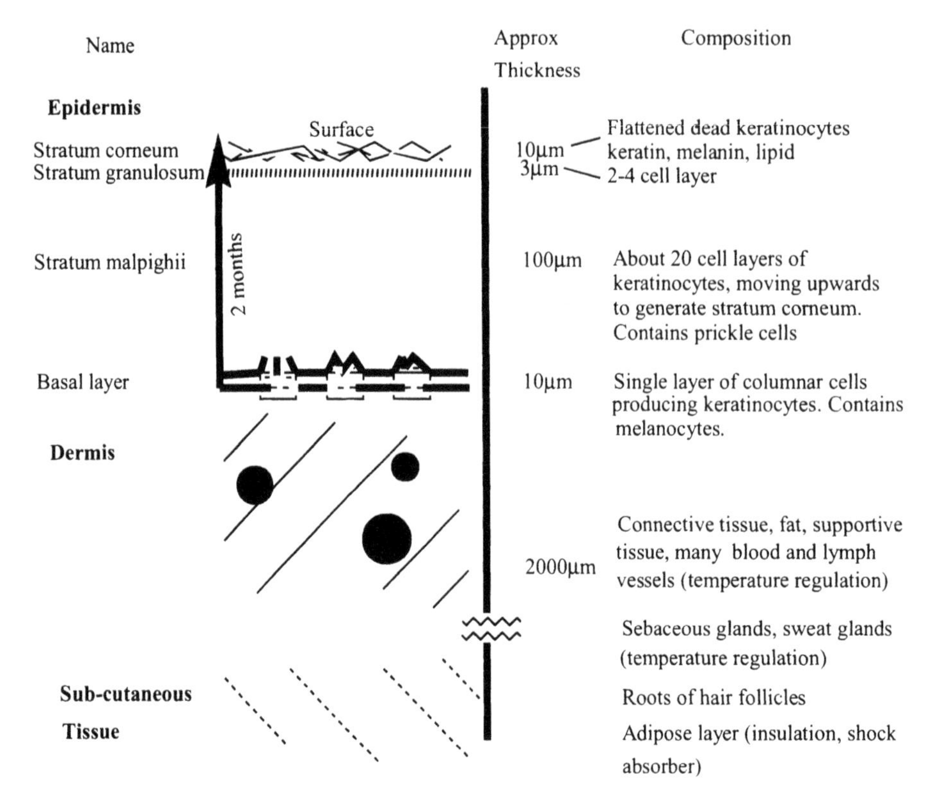

FIGURE 5.1 SCHEMATIC DIAGRAM OF THE STRUCTURE OF HUMAN SKIN.

W m^{-2}) falling in the wavelength range 220 nm to 11000 nm (Figure 5.2). Fortunately the light reaching the skin at sea-level has been filtered through the atmosphere. Carbon dioxide and water vapour absorb various bands in the near infrared (see the various absorption bands — pointing downwards — between *ca.* 700 nm and 2000 nm in Figure 5.2). However, from the biological point of view the most significant absorption is that of the ozone layer (O_3, λ_{max} 251 nm), which essentially cuts out the very damaging UV-C. Thus, solar radiation at sea level starts at about 290 nm (rather than at ~220 nm, where it would start if there were no ozone layer, Figure 5.2) and stretches, through the visible, well into the infrared.

Solar irradiation of the human skin causes three effects: erythema, tanning and thickening. The irradiation between about 290 nm and 320 nm (ie in the UV-B region) is responsible for the erythema which is commonly called sunburn. The effect varies from one individual to another, but is generally biphasic: an immediate reddening appears during exposure and diminishes shortly after, only to reappear after about 2 hours and peak at about 15 hours, finally with macroscopic flaking of the dead surface layer. UV-A by itself does not appear to cause the effect, but enhances the biological response of UV-B when both are present (as they are in sunlight).

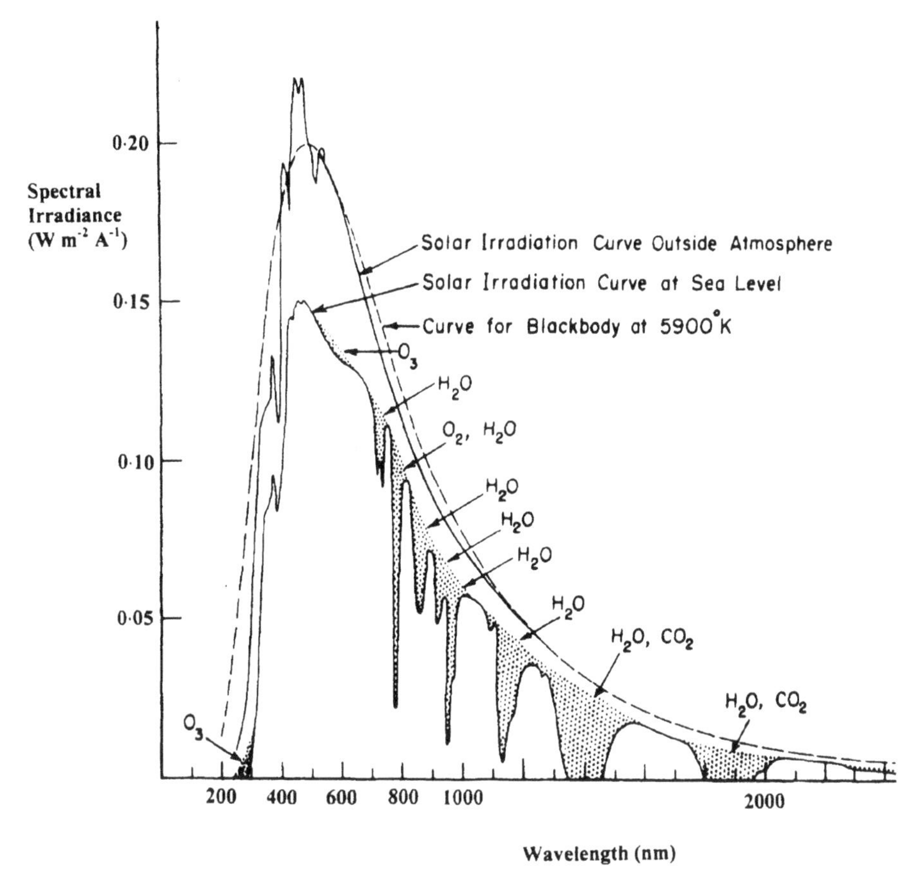

FIGURE 5.2 SOLAR RADIATION ABOVE THE ATMOSPHERE AND AT SEA LEVEL. SHADED AREAS INDICATE ATMOSPHERIC ABSORPTION AT SEA LEVEL DUE TO THE MOLECULES SHOWN. REPRODUCED BY PERMISSION OF MCGRAW-HILL FROM P. R. GAST, "SOLAR ELECTROMAGNETIC RADIATION", SECTION 16.1, IN *HANDBOOK OF GEOPHYSICS AND SPACE ENVIRONMENTS,* (EDITED BY S.L. VALLEY), MCGRAW-HILL, NEW YORK, 1965.

Sunburn arises by damage to the epidermis, and to a lesser extent, the dermis (vasodilation). Although the action spectrum (Figure 5.3) is well documented, the basic photochemistry of the effect is not really understood. It is supposed that substances absorbing in the UV-B region [end absorption by DNA (eg guanine) and protein (eg tryptophan, tyrosine)] trigger the effect.

Light in the sunburn region, and also at longer wavelengths though less efficiently, stimulates a protective process, the increase in melanin synthesis in the melanocytes. Melanin is a polymer produced by the enzymatic oxidative condensation of 3,4-dihydroxyphenylalanine (DOPA). It is produced initially at or near the basal layer, but is delivered into the stratum malpighii. The production of effective new melanin is a slower process (2–20 days) than is sunburn, so it essentially helps to protect against second and subsequent exposures, and builds up further with each incident.

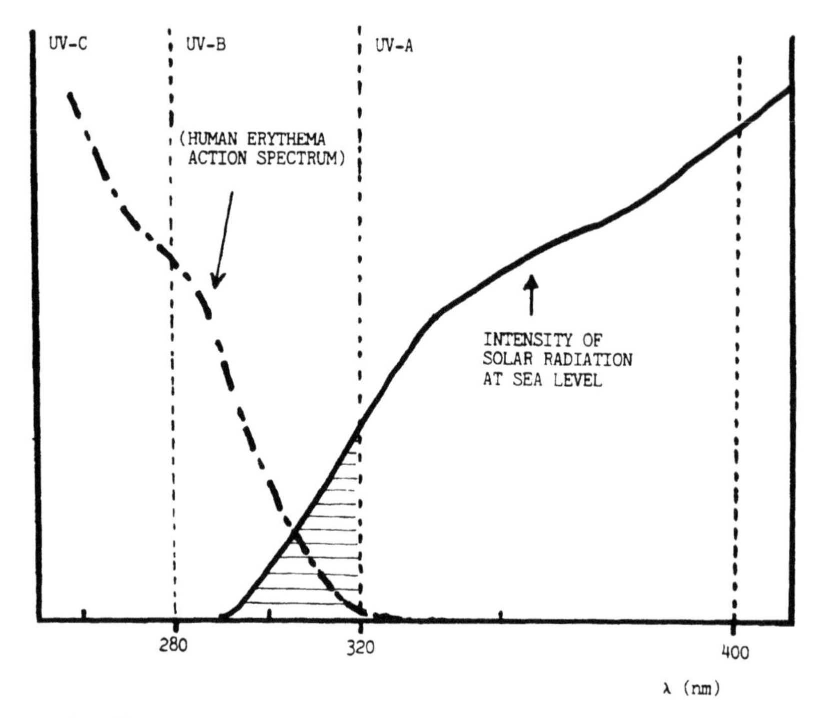

FIGURE 5.3 DIAGRAM TO SHOW ACTION SPECTRUM FOR SKIN ERYTHEMA (SUNBURN) –·–·–·; AND THE INTENSITY OF SOLAR RADIATION AT SEA LEVEL. AN EFFECTIVE SUNSCREEN HAS TO REMOVE THE LIGHT IN THE SHADED AREA (MAINLY UV-B).

5.2.2 Artificial Sunscreens

Melanin is a natural sunscreen. Artificial sunscreens seek to prevent sunburn by removing the incident radiation in the 290–320 nm region — the shaded area in Figure 5.3.

Artificial sunscreens are an important commercial item. They are basically of two sorts:

(i) *Physical sunscreens.* These employ finely divided inert particles principally to reflect and scatter the incident radiation. (Melanin is formed on particulate structure called melanosomes which essentially act in this way). Artificial examples are usually opaque oxides, such as TiO_2, ZnO or talc applied in a cream.

(ii) *Chemical sunscreens.* These act principally by absorbing the radiation, and converting it to heat. They are generally organic substances with strong absorption in the UV-B/A such as anthranilates, benzophenones, cinnamates, salicylates, and dibenzoylmethanes. In many cases, ϕ_f is low, and ϕ_{isc} and ϕ_T are high: luminescence is not observed since at room temperature the T_1 state reverts to S_0 with the liberation of heat.

Many of these substances, or combinations of substances, are designed so that while the UV-B is absorbed, enough UV-A gets through to start the photogeneration of melanin ie tanning. In some cases the process is facilitated by photoisomerisation.

Thus 2-hydroxybenzophenone (**37**) does not fluoresce or phosphoresce: the S_1 state undergoes efficient intersystem crossing to give the triplet, which as a carbonyl $n \rightarrow \pi^*$ excited state rapidly abstracts a hydrogen atom from the hydroxyl group, giving the photoenol (**38**). This is an energy-rich tautomer of the original compound, and it slowly reverts to it, liberating heat, and regenerating the sunscreen molecule for another cycle.

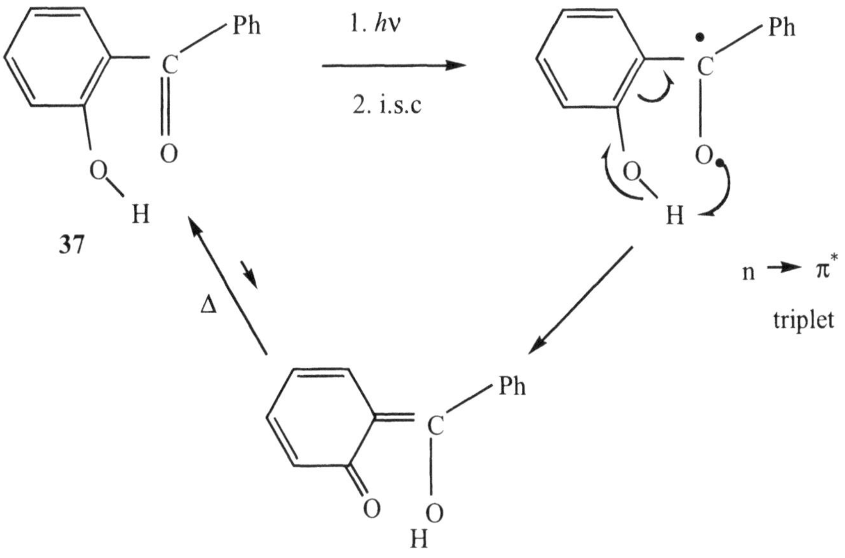

1. $h\nu$
2. i.s.c

$n \rightarrow \pi^*$
triplet

Δ

37

38

Urocanic acid (**39**, λ_{max} 277 nm) occurs naturally in the skin, and is thought to play a photoprotective role using a *trans-cis* photoisomerisation to absorb light, followed by a slow, dark isomerisation to regenerate the thermodynamically more stable *trans*-isomer.

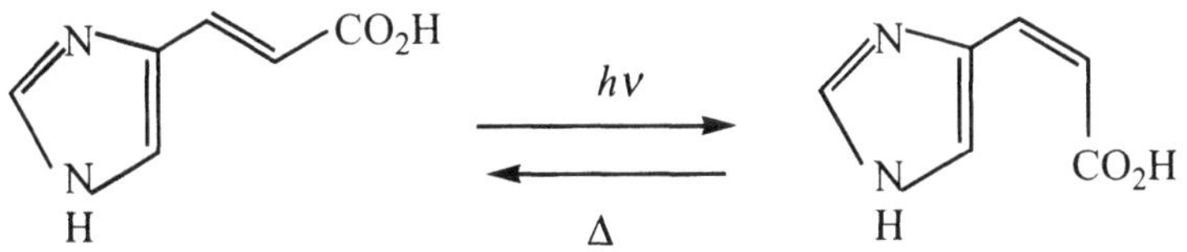

$h\nu$

Δ

39 Urocanic acid

5.3 SKIN CARCINOGENESIS

Three skin cancers are relevant here: basal cell carcinoma (BCC), squamous cell carcinoma (SCC), both arising from keratinocytes, and cutaneous melanoma, arising from malignant transformations of the melanocytes. There is strong epidemiological evidence that exposure to sunlight causes skin cancer in man. Thus, skin cancers occur predominantly on the exposed parts of the body, they are more common in those parts of the world that have the most sunshine, and light-skinned people are more susceptible than are dark-skinned. The effect of sunlight seems to be additive, and again the 290–320 nm region of sunlight is most effective. Sunlight seems to be responsible, in part at least, for squamous cell carcinoma and basal cell carcinoma: the photochemical generation of melanoma is more controversial, as is the extent to which sunscreen agents (section 5.2) give protection against the onset of skin cancer.

In experimental animals, it has been clearly shown that both solar and artificial UV irradiation causes cancer of the skin.

The principal substances which occur naturally in the skin and which absorb (including end absorption) in the UV-B region include the purine and pyrimidine bases of DNA (λ_{max} ~ 260 nm), certain α-amino acid side chains (eg tyrosine, tryptophan, histidine) in protein (λ_{max} ~ 275 nm), melanin, urocanic acid (**39**, λ_{max} ~ 277 nm), haemoglobin and the provitamins D (section 5.4). Melanin is a natural physical sunscreen (section 5.2): and urocanic acid probably acts as a natural chemical sunscreen (section 5.2.2).

It appears that the photochemical processes responsible for UV-induced skin carcinogenesis involve DNA photoreactions in some way. Absorption by cells at ~260 nm includes $\pi \rightarrow \pi^*$ absorption by purines and pyrimidines in DNA, and the end absorption of this band extends into the UV-B. Panel 5.1 shows some of the photoreactions of the DNA bases. Phosphorescence studies show emission from the triplet of thymine, and irradiation of DNA leads to cyclobutane dimers and to oxetane dimers (Panel 5.1) causing irreversible chemical change to the genetic material. Mechanisms exist in living things for the repair of damaged DNA: in humans a dark repair mechanism (excision repair) is the most important. Where this repair mechanism is inefficient, the individual is particularly susceptible to skin cancer (xeroderma pigmentosum).

5.4 PHOTOTHERAPY OF RICKETS (OSTEOMALACIA)

The background to this subject has already been outlined in section 1.2.1.

Amongst the family of D vitamins, two compounds are of especial importance and may be taken to exemplify the characteristic photochemical generation of vitamin D activity, which in ricketic patients leads to an increase in calcium concentration in plasma and a stimulation of apatite deposition in bone. These two compounds are ergosterol (**40**), a steroid which occurs in yeasts and fungi, and to a lesser extent in higher plants; and 7-dehydrocholesterol (**41**) which occurs in fish oils. In nutritional terms, the recommended daily intake of vitamin D is 5–10 µg: herring contains about 25 µg/100g, and milk about 0.03 µg/100g. Vitamin D_2 (from the irradiation of ergosterol) is used to enrich foodstuffs (eg milk, margarine) where the diet is deficient. However 7-dehydrocholesterol (**41**) (but not vitamin D_2, also called ergocalciferol) is also biosynthesised by humans, and to the extent that the nutritional need is met in this way, it is not a true vitamin. 7-Dehydrocholesterol is an intermediate on the biosynthetic pathway to cholesterol: it is formed in the sebaceous glands, exudes on to the skin surface, and is then taken up by the epidermis and dermis. It does not appear to accumulate elsewhere.

Panel 5.1 Photochemistry of DNA bases

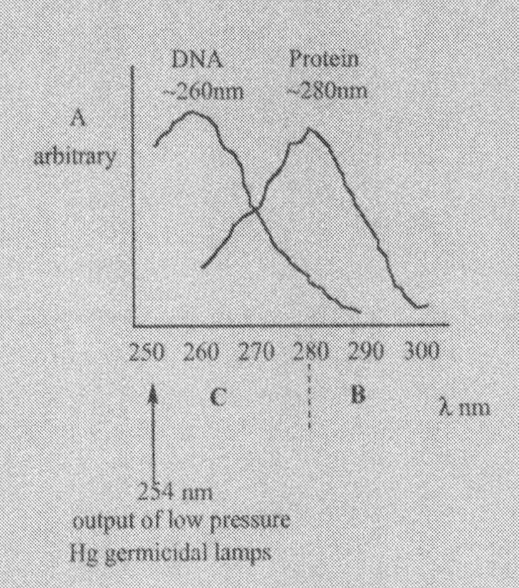

DIAGRAM TO SHOW ABSORPTION OF DNA AND PROTEIN IN THE UV REGION

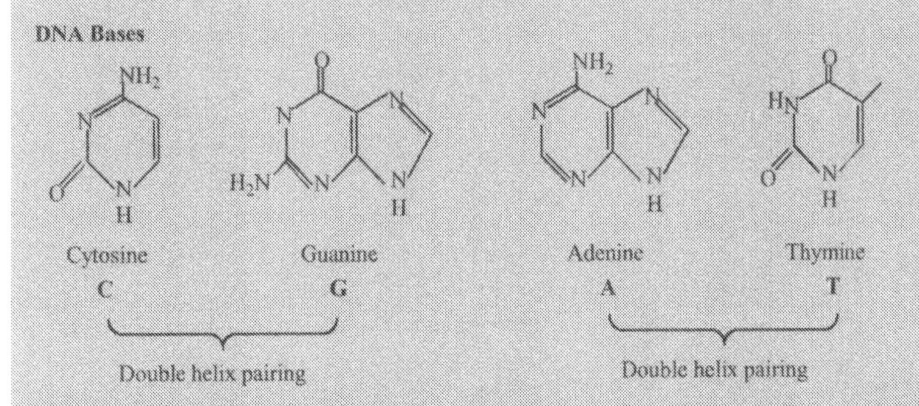

Triplet energies decrease in the series:

C > G > A > T

and as a consequence phosphorescence is usually detected from thymine triplets. Much of the photochemistry that is known involves pyrimidines: photoadducts and pyrimidine dimers, as shown next.

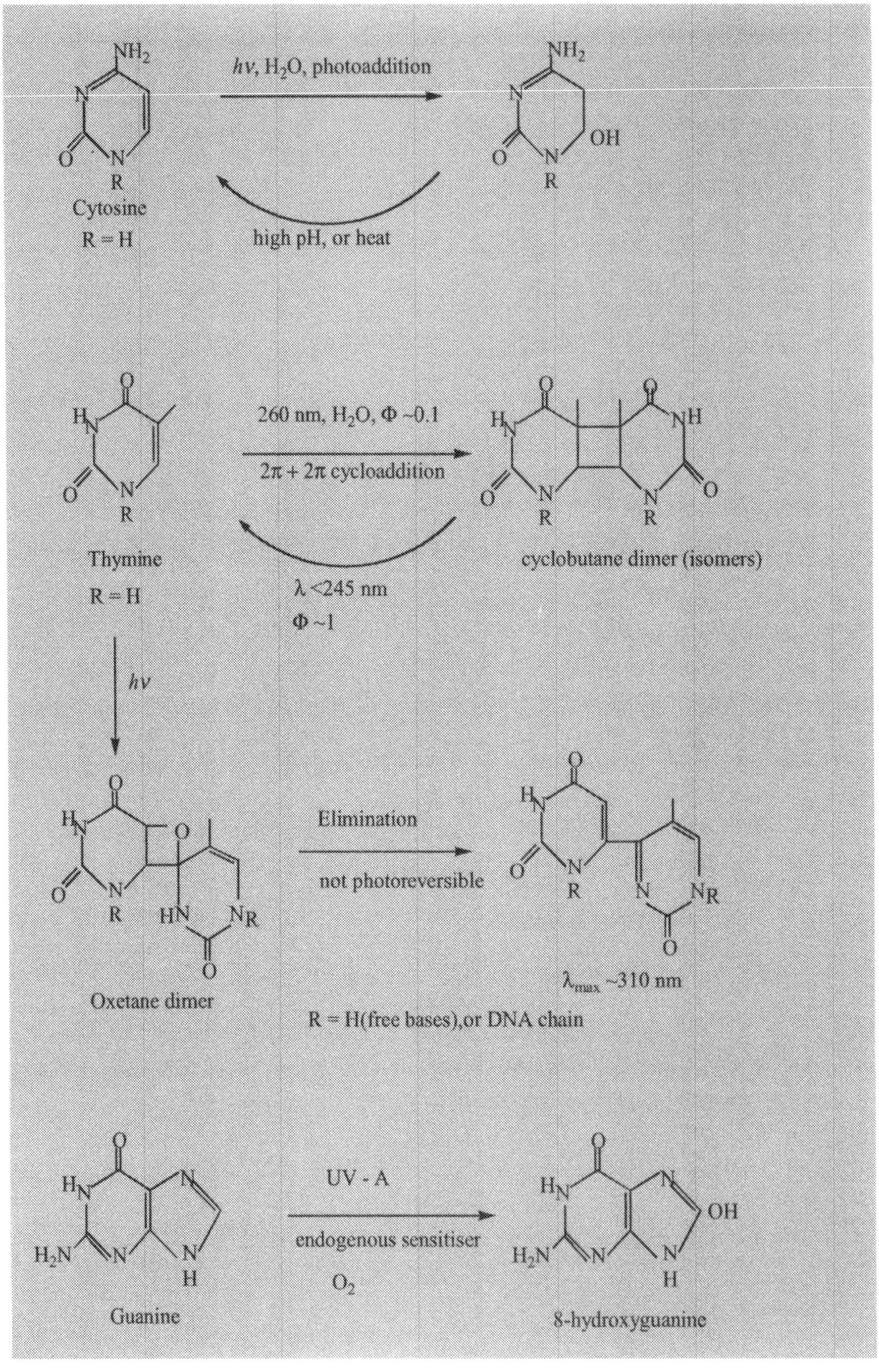

Cytosine
R = H

$h\nu$, H_2O, photoaddition

high pH, or heat

Thymine
R = H

260 nm, H_2O, Φ ~0.1

$2\pi + 2\pi$ cycloaddition

λ <245 nm
Φ ~1

cyclobutane dimer (isomers)

$h\nu$

Oxetane dimer

Elimination

not photoreversible

λ_{max} ~310 nm

R = H(free bases),or DNA chain

Guanine

UV - A

endogenous sensitiser

O_2

8-hydroxyguanine

Both ergosterol (**40**) and 7-dehydrocholesterol (**41**), which are also known as provitamins D_2 and D_3, respectively, on irradiation at about 280–310 nm undergo photochemical processes leading to vitamin D. Ergosterol gives vitamin D_2, while 7-dehydrocholesterol gives vitamin D_3. The two compounds have the same chromophore, and differ only in the nature of the alkyl substituent at C-17, and hence the photochemical and dark reactions will be discussed here in terms of 7-dehydrocholesterol **(41).**

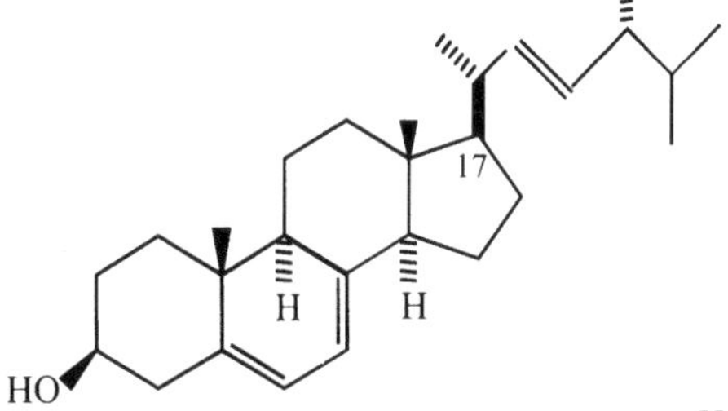

40 Ergosterol

(Provitamin D_2)

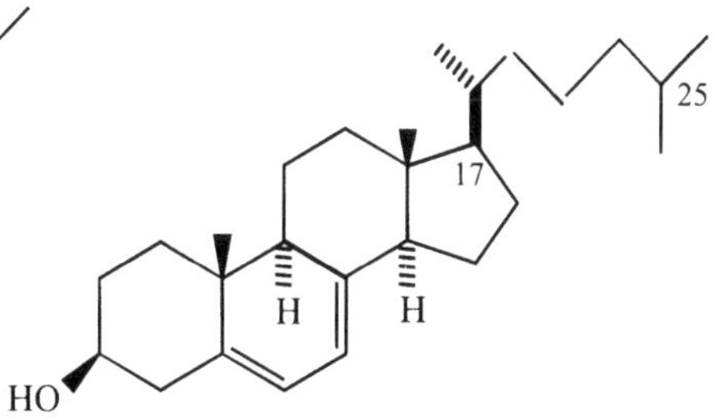

41 7-Dehydrocholesterol

(Provitamin D_3)

In chemical terms 7-dehydrocholesterol (**41**) undergoes a series of photochemical reactions, shown in Figure 5.4, when irradiated with light at about 280–305 nm (optimum 296 nm). The photochemical reaction is of the direct type ie no photosensitiser is involved. Pre-vitamin D_3 (**42**) is central to this process: according to the Woodward-Hoffmann rules on the conservation of orbital symmetry in concerted reactions, the reversible concerted cyclisation of the 6π-hexatriene system (**42**) is an allowed process photochemically and is constrained to occur in the conrotatory mode. Thus reclosure by reverse reaction gives (**41**) back again, while electrocyclisation in the alternative conrotatory sense gives lumisterol$_3$ (**43**). This is shown in terms of the symmetry of the controlling molecular orbital (ψ_4) in Figure 5.5. A photochemical *cis-trans* isomerisation leads reversibly from previtamin D_3 to tachysterol$_3$ (**44**). However, the vitamin itself, vitamin D_3 (**45**), also called cholecalciferol, is formed from pre-vitamin D_3 (**42**) in a thermal step, which in chemical terms is a sigmatropic [1,7] hydrogen shift. As mentioned earlier, besides being formed in this way photochemically, vitamin D_3 also occurs as such in fish liver oil (eg cod, tuna), and so there is a direct dietary input of this vitamin.

The situation in biological terms is described in Figure 5.6, which should be considered in conjunction with the more detailed diagram of skin structure given earlier (Figure 5.1). 7-Dehydrocholesterol in the epidermal layer is converted by ultraviolet light (UV-B region) to previtamin D_3: this is changed in a slow thermal reaction, mainly in the malpighian and basal layers, into vitamin D_3. Diffusing to the vascular system, the vitamin is then carried around the blood stream by a specific vitamin D_3 binding protein in the plasma. Experiments *in vivo* (rat skin irradiated with 0.2 J cm^{-2} at 253–400 nm) have shown that the major photoproduct is

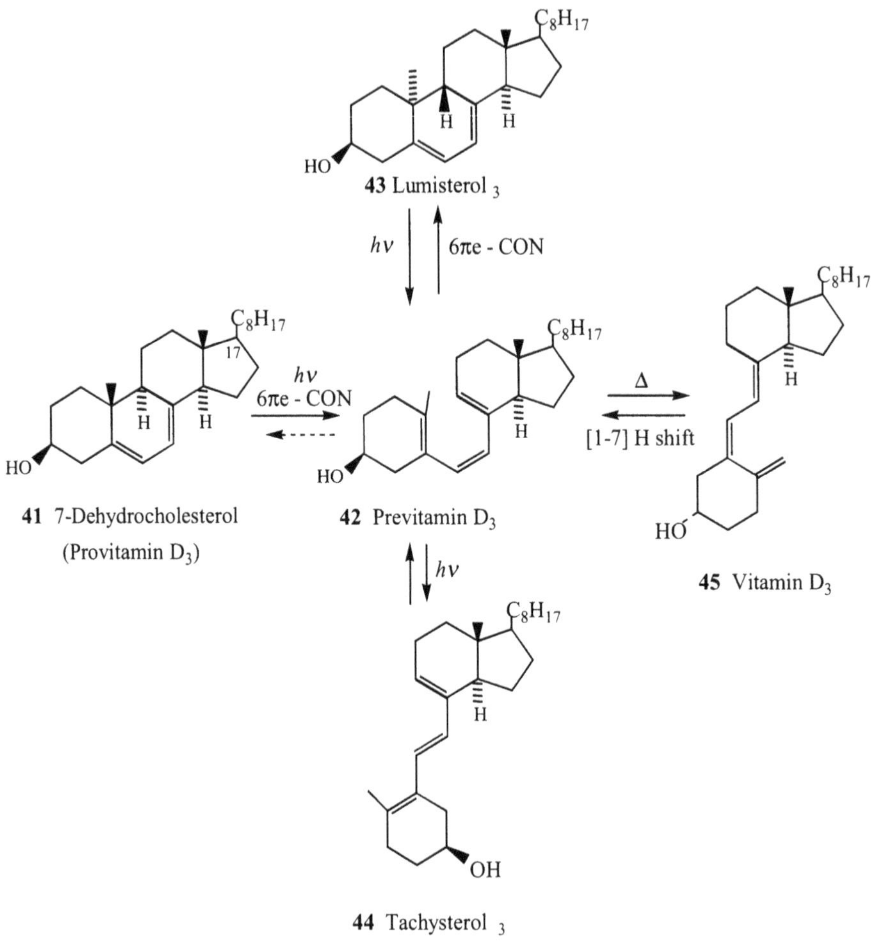

C_8H_{17}

HO

43 Lumisterol $_3$

$h\nu$ 6πe - CON

41 7-Dehydrocholesterol (Provitamin D_3)

$h\nu$ 6πe - CON

42 Previtamin D_3

Δ

[1-7] H shift

45 Vitamin D_3

$h\nu$

44 Tachysterol $_3$

FIGURE 5.4 INITIAL PHOTOCHEMICAL REACTIONS OF 7-DEHYDROCHOLESTEROL (41). ERGOSTEROL (40) UNDERGOES ANALOGOUS CHANGES LEADING TO THE D_2 SERIES OF COMPOUNDS.

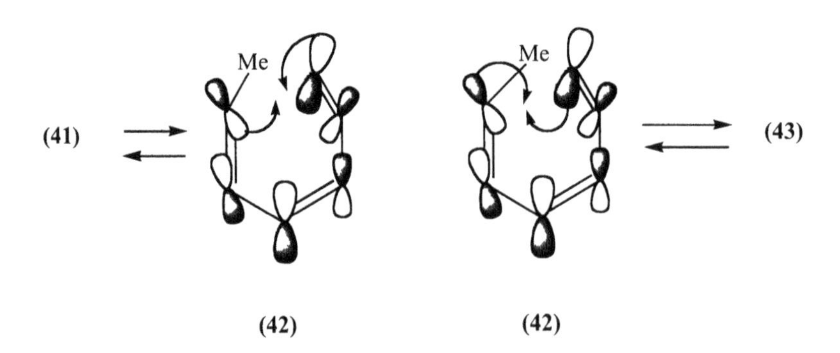

FIGURE 5.5 CONCERTED PHOTOCYCLISATION OF PREVITAMIN D_3 (42) SHOWING CONROTATION OF THE HOMO OF THE EXCITED STATE OF THE TRIENE TO FORM A NEW σ-BOND. CONROTATION IN ONE SENSE GIVES 7-DEHYDROCHOLESTEROL (41); AND, IN THE OTHER SENSE, ITS DIASTEREOISOMER, LUMISTEROL$_3$ (43).

previtamin D_3, neither lumisterol$_3$ nor tachysterol$_3$ being detected under these conditions. The molecular mechanism of this streamlining of the pathway *in vivo* is not obvious, since the photoisomers (**43**) and (**44**) are readily detected *in vitro.*

Since vitamin D_3 (like previtamin D_3) is a conjugated triene, it will also undergo photochemical rearrangements: amongst the products are the isomeric suprasterols$_3$ I and II in which electrocyclisation reactions have occurred to give compounds which, having only one double bond, do not absorb in the near UV region. They do not have vitamin D reactivity. Hence in commercial photochemical syntheses of the D vitamins it is important to carry out the photochemical reaction of the

(**45**) Vitamin D_3

(drawn as 6,7-cis conformation)

antarafacial addition to 7,8-double bond

Suprasterol $_3$ I

Suprasterol $_3$ II

provitamin in the cold, so that thermal formation of the vitamin can follow as a separate dark step. And in the biological situation (Figure 5.6) it is necessary for the previtamin D_3 to be transported away from the epidermis and into the dermal layer for the thermal (38°C) transformation to vitamin D_3 in a region of reduced light intensity.

Vitamin D_3 is not the end of the story. Subsequent hydroxylation involving a cytochrome P_{450} enzyme occurs in the liver to give 25-hydroxy vitamin D_3 which is further hydroxylated in the kidney to give 1α, 25-dihydroxy vitamin D_3 (**46**) and 1α, 24(*R*),25-trihydroxy vitamin D_3 (**47**). These both stimulate calcium absorption by the intestine, while the former, which appears to be the key metabolite, serves as the calcium mobilising hormone and phosphate transport hormone in bone formation.

46 1α,25-Dihydroxyvitamin D_3 (Calcitriol)

47 1α, 24(*R*), 25-Trihydroxy vitamin D_3

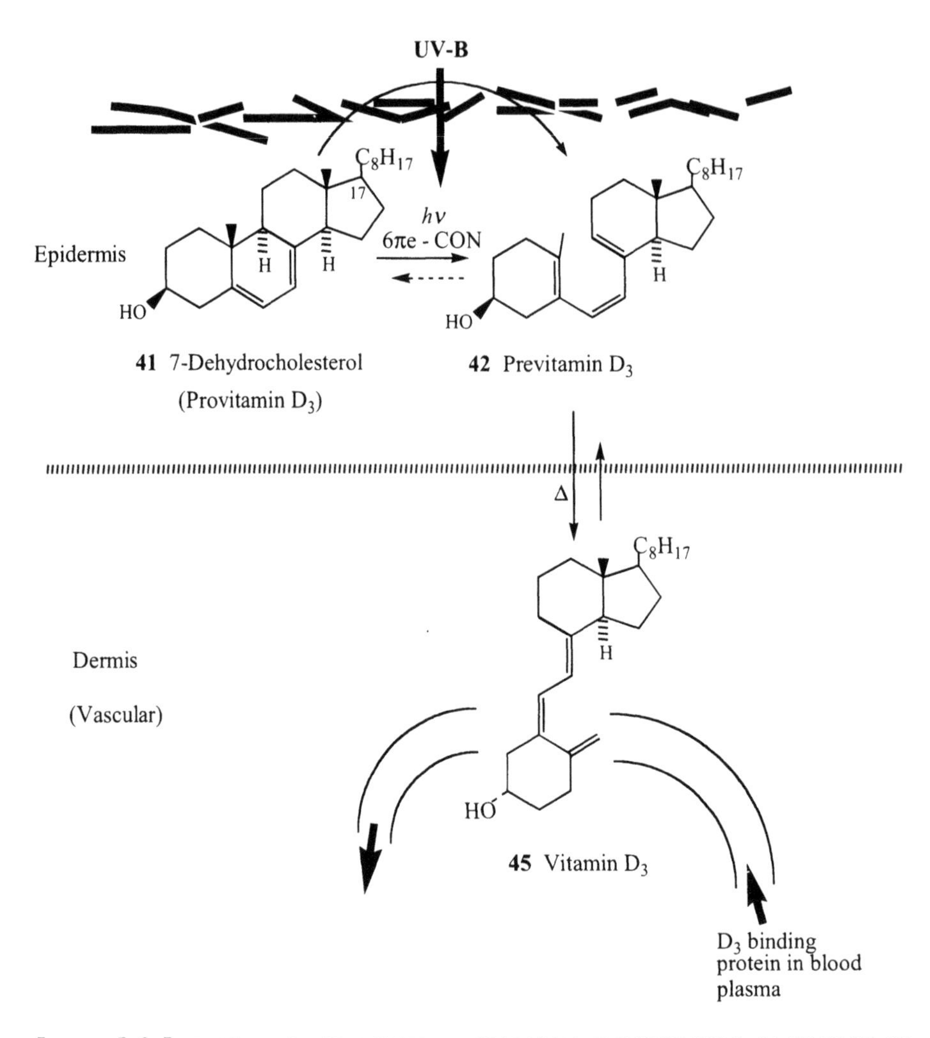

FIGURE 5.6 PHOTOTHERAPY OF RICKETS: MACROSCOPIC INTERPRETATION IN RELATION TO MOLECULAR CHANGES.

5.5 PSORIASIS

A general description of psoriasis and its photochemotherapy, with a brief historical background, has been provided in section 1.2.2. Because of the care needed with this treatment to safeguard against the problems involved (short term severe erythema, ageing of the skin, skin carcinogenesis) it is usually available only in specialised centres.

5.5.1 Photosensitisers for PUVA

The photosensitisers used in treating psoriasis are typically furocoumarins, and they fall into two classes, linear and angular.

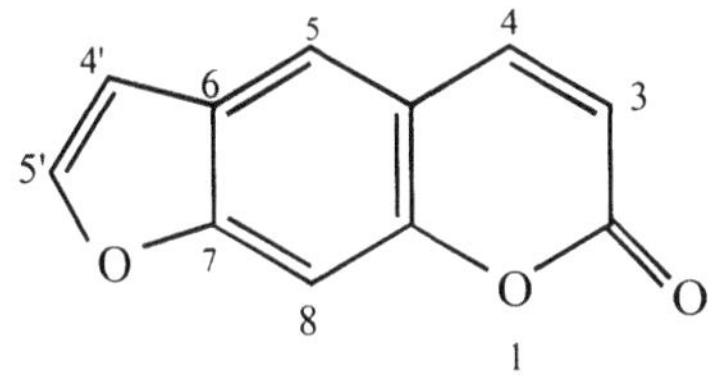

Psoralen
(linear)

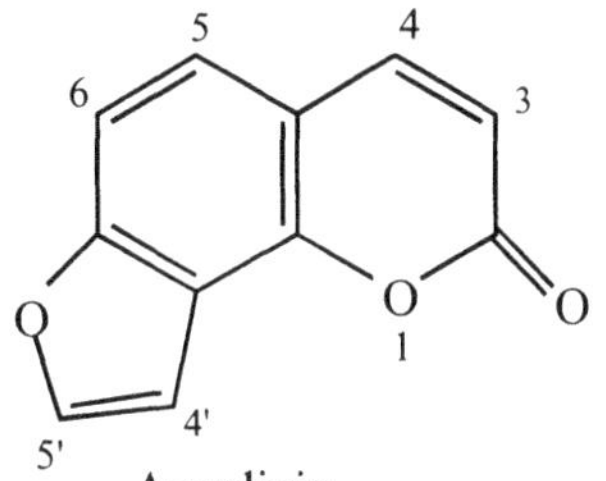

Angelicin
(angular)

(In its wisdom, Chemical Abstracts uses a different numbering scheme for psoralen to that used by scientists in the field, so be warned. It is:

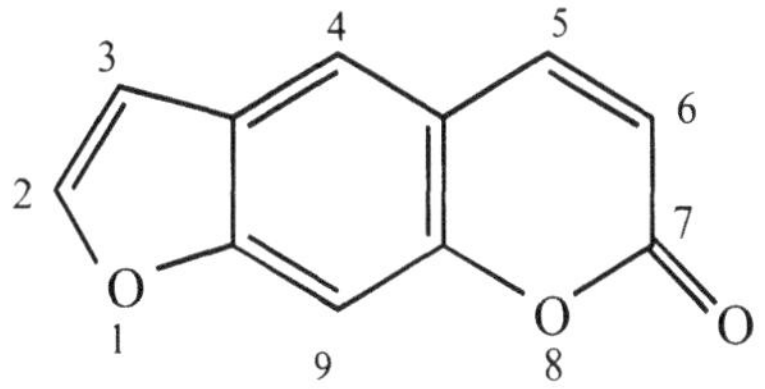

In psoriasis phototreatment,examples of the linear class of compound are the most commonly employed, especially 8-methoxypsoralen (**48**), 5-methoxypsoralen (**49**), and 4,5′,8-trimethylpsoralen (**50**).

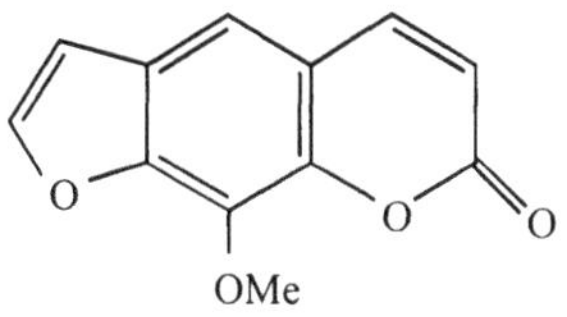

48 8-MOP

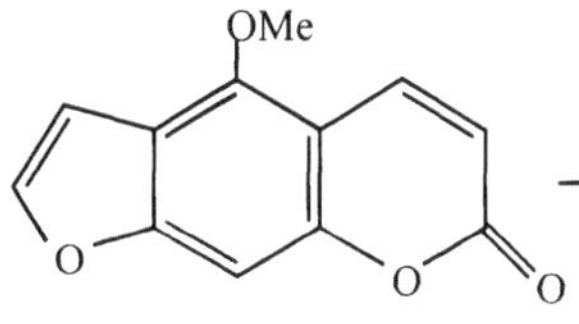

49 5-MOP

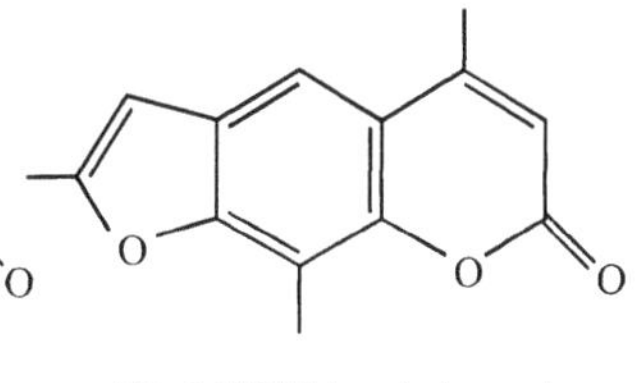

50 4,5',8-Trimethylpsoralen

These substances occur naturally in the *Umbelliferae* and *Rutaceae* families. For example psoralen, 8-methoxypsoralen and 5-methoxypsoralen occur in the *Umbelliferae* (eg parsnips, celery), while angelicin occurs in angelica. There have been occasional reports of photosensitivity of the skin following prolonged handling of plant products of these families. Historically, 8-methoxypsoralen (8-MOP) was called xanthotoxin, and 5-methoxypsoralen (5-MOP) was called bergapten (isolated from bergamot fruit), but these names have fallen into disuse.

Because of the small amounts required, these photosensitisers are generally isolated from nature. The substances have been prepared by synthesis, but the syntheses are rather tortuous with only modest overall yields. Figure 5.7 shows an example, the synthesis of 8-methoxypsoralen.

5.5.2 Mechanism

8-Methoxypsoralen can act as a sensitiser of singlet oxygen formation, but the quantum yield is low (0.009, Table 2.1), and the principal photobiological effect is not regarded as being due to singlet oxygen.

Following the work particularly of the Italian School (Musajo) in the 1960s, it is now recognised that the principal physiological process is an orbital symmetry allowed $2\pi + 2\pi$ photoadditon between the furocourmarin and the bases, particularly the pyrimidine bases, of DNA and RNA. The reaction has been studied with the bases themselves, eg:

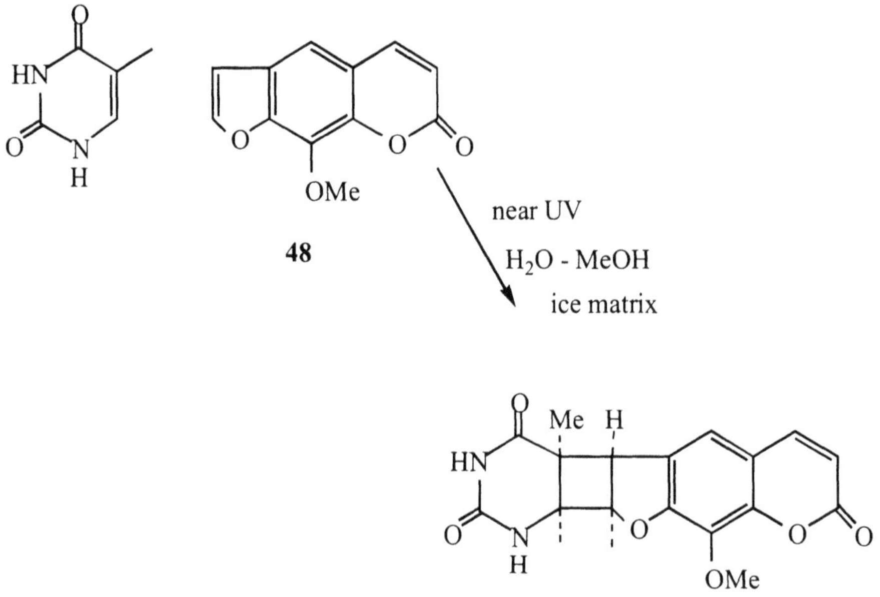

48

near UV
H_2O - MeOH
ice matrix

In this particular example, conducted in a frozen water-methanol matrix, X-ray analysis of the crystals isolated showed the regio and stereoisomerism indicated. In general, however, in fluid solution mixtures of isomers are formed. Addition can occur at the 4′, 5′ (furan) bond, as shown above; at the 3,4 (δ-lactone) bond; (where monoaddition occurs in nucleic acids, this is the major adduct) or at both, thus:

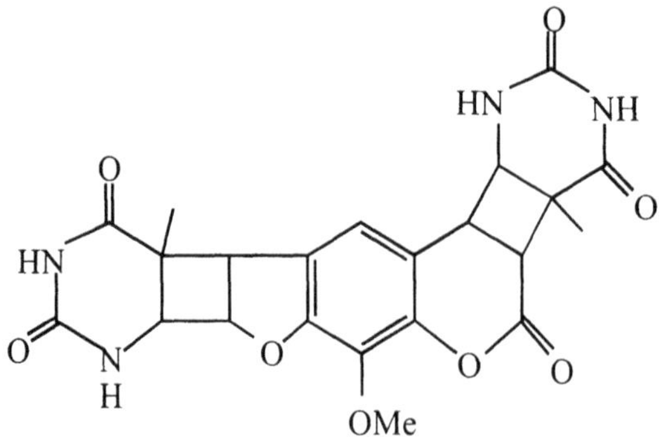

(isomers)

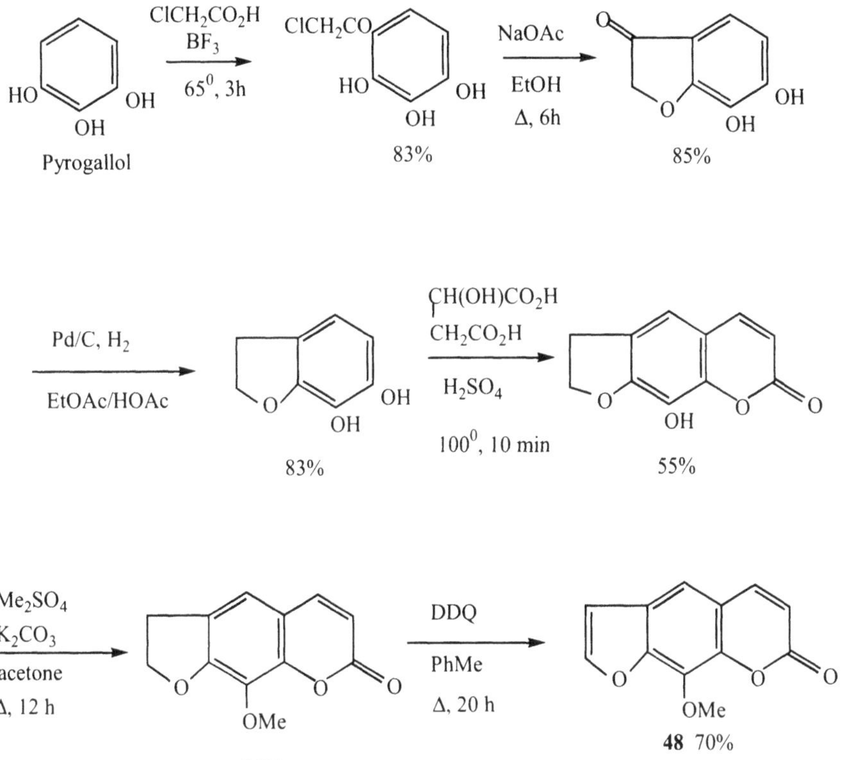

FIGURE 5.7 SYNTHESIS OF 8-METHOXYPSORALEN (8-MOP) FROM PYROGALLOL ACCORDING TO L.J. GLUNZ AND D.E. DICKSON, US PATENT 4,129,576 (DEC 12 1978). SEE ALSO E. SPÄTH AND M. PAILER, *BER. DEUTSCH. CHEM. GESELL.*,1936, **69**, 767–770.

It is this possibility of tying two thymine residues together with covalent linking that appears to be important in interrupting the overproduction of DNA and of cells in psoriasis.

The interaction of furocoumarins with DNA and RNA has also been studied extensively. The first step appears to be intercalation of the planar furocoumarin into the DNA helix to give a molecular complex in which the photosensitiser is located between two planar heterocyclic rings (purine or pyrimidine). There are about 5 complexing sites for every 100 nucleotide units in DNA; and single stranded RNA and DNA undergo intercalation less avidly than does double stranded DNA. This interaction is, of course, a dark process, but it places the furocoumarin close to the heterocyclic bases of the nucleic acids, and, especially where these bases are thymines, on irradiation with UV-A photoaddition and cross linking are facilitated.

5.6 PHOTOTHERAPY OF NEONATAL HYPERBILIRUBINEMIA

The background and general description of this topic have been given in section 1.2.3. Here we shall look at mechanistic aspects.

The condition arises from a defect in haem catabolism (Panel 5.2). During this process haem is degraded to bilirubin. It turns out that bilirubin in spite of its solubilising functions (lactam, propionic acid) is virtually insoluble in water at pH 7. In the normal course of events it is solubilised as the diglucuronide. When the enzymatic capacity to do this is low, bilirubin is retained in the body, and bilirubinemia results. On irradiation of the tissues with visible light, bilirubin concentrations fall, that is, *photobleaching* occurs. At least two mechanisms are responsible for this: *photofragmentation*, which breaks the molecule of bilirubin into small, colourless, water soluble and readily excretable products of oxidative degradation; and *photoisomerisation / photosolubilisation*, in which the bilirubin molecule is (temporarily, in some processes) converted into an isomer which is more soluble in water, and which can pass from the liver to the bile duct without conjugation to D-glucuronic acid.

5.6.1 Photofragmentation

Photooxygenation of bilirubin occurs to give a complex mixture of products. Although the triplet energy of bilirubin is sufficient (*ca.* 150 kJ mol^{-1}) it emerges that bilirubin (**51**, bilirubin IXα, Panel 5.3) is a rather poor sensitiser of singlet oxygen formation, but rapidly interacts with singlet oxygen both to quench it physically (k_q ~10^9 l mol^{-1} s^{-1}) and to react with it chemically (k_r ~ 10^8 l mol^{-1} s^{-1}). The latter process destroys the chromophore. The products which have been identified include imides (**53**, **54**) and propentdyopent adducts (**55**), both of which are water soluble and colourless. The structures are shown in Figure 5.8. Propentdyopent adducts have been encountered earlier as products of the reaction of singlet oxygen with pyrromethenones (Panel 3.3). The photochemical oxidation has been studied chemically in methanol solution, under which conditions the propentdyopents are obtained as major products (60% of mass balance) as methanol adducts. *In vivo* it is the water adducts that are produced: they have been detected in the urine of jaundiced infants following phototherapy.

The evidence for the formation and involvement of singlet oxygen (section 4.4.3) includes the following. Bilirubin has a triplet energy of 150 kJ mol^{-1}, although both ϕ_T and ϕ_Δ are low. Photobleaching occurs to give the same products when known singlet oxygen photosensitisers (such as rose bengal) are used, and, indeed, the photobleaching of bilirubin has been adapted to provide a simple method for comparing the relative efficiencies of different photosensitisers. Photobleaching also occurs when singlet oxygen is produced by direct laser excitation of solutions of bilirubin and oxygen in fluorinated solvents. The rate of photobleaching is decreased by singlet oxygen quenchers (eg β-carotene) and chemical traps (eg 2,5-dimethylfuran): and the reaction goes faster in deuteriated solvents than in protiated ones. This evidence points to the involvement of singlet oxygen, but attention should be paid to the caveats in section 4.4.3.

Figure 5.9 shows how singlet oxygen reactions may be invoked to account for the cleavage products observed.

Panel 5.2 Haem catabolism

Monoglucuronides and other sugar conjugates are also detected: bilirubin diglucuronide appears to be the predominant pathway. The cleavage of the haem occurs at the α position (C-5) of protohaem (iron(II) protoporphyrin IX) and hence when it is necessary to discuss positional isomers, natural biliverdin and bilirubin are referred to as the IXα isomers.

Panel 5.3 Positional isomers of bilirubin

Dipyrrylmethanes undergo the pyrrole exchange reaction, for example under acid conditions. This involves acid-catalysed cleavage and reassembly, thus:

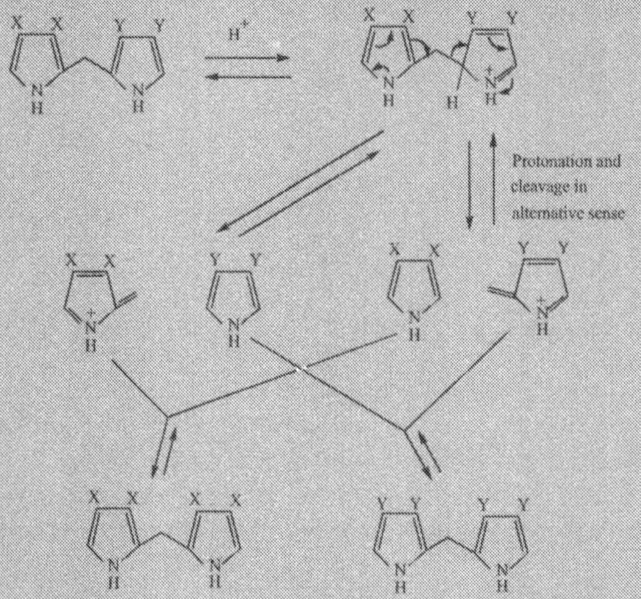

Protonation and cleavage in alternative sense

So a mixture of three dipyrrylmethanes is produced. As may be imagined, this needs to be watched when dipyrrylmethanes (especially those without electron withdrawing substituents on the π-system) are used in stepwise synthesis.

The reaction occurs with the dipyrrylmethane unit (rings B and C) when bilirubin (the common natural isomer, IX α) is subjected to acid conditions.

Two new products are formed, called bilirubin IIIα and bilirubin XIIIα, which can be separated from one another by preparative TLC or by HPLC.

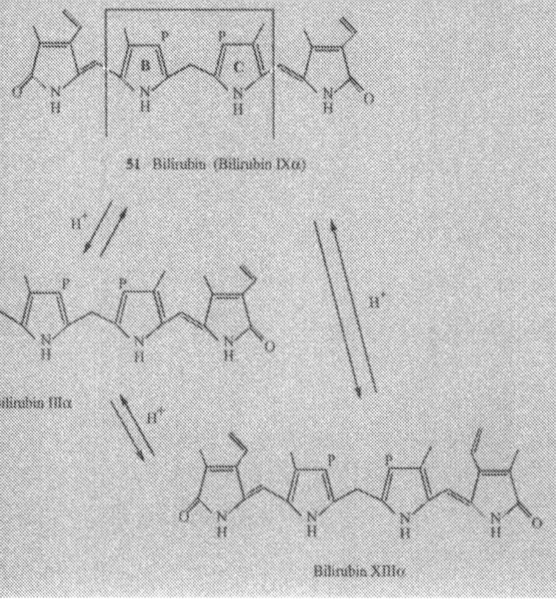

51 Bilirubin (Bilirubin IXα)

Bilirubin IIIα

Bilirubin XIIIα

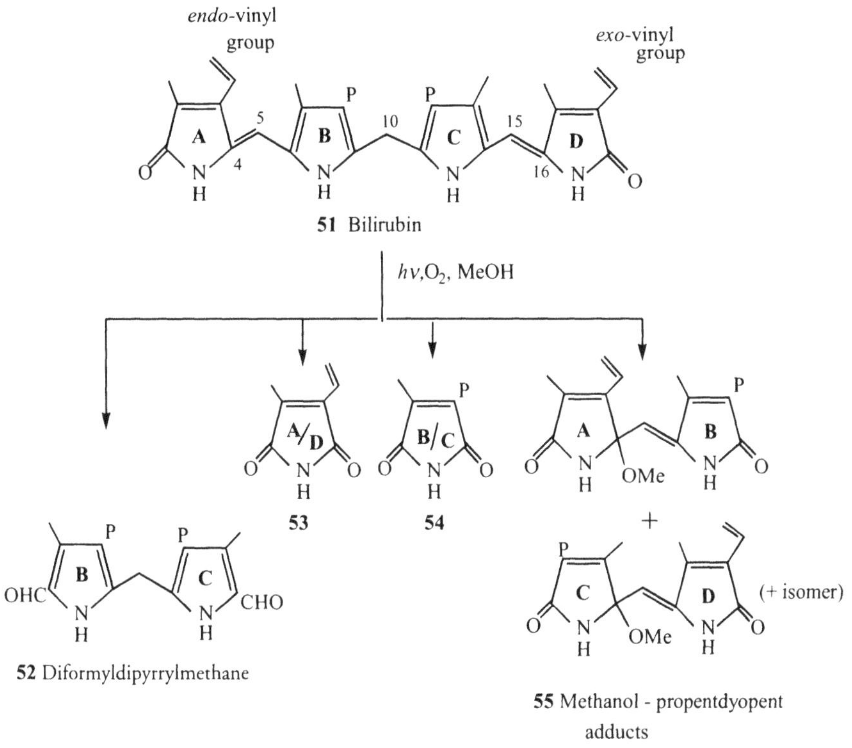

endo-vinyl group

exo-vinyl group

51 Bilirubin

$h\nu$, O_2, MeOH

52 Diformyldipyrrylmethane

(+ isomer)

55 Methanol - propentdyopent adducts

53 Methylvinylmaleimide

54 Haematinic acid imide

FIGURE 5.8 PHOTOFRAGMENTATION OF BILIRUBIN IN METHANOL (THE CURIOUS NAME OF THE PROPENTDYOPENT ADDUCTS ARISES BECAUSE HISTORICALLY THEY WERE COMMONLY IDENTIFIED BY A SPECIFIC COLOUR TEST WITH ALKALINE SODIUM DITHIONITE, WHICH GIVES A STRIKING PINK COLOUR, λ_{MAX} 525 NM).

5.6.2 Photoisomerisation / Photosolubilisation

Bilirubin is very insoluble in water. If the orange solid is shaken with distilled water the liquid does not become appreciably coloured to the naked eye. However the substance is soluble in polar aprotic solvents (such as DMSO) and in some non-polar organic solvents (such as CS_2).

The reason for the insolubility in water becomes clear from the X-ray crystal structure. The molecule takes the form of a ridge tile (Figure 5.10) because of intramolecular hydrogen bonding between the four imino hydrogens and the two carboxylic acid functions. The bridge double bonds (at C-4 and C-15) have the *Z* configuration. These structural features are seen most clearly in structure (**51A**) in Figure 5.11 where the ridge tile is being looked at in plan view.

5.6.2.1 Photochemical configurational isomerisation (Photobilirubins I)

Photochemical geometrical isomerisation of these bridge double bonds opens out the structure (shown for the 4*Z*, 15*E* isomer (**51B**) in Figure 5.11) to give the 4*Z*, 15*E* (**51B**), 4*E*, 15*Z* and 4*E*, 15*E* isomers. In these, some of the imino hydrogens and propionic acid groups are redisposed so that they can no longer interact intramolecularly, and they are free to break up the hydrogen bonding system in water.

$$\text{Bilirubin }(S_0) \xrightarrow{h\nu} (S_1) \xrightarrow[\phi_{isc} < 0.01]{\text{isc}} (T_1)$$

$$\text{Bilirubin }(T_1) + {}^3O_2 \longrightarrow \text{Bilirubin }(S_0) + {}^1O_2$$

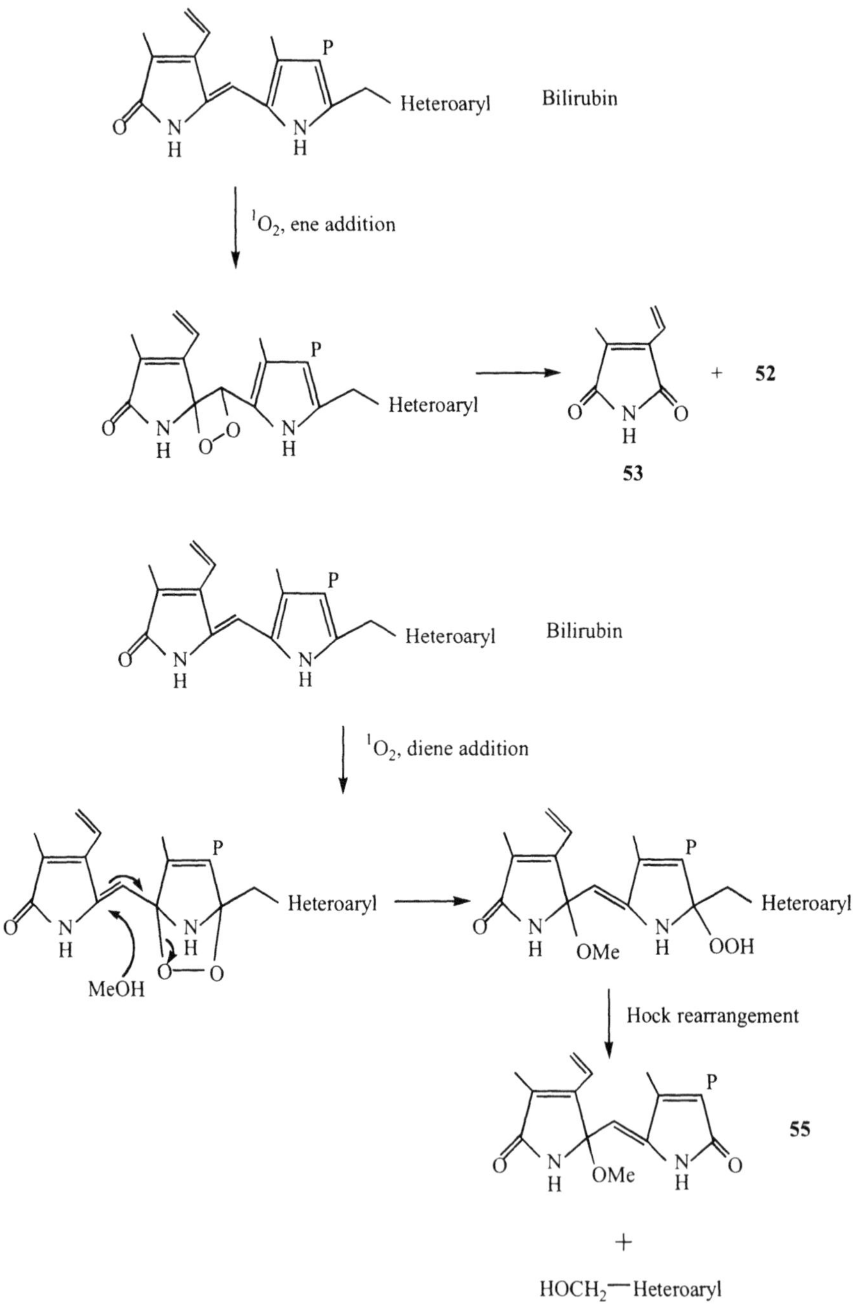

P
O N H N H Heteroaryl Bilirubin

${}^1O_2$, ene addition

P
O N H O O N H Heteroaryl → O N H O + 52

53

P
O N H N H Heteroaryl Bilirubin

${}^1O_2$, diene addition

P
O N H MeOH N H O O Heteroaryl → P
O N H OMe N H OOH Heteroaryl

Hock rearrangement

P
O N H OMe N H O 55

+

$HOCH_2$—Heteroaryl

FIGURE 5.9 RATIONALISATION OF FRAGMENTATION ON PHOTOOXYGENATION OF BILIRUBIN.

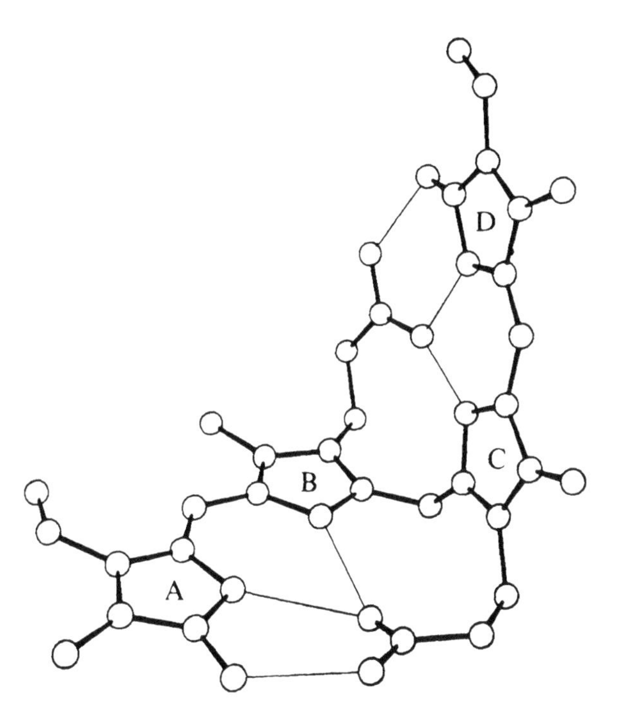

FIGURE 5.10 *X*-RAY STRUCTURE OF BILIRUBIN, SHOWING THE RIDGE-TILE STRUCTURE. THE FINE LINES REPRESENT HYDROGEN BONDS. (REPRINTED WITH PERMISSION FROM *NATURE*, R. BONNETT, J.E. DAVIES AND M.B. HURSTHOUSE, *NATURE*, 1976, **262**, 326–328. COPYRIGHT 1976 BY MACMILLAN MAGAZINES LIMITED).

Consequently these geometrical isomers of bilirubin are more soluble in water than is bilirubin itself , and they can be secreted into the bile without glucuronidation. However, the *E* photoisomers are less stable thermodynamically than the *Z,Z* isomer, and in the bile they gradually revert back to it. Thus bilirubin appears in the bile after phototherapy.

The quantum yield for the photoisomerisation $Z \rightarrow E$ is about 0.2. With blue fluorescent light (as used in phototherapy of neonatal jaundice) in ammoniacal methanol solution the distribution of isomers is approximately 80%, 4*Z*, 15*Z*; 14% 4*Z*, 15*E*; 6% 4*E*, 15*Z* and 1% 4*E*, 15*E*. Binding to serum albumin does not prevent photoisomerisation, although it does influence the regioselectivity of the process. Thus on irradiation of bilirubin bound to human serum albumin the 4*E*, 15*Z* photoisomer is formed preferentially. Protein binding also enhances the thermal stability of the *E* isomers.

5.6.2.2 Photocyclisation (Photobilirubins II, lumirubins)

Photocyclisation proceeds from the 4*E*, 15*Z* isomer, and gives products called photobilirubins IIA and IIB (or lumirubin and isolumirubin). Two products arise, and may be separated by chromatography; they are diastereoisomers formed because two new chiral centres are created. Photobilirubin II has a much enhanced solubility in water compared with bilirubin. Partition coefficient measurements between

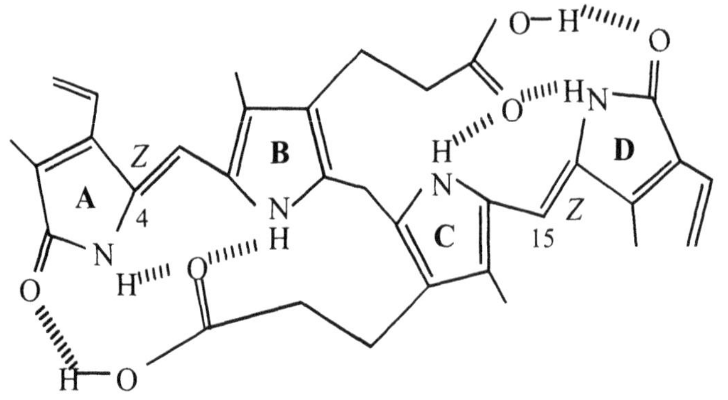

51A 4*Z*,15*Z* - Bilirubin

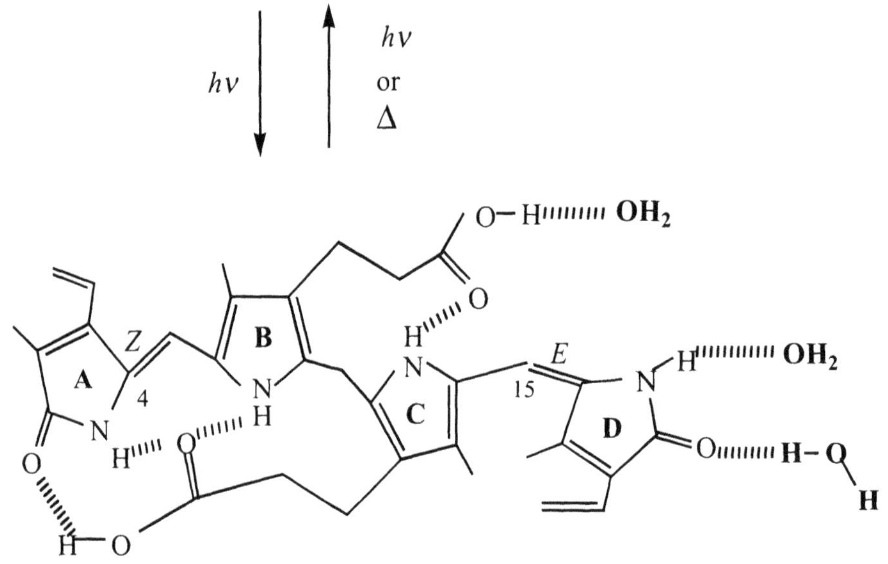

$h\nu$

$h\nu$ or Δ

51B 4*Z*,15*E* - Bilirubin

FIGURE 5.11 PHOTOISOMERISATION OF BILIRUBIN (4*Z*, 15*Z*) TO GENERATE CONFIGURATIONAL PHOTOISOMERS (4*Z*, 15*E* SHOWN; 4*E*, 15*Z*; 4*E*, 15*E*) WHICH ARE NOW FREELY SOLUBLE IN WATER.

phosphate buffer (pH7.4) and chloroform give $[PBII]_{CHCl3} / [PBII]_{buffer} = 0.67$. In a similar experiment with bilirubin, the concentration in the aqueous layer is too low to measure. The photocyclisation can be understood by reference to the model reaction with *Z*-vinylneoxanthobilirubinic acid (**56**) which, on irradiation with visible light, is transformed to the *E*-isomer (**57**). This is a well-known reversible photoreaction of simple pyrromethenones: it is the reaction described above in section 5.6.2.1, since bilirubin (**51**) is comprised of two slightly different pyrromethenone chromophores. However, in the case of *E* isomer (**57**) there is an alternative to photoreversal , which is photocyclisation involving the β-vinyl group to give the seven-membered ring system (**58**).

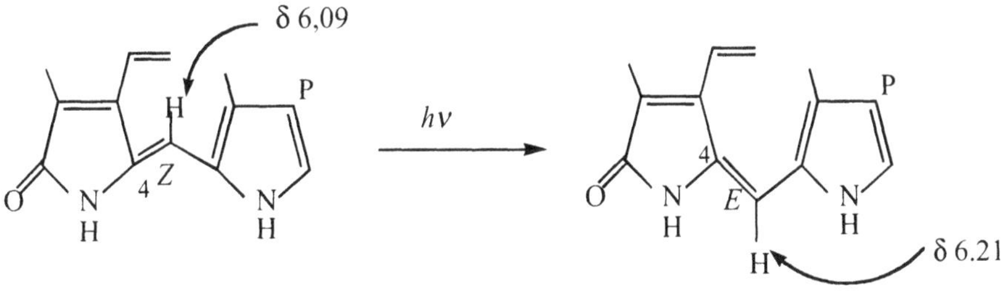

δ 6,09

hν

δ 6.21

56 ***Z*** - Vinylneoxantho bilirubinic acid

57 *E* - Vinylneoxantho bilirubinic acid

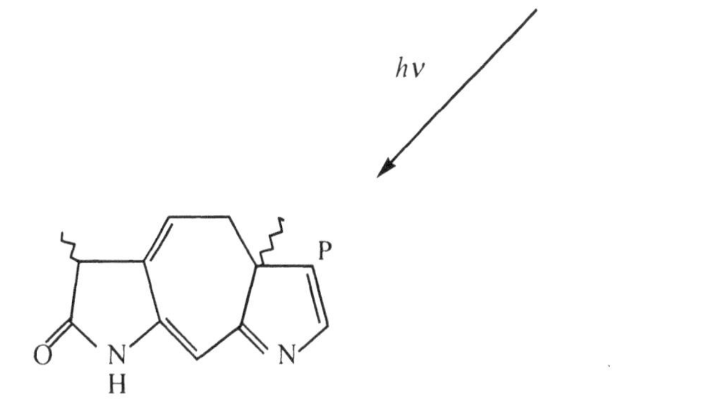

58 Photoneoxanthobilirubinic acid

A possible mechanistic interpretation of the cyclisation is shown in Figure 5.12

The structure of photobilirubin II, in which photocyclisation of only the *endo*-vinyl group has occurred, is shown at (**59**). As in (**58**) two chiral centres have been introduced (at C-2 and C-7) leading to the two diastereoismers (IIA and IIB) already mentioned. The photocyclisation presumably starts from 4*E*, 15*Z*-bilirubin, (as with the model reaction **56** → **57** → **58**). At 450 nm, the quantum yield for photocyclisation of 4*E*, 15*Z*- bilirubin has been estimated to be 0.12, and for configurational isomerisation back to 4*Z*, 15*Z*-bilirubin it is lower at 0.03. Further irradiation of **59**

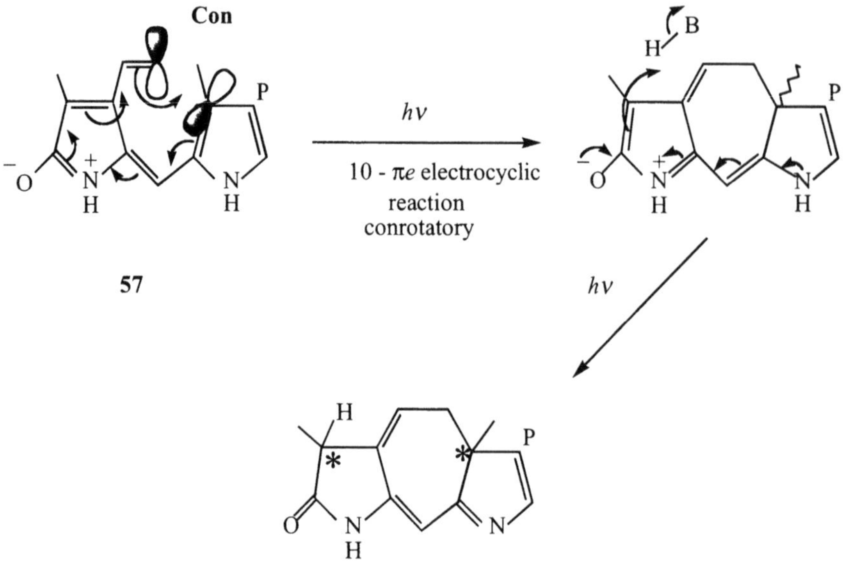

Con

57

$h\nu$

10 - πe electrocyclic reaction conrotatory

$h\nu$

58 Photoneoxanthobilirubinic acid

FIGURE 5.12 PROPOSED MECHANISM OF THE PHOTOCYCLISATION OF THE 3-VINYLPYRROMETHANONE SYSTEM.

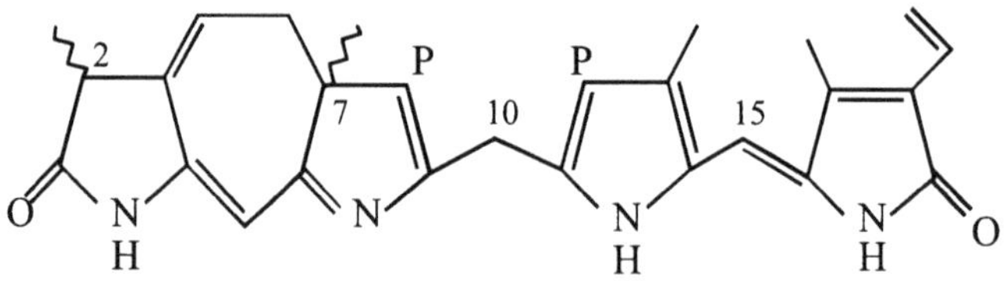

59 Photobilirubin II

(diastereoisomers IIA and IIB, lumirubin and isolumirubin)

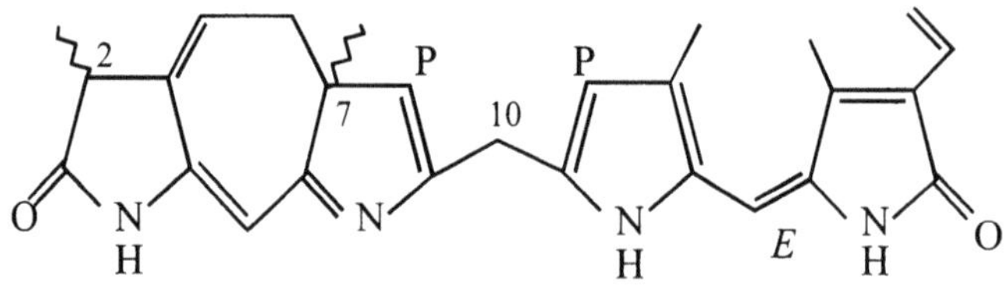

60 Photobilirubin III (diastereoisomers)

gives the 15-*E* isomer (which has been called photobilirubin III), as well as some bilirubin. Hence the photocyclisation is reversible.

The phototransformations are summarised in Panel 5.4. The relative importance of the three processes — photooxidation, photoconfigurational isomerisation, and

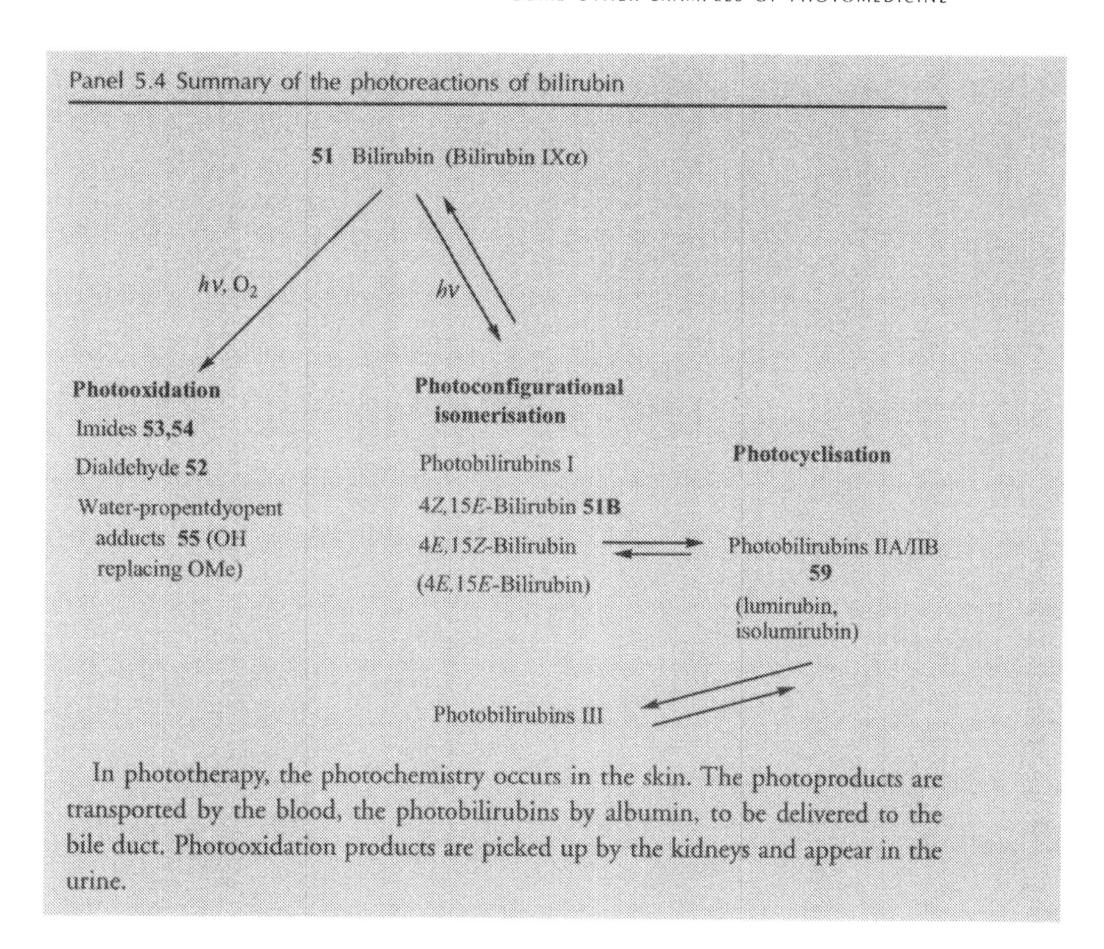

photocyclisation — *in vivo* is not known with certainty, but it appears that, in human infants undergoing phototherapy, effectiveness in removing bilirubin from the system increases along this series.

BIBLIOGRAPHY

United Nations Environment Programme, 'Environmental effects of ozone depletion: 1998 assessment', *J. Photochem. Photobiol. B Biol.*, 1998, **46**, 1–108. Produced by the UN for governments, this document carries much of general photobiological interest, and sections dealing with solar UV and health, including sunburn and skin cancer.

A.F. DeLuca, 'Vitamin D: the vitamin and the hormone', *Federation Proceedings*, 1974, **33**, 2211–2219.

D.E.M. Lawson and M. Davie, 'Aspects of the metabolism and function of vitamin D', *Vitamins and Hormones*, 1979, **37**, 1–67. A clear account of vitamin D biochemistry with a section on ultraviolet light in the management of vitamin D deficiency.

H. Jones and G.M. Rasmusson, 'Recent advances in the biology and chemistry of vitamin D', *Progress in the Chemistry of Natural Products*, 1980, **39**, 63–121. Emphasis on chemical and photochemical aspects.

P-S. Song and K.J. Tapley, 'Photochemistry and photobiology of psoralens', *Photochem. Photobiol.*, 1979, **29**, 1177–1197.

A.R. Young 'Photochemotherapy and skin carcinogenesis: a critical review', in *The Fundamental Bases of Phototherapy*, (eds H. Hönigsmann, G. Jori, A.R. Young), OEMF, Milan 1996. ISBN 88-7076-206-8. pp.77–87. A carefully argued presentation of the evidence on the risk of skin cancer in PUVA treatment.

D.A. Lightner and A.F. McDonagh, 'Molecular mechanisms of phototherapy for neonatal jaundice', *Acc. Chem. Research*, 1984, **17**, 417–424.

CHAPTER 6

THE CHEMISTRY OF HAEMATOPORPHYRIN DERIVATIVE (HpD)

"For this end I caused Twelve Ounces of dry'd Blood to be carefully distill'd by an expert Laborant, well admonished of the Difficulties of his Task, and the exactness he was to aim at in performing it. ...... The four Ounces and Two Drams of *Caput Mortuum* being diligently calcin'd, afforded but Six Drams and a half of Ashes. .... These Ashes were not White or Gray, as those of other Bodies use to be, but of a Reddish Colour, much like that of Bricks."

Robert Boyle, *Memoirs for the Natural History of Human Blood*, London, 1683/4

"*Preparation of Hematoporphyrin Derivatives.* — Recrystallized hematoporphyrin hydrochloride purchased from the Nutritional Biochemicals Corporation was used. The crude hematoporphyrin was dissolved in a mixture of 19 parts glacial acetic acid and one part concentrated sulfuric acid and allowed to stand for 15 minutes. The mixture was then filtered to remove the undissolved residue. Next the porphyrin solution was neutralized by the addition of 15 to 20 volumes of approximately 3% sodium acetate solution. Congo red paper was used as the indicator. With neutralization, the derivative was precipitated out and it was removed by filtration.

The precipitate was washed several times with distilled water and allowed to dry in air, at room temperature, in the dark.[a] The derivative was then dissolved in saline made completely alkaline by the addition of normal sodium hydroxide. The final pH was adjusted to pH 7.4 by the addition of hydrochloric acid.[b] Solutions containing 10mg per millilitre of derivative were used throughout the experimental studies. They were kept in brown bottles and refrigerated when not in use."

R.L. Lipson and E.J. Baldes, *Arch. Dermatol.*, 1960, **82**, 508

[a] HpD Stage I
[b] HpD Stage II

6.1 INTRODUCTION

The subject has already been introduced in a descriptive way in section 1.2.4. Haematoporphyrin (**2**) has two secondary benzylic-type alcohol functions at C-3^1 and C-8^1. Diastereoisomerism arises as a result of these two positions; and because of their benzylic reactivity it is one of the most difficult of the naturally-derived

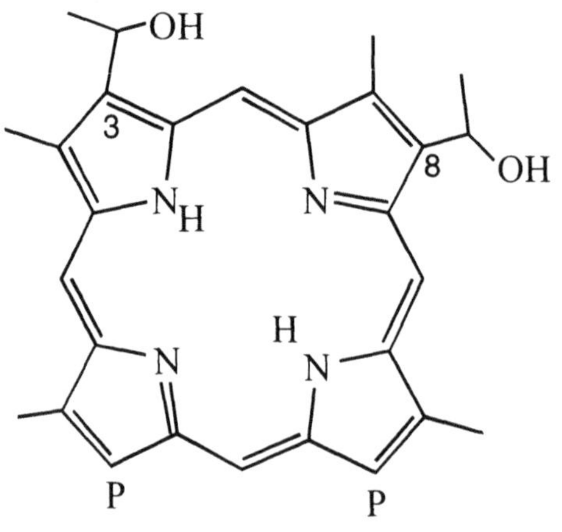

2 Haematoporphyrin (P = $-CH_2CH_2CO_2H$)

porphyrins to get pure. (Schwartz found that various commercial samples contained up to 15 components). Haematoporphyrin is readily available in crude form from slaughterhouse blood (or from haemoglobin from that source). Treatment with mineral acid (H_2SO_4 or HBr-HOAc) followed by hydrolytic work-up removes the protein and the iron to generate haematoporphyrin (generally isolated as its dihydrochloride) in one operation. The procedure was first described by Scherer in 1841: although he did not characterise the product, he did show that it was iron-free, and was thus the first to demonstrate that the red colour of haemoglobin was not primarily due to the iron which it contained. Haematoporphyrin was first properly described in 1867 by Thudichum (who called it "cruentine"): however, the name which has stuck, and which has led to the series name (Greek: *porphuros* = purple) was provided by Hoppe-Seyler in 1871.

In the early 1960s Lipson and his colleagues at the University of Minnesota were searching for a more soluble photosensitiser than the common porphyrins such as haematoporphyrin (**2**) and protoporphyrin (**61**). Following advice from Schwartz they treated haematoporphyrin dihydrochloride with sulphuric acid in acetic acid at room temperature for 15 minutes, precipitated it out with aqueous sodium acetate, then treated it with alkali before neutralising to pH 7.4 and preparing it for injection. The preparation was called Haematoporphyrin Derivative (HpD), and was found to be a powerful photosensitiser. Moreover, it localised preferentially in tumour tissue, allowing the tumour to be visualised by the bright red fluorescence which occurs on irradiation with UV light (366 nm). Added to this diagnostic application, it eventually became clear that by changing to red light the photodynamic effect could be brought in to play, leading to photodynamic therapy of cancer. Commercial development of HpD led to various proprietary photosensitisers, of which Photofrin (QLT Phototherapeutics Inc, Vancouver) received the first regulatory approval in Canada in 1993 (section 1.2.4).

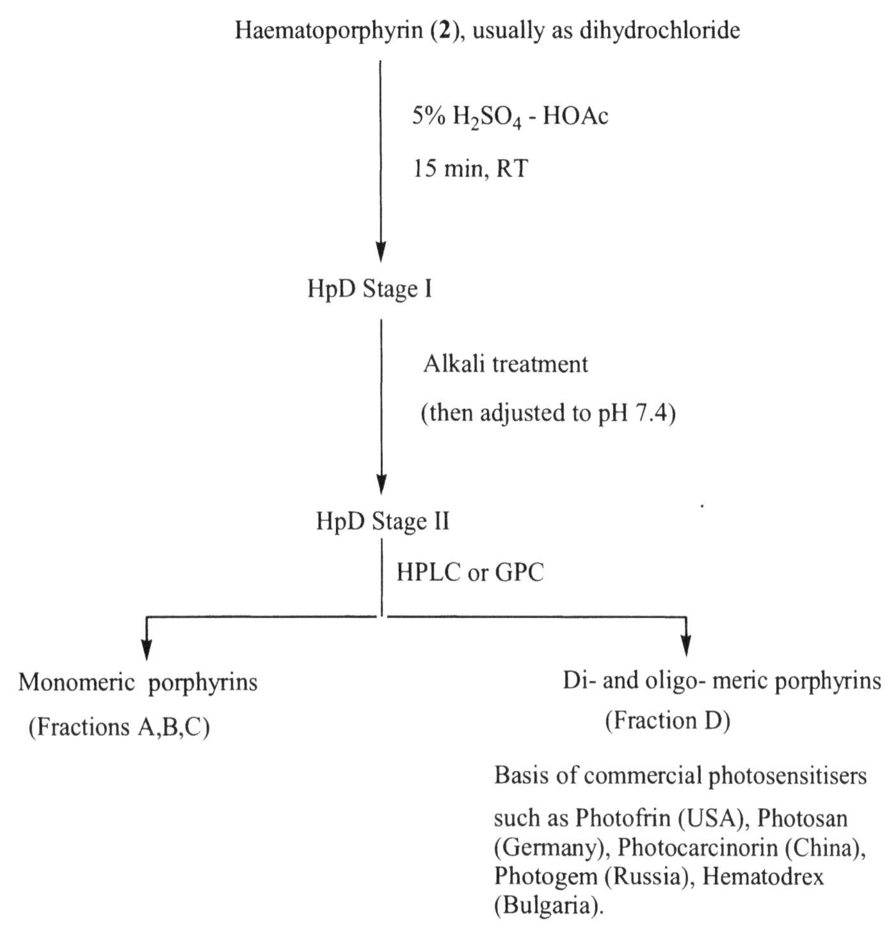

FIGURE 6.1 OUTLINE SCHEME FOR THE PREPARATION OF HPD AND ITS COMMERCIAL VARIANTS.

The preparation of HpD, as first described by Lipson and Baldes, is given as a quotation at the head of this chapter. When the work was originally done, there was little understanding of the chemistry involved, and, in truth, the chemistry remains something of a minefield. In order to understand it better, it makes sense to identify the two types of material involved (both of which have been called HpD in the literature). Here we shall refer to the product from the sulphuric acid/acetic acid treatment of haematoporphyrin as HpD Stage I, and the product from its subsequent treatment with alkali, followed by neutralisation, as HpD Stage II as shown in Figure 6.1. It is HpD Stage II (or a processed version of it) that is actually administered in PDT.

6.2 CHEMISTRY OF HPD STAGE I.

Examination of the black solid precipitate (HpD Stage I, Figure 6.1) by HPLC reveals a mixture of about 10 components (Figure 6.2).

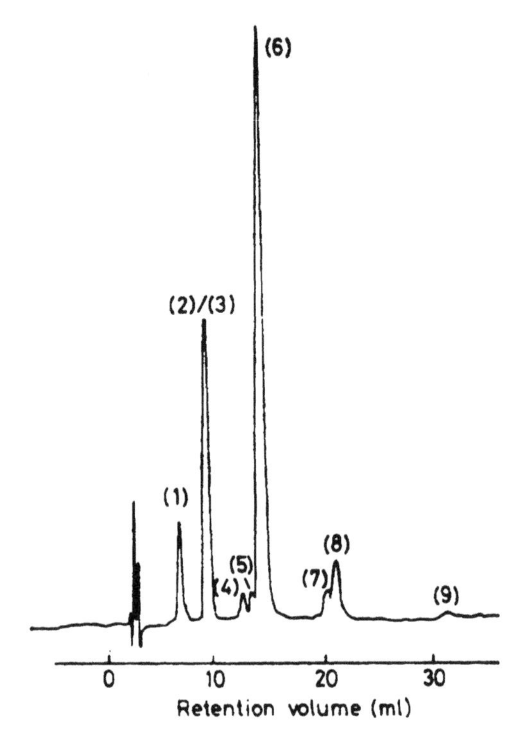

FIGURE 6.2 SEPARATION OF COMPONENTS OF HpD STAGE I BY ANALYTICAL HPLC ON A REVERSE PHASE COLUMN ELUTED WITH AQUEOUS TETRAHYDROFURAN — ACETATE BUFFER. (DIASTEREOISOMERS CAN BE SEPARATED, BUT ARE NOT SEPARATED HERE). THE NUMERALS ON THE PEAKS REFER TO THE FIRST COLUMN IN TABLE 6.1. (REPRODUCED FROM R. BONNETT, R.J. RIDGE, P.A. SCOURIDES AND M.C. BERENBAUM, *J. CHEM. SOC., PERKIN TRANS. I*, 1981, 3135–3140 WITH PERMISSION).

Preparative separation and characterisation of these components shows that esterification and elimination reactions have occurred, the products being identified as shown in Table 6.1. The main product is the diacetate of haematoporphyrin (**62**), the next most abundant components being the monoacetate fraction (**63a,b**). Although the mixture is complex, the products are those expected from acid-catalysed acylation and elimination reactions. Elimination occurs under the mild condition employed because of the ready formation of, and proton loss from, the benzylic carbocations at C-3^1 and C-8^1.

6.3 CHEMISTRY OF HPD STAGE II

The second stage of the preparation of HpD involves treatment of the mixture just described (largely mono and di acetates of haematoporphyrins) with alkali at room temperature. As would be expected, this leads to the formation of hydrolysis products, and at the same time a certain amount of elimination occurs, so that haematoporphyrin, mono(1-hydroxyethyl)monovinyldeuteroporphyrins, and protoporphyrin are all detected (fractions A, B and C). At the same time higher

TABLE 6.1 TYPICAL COMPOSITION OF HPD STAGE I.

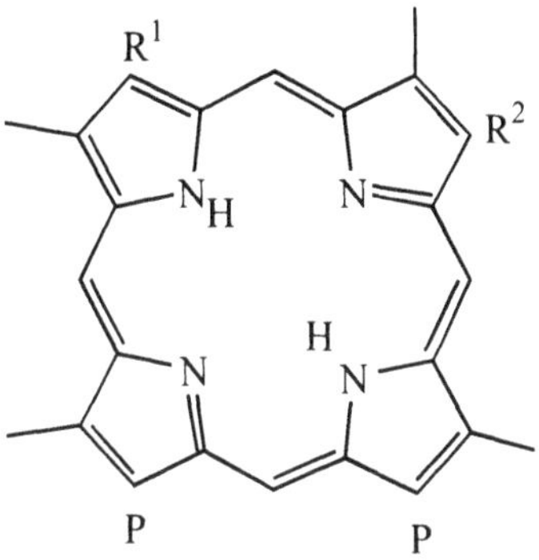

Key to Fig 6.2				Typical relative abundance %
1	**2**	$R^1 = R^2 = CHOHMe$	Haematoporphyrin	5.1
2/3	**63a**	$R^1 = CH(OH)Me$, $R^2 = CH(OAc)Me$	O-Acetyl haematoporphyrin isomers	22.4
	63b	$R^1 = CH(OAc)Me$, $R^2 = CH(OH)Me$		
4/5	**64a**	$R^1 = CH = CH_2$, $R^2 = CH(OH)Me$	8-(1-Hydroxyethyl)-3 - vinyl deuteroporphyrin	1.4
	64b	$R^1 = CH(OH)Me$, $R^2 = CH = CH_2$	3-(1-Hydroxyethyl)-8 - vinyl deuteroporphyrin	1.8
6	**62**	$R^1 = R^2 = CH(OAc)Me$	O,O′-Diacetyl haematoporphyrin	60.3
7/8	**65a**	$R^1 = CH = CH_2$, $R^2 = CH(OAc)Me$	8-(1-Acetoxyethyl) 3-vinyldeuteroporphyrin	2.7
	65b	$R^1 = CH(OAc)Me$, $R^2 = CH=CH_2$	3-(1-Acetoxyethyl) 8-vinyldeuteroporphyrin	4.8
9	**61**	$R^1 = R^2 = CH = CH_2$	Protoporphyrin	1.4

molecular weight material is formed (Fraction D), and it is in this fraction that the principal biological activity resides (Figure 6.3).

Fraction D is a complex mixture of polymers mainly in the range dimer to hexamer. Although this fraction is biologically active, attempts to isolate a "magic bullet" with very high activity have not been successful. It seems that it contains a number (probably a very large number) of components, many of which have modest PDT activity. The commercial preparations (eg Photofrin) are based on this higher molecular weight fraction, which is separated out by either high pressure

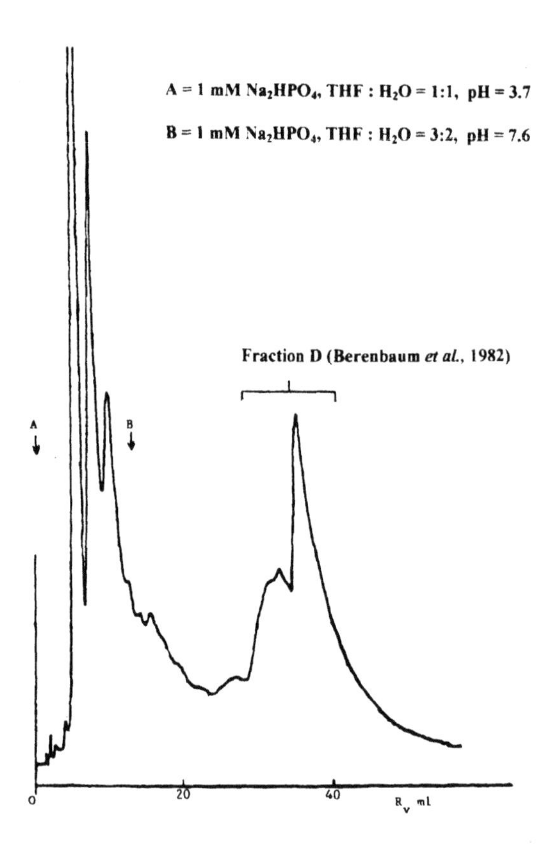

FIGURE 6.3 HPLC OF HPD STAGE II, PREPARATIVE RP-18 COLUMN ELUTED WITH THE SEQUENCE OF SOLVENTS SHOWN. MONOMERS EMERGE WITH SOLVENT SYSTEM A; THE MAJORITY OF THE BIOLOGICAL ACTIVITY RESIDES IN THE OLIGOMERIC FRACTION D. (M.C. BERENBAUM, R. BONNETT AND P.A. SCOURIDES, *BR. J. CANCER,* 1982, **45**, 571–581).

liquid chromatography or by gel permeation chromatography. This procedure increases the relative amount of the higher molecular weight material, but does not eliminate monomeric porphyrins. Thus an analysis by Russian workers gives the ratio of monomer: dimer: oligomer as 22: 23: 55 for HpD Stage II and 14: 19: 67 for Photofrin. Capillary electrophoresis has revealed 60 components in Photofrin.

Why so many? The explanation here rests on the benzylic reactivity of haematoporphyrin at C-3^1 and C-8^1, which also accounts for the difficulty in getting this substance pure (section 6.1). On treatment with alkali, three types of internuclear linkage are formed, viz:

Ester

or S_N2

Ether

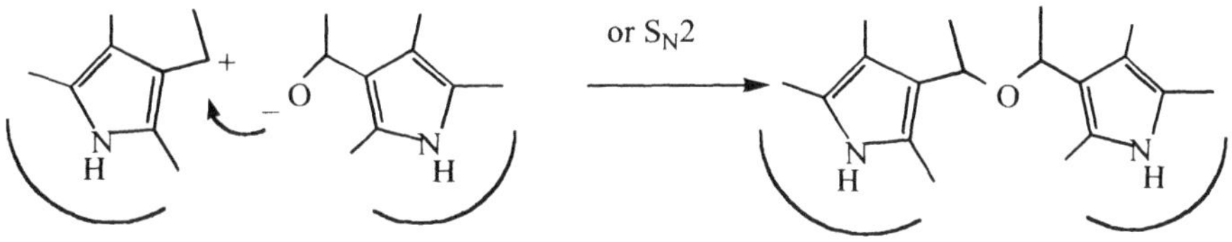

or S_N2

Carbon-carbon

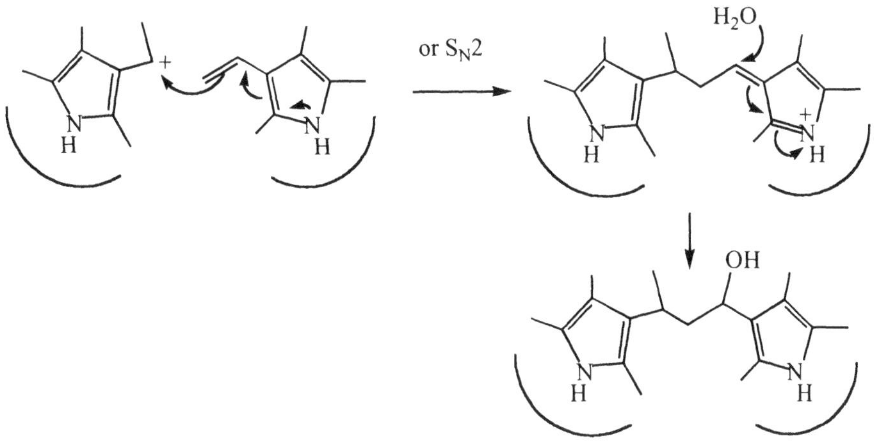

or S_N2

H_2O

OH

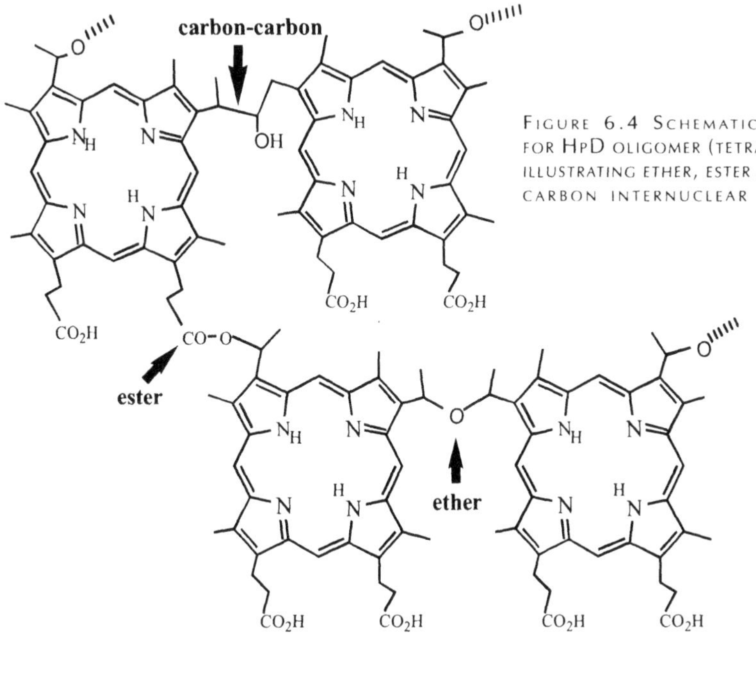

carbon-carbon

ester

ether

FIGURE 6.4 SCHEMATIC STRUCTURE FOR HPD OLIGOMER (TETRAMER DRAWN) ILLUSTRATING ETHER, ESTER AND CARBON-CARBON INTERNUCLEAR LINKAGES.

The internuclear linkages have been identified by a variety of methods. The ester linkage is cleaved under mild conditions of acid or alkaline hydrolysis, and also by lithium aluminium hydride; the ether is cleaved by highly acidic conditions, but not by alkaline hydrolysis; and the carbon-carbon bond linkage resists both acid and alkaline hydrolysis (see also Panel 6.1). There is some evidence that the ester may be converted gradually to the ether. However, chromic acid oxidation of HpD under mild conditions, followed by GC-MS separation and identification, gives *bis*-maleimide fragments (**66**, **67**) which reveal the two types of linkage, thus:

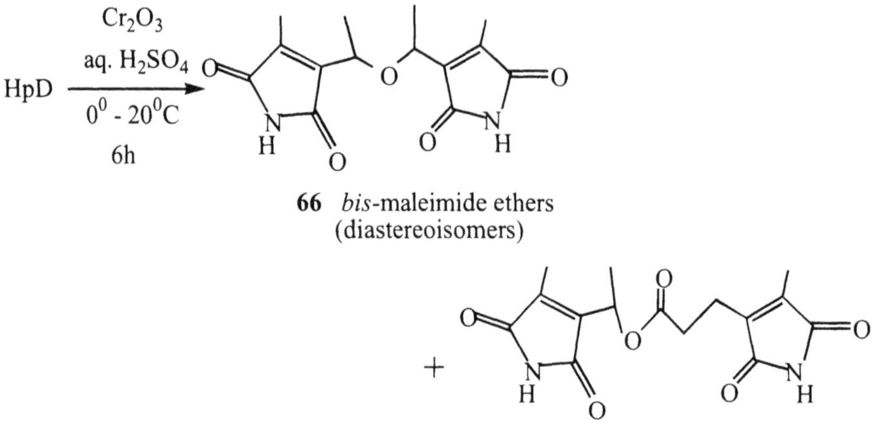

HpD → (Cr_2O_3, aq. H_2SO_4, 0^0 - 20^0C, 6h)

66 *bis*-maleimide ethers (diastereoisomers)

+ **67** *bis*-maleimide ester

Authentic compounds were made for comparison: **67** by chromic acid oxidation of the model ester **68** (Panel 6.1), and **66** in low yield from 1-methoxyethylmethylmaleimide, thus:

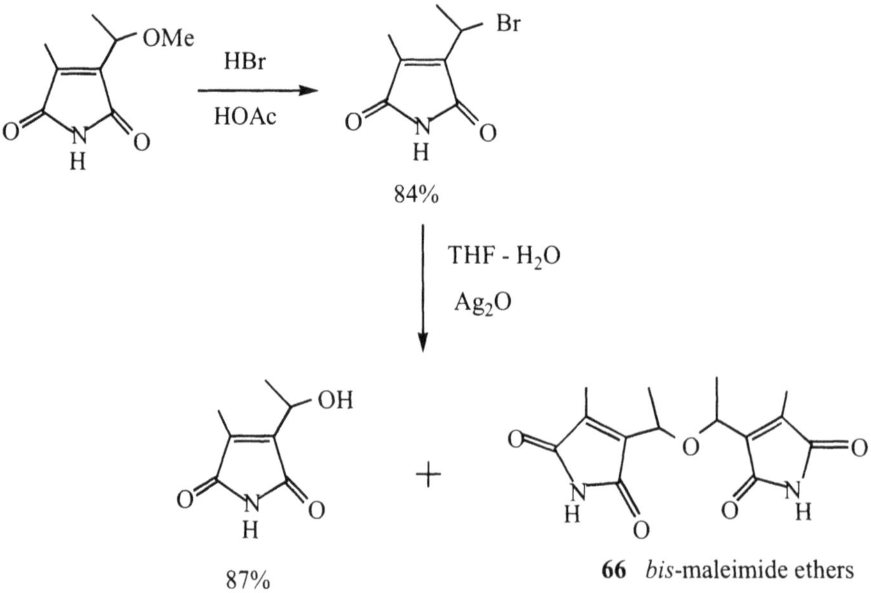

→ (HBr, HOAc) 84%

↓ ($THF - H_2O$, Ag_2O)

87% + **66** *bis*-maleimide ethers 5%

Panel 6.1 Some model dimer systems

A. Ester

(N. Risch, U.Hesse, A. Josephs and A. Gauler, *Liebigs Ann.*,1996, 1871-1874)

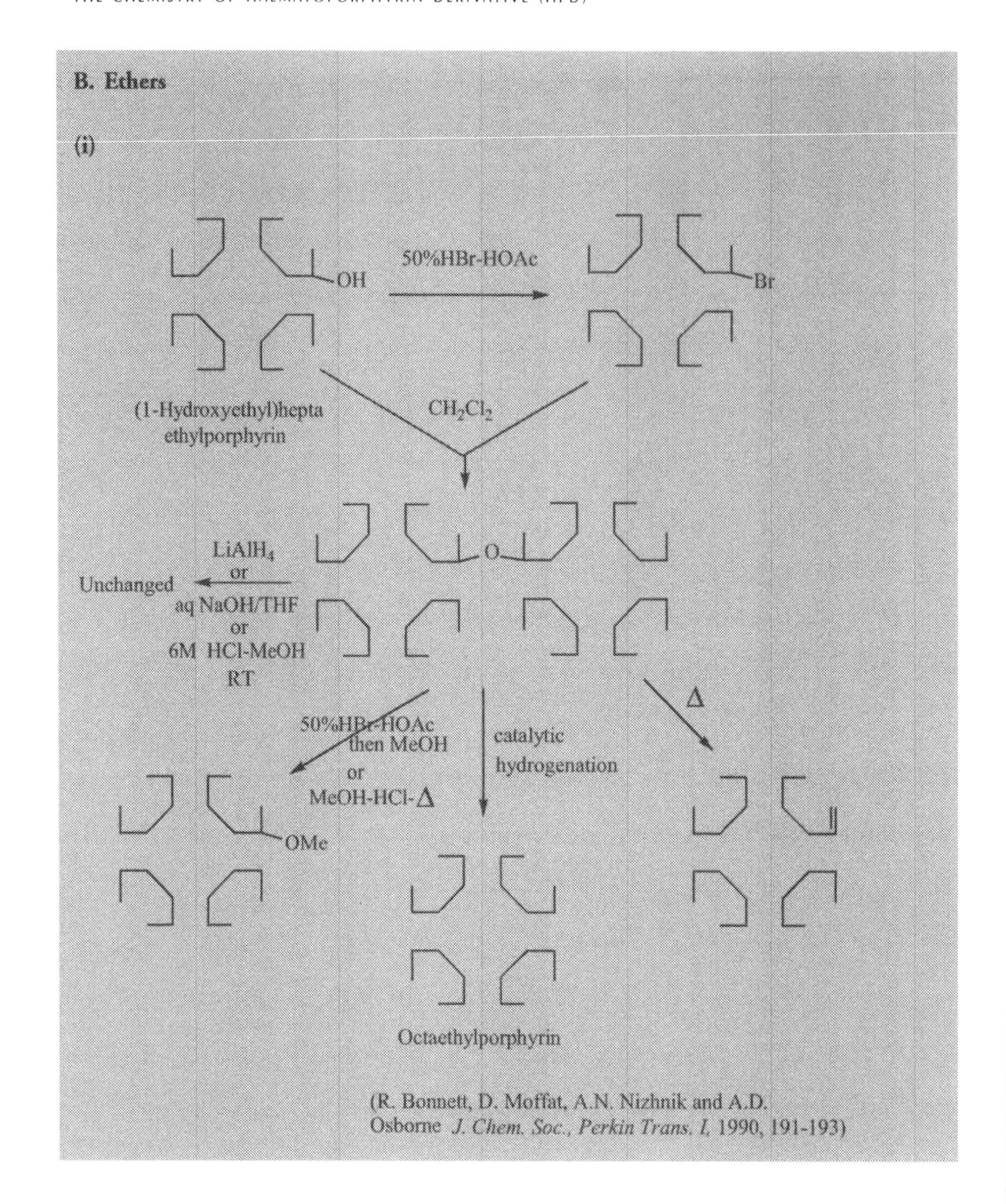
B. Ethers
(i)
OH
50%HBr-HOAc
Br
(1-Hydroxyethyl)hepta ethylporphyrin
CH_2Cl_2
O
$LiAlH_4$
or
Unchanged
aq NaOH/THF
or
6M HCl-MeOH
RT
50%HBr-HOAc
then MeOH
or
MeOH-HCl-Δ
catalytic hydrogenation
Δ
OMe
Octaethylporphyrin
(R. Bonnett, D. Moffat, A.N. Nizhnik and A.D. Osborne J. Chem. Soc., Perkin Trans. I, 1990, 191-193)

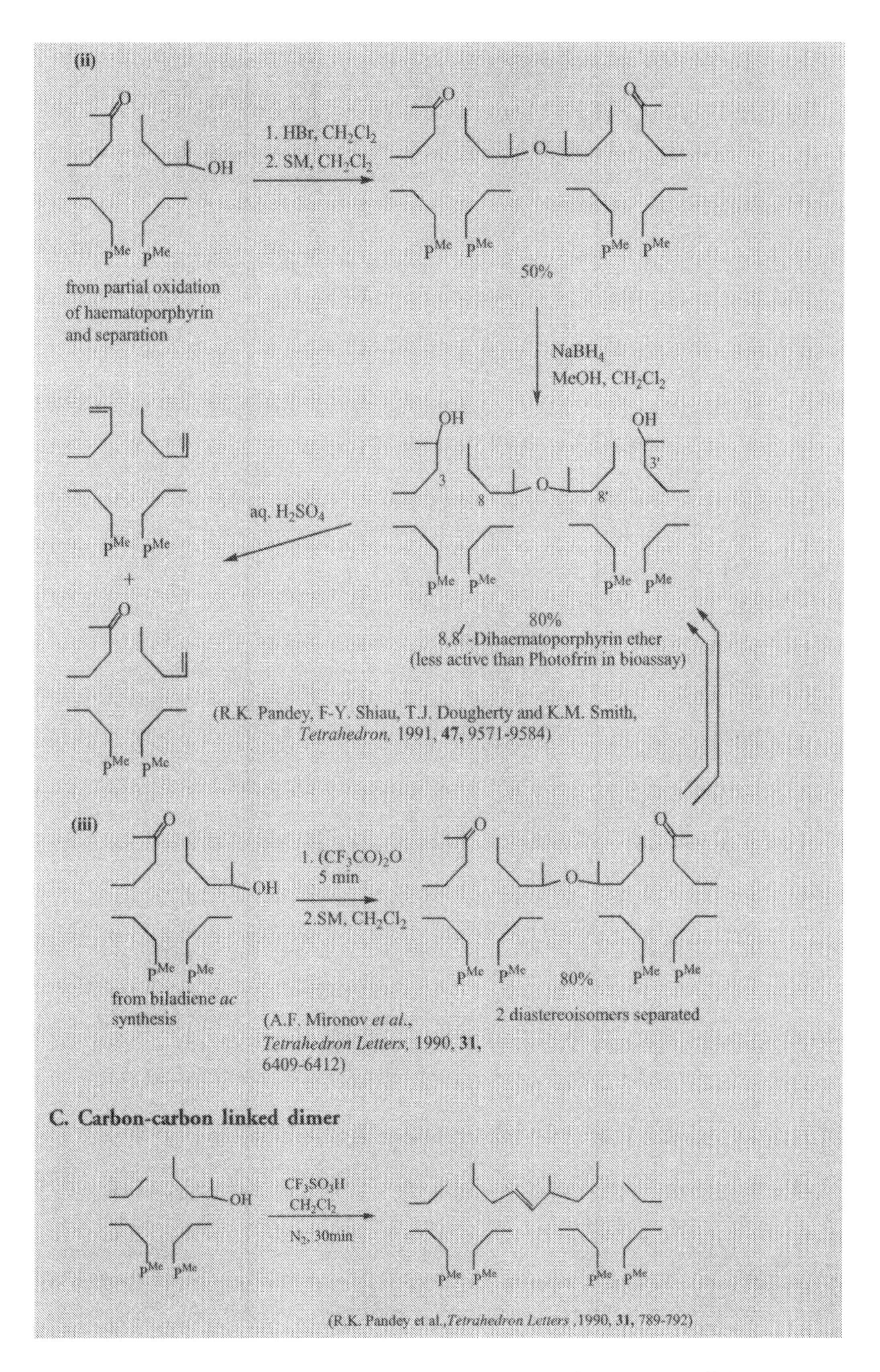

(ii)
1. HBr, CH_2Cl_2
2. SM, CH_2Cl_2
OH
O
P^{Me}
50%
from partial oxidation of haematoporphyrin and separation
$NaBH_4$
MeOH, CH_2Cl_2
aq. H_2SO_4
3
8
8'
3'
+
80%
8,8'-Dihaematoporphyrin ether
(less active than Photofrin in bioassay)
(R.K. Pandey, F-Y. Shiau, T.J. Dougherty and K.M. Smith, *Tetrahedron*, 1991, **47**, 9571-9584)
(iii)
1. $(CF_3CO)_2O$
5 min
2.SM, CH_2Cl_2
80%
from biladiene *ac* synthesis
2 diastereoisomers separated
(A.F. Mironov *et al.*, *Tetrahedron Letters*, 1990, **31**, 6409-6412)
C. Carbon-carbon linked dimer
CF_3SO_3H
CH_2Cl_2
N_2, 30min
(R.K. Pandey et al., *Tetrahedron Letters*, 1990, **31**, 789-792)

The synthesis of simple model porphyrin dimer systems is outlined in Panel 6.1. They have also played a part in elucidating the general nature of the polymeric linkages present in HpD. A schematic structure is shown at Figure 6.4.

The cause of the complexity becomes apparent. We have three types of internuclear linkage involving benzylic positions, and two different benzylic positions in haematoporphyrin. If we confine ourselves to a consideration of ether dimers (once thought to be the principal constituent of Photofrin and named dihaematoporphyrin ether, DHE, which acronym may be encountered in the older literature) then as Panel 6.2 shows, we have three types of positional isomer ($3, 3^1$; $3, 8^1$; $8, 8^1$) each of which has four chiral centres! And the situation becomes more complex still when trimers, tetramers etc., and the various types of inter-porphyrin linkage, are taken into account.

It is thus clear that HpD is a **very** complex mixture. It is also now known that crude haematoporphyrin contains oligomeric impurities which are active in PDT. To emphasise a point made earlier, these complexities can all be attributed to one structural feature, namely that haematoporphyrin is a bis-benzylic alcohol, and that the [porphyrinyl-$CHCH_3$]+ carbocation is readily formed and highly reactive.

6.4 CLINICAL DEVELOPMENT

The first published account of clinical PDT with HpD was a single case of bladder carcinoma reported in 1976 by Kelly and Snell (London). Dougherty and his colleagues (Buffalo, USA) reported the first extensive clinical survey with HpD in 1978, involving 25 patients with cutaneous or sub-cutaneous malignancies. The results were very promising: to quote from the 1978 report: "*Administration of haematoporphyrin derivative i.v. followed by local exposure to red light has resulted in complete or partial response in 111 of 113 cutaneous or s.c. malignant lesions. Tumors treated have included carcinomas of the breast, colon, prostate, squamous cell, basal cell, and endometrium; mycosis fungoides; chondrosarcoma; and angiosarcoma. No type has been found to be unresponsive*".

However, it took another 15 years for the first regulatory approval to come through. During this time HpD became commercialised, the commercial products being "purified" HpD, purified in the sense that some of the monomeric porphyrin fraction was removed (Figure 6.1). The main proprietary product was Photofrin, originally developed by Dougherty, but more recently commercialised by QLT Phototherapeutics Inc (Vancouver): but other commercial variants emerged in various countries, for example in Bulgaria (Hematodrex), China (Photocarcinorin), Germany (Photosan) and Russia (Photogem).

During the 1980s there was considerable clinical activity worldwide. For example, in Japan, Hayata and Kato and their teams reported clinical results on the PDT of lung cancer in 1982. These workers used HpD and later Photofrin as the sensitiser. In the initial investigation, 299 lesions at various stages were treated, of which 134 were judged to show complete remission. Subsequent studies were made by this group on early stage squamous cell carcinoma, and on intraoperative and palliative applications. A multicentre clinical trial for early stage lung cancer was conducted.

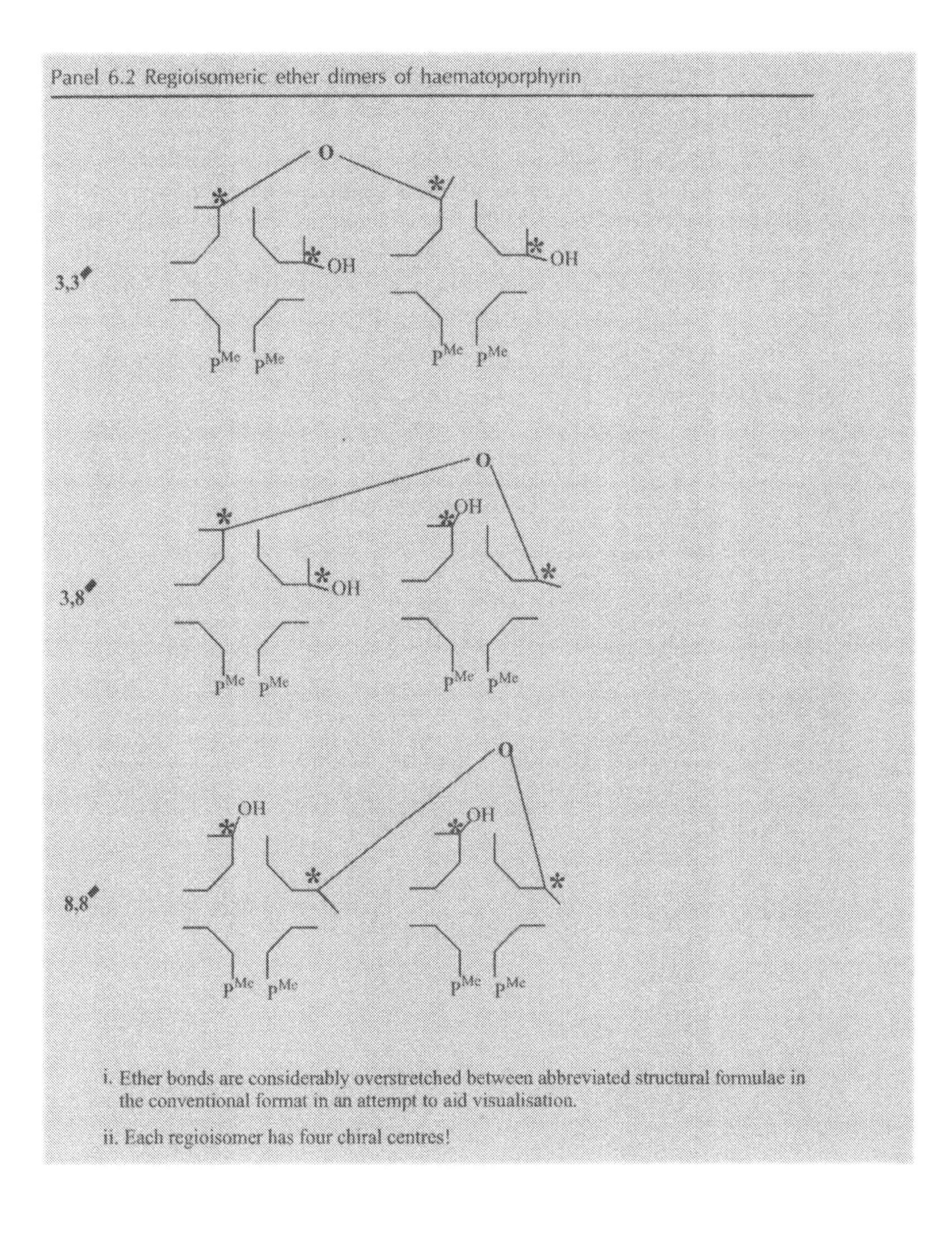

Panel 6.2 Regioisomeric ether dimers of haematoporphyrin

i. Ether bonds are considerably overstretched between abbreviated structural formulae in the conventional format in an attempt to aid visualisation.

ii. Each regioisomer has four chiral centres!

In other centres, superficial bladder cancer, gynaecological cancer, skin cancer, and cancer of the aerodigestive tract were treated with PDT, generally with positive results. There were problems, a significant one being long term (several weeks) skin photosensitisation: but there were also benefits. One of these was the satisfactory to excellent long term cosmetic result in sensitive areas such as the head and neck, which appears to be due to the resistance of collagen to destruction by PDT, allowing tissue to reform around it. This good recovery of normal appearance seems to be a general (and unexpected) feature of PDT treatment, and is in contrast to the outcome of surgery in these sensitive areas.

In 1993 the Canadian authorities granted the first regulatory approval for cancer PDT using Photofrin. The approval was strictly limited (papilloma of the bladder) but this was nevertheless a turning point. Subsequent approvals for use have included France (oesophagus, bladder), Germany (lung), Japan (lung, oesophagus, stomach, cervix), the Netherlands (lung, oesophagus) and the USA (oesophagus). Approval for palliative use (bronchus, oesophagus) was given in the UK in 1999. This was the first regulatory approval for PDT in the UK.

BIBLIOGRAPHY

J.F. Kelly and M.E. Snell, 'Hematoporphyrin derivative: a possible aid in the diagnosis and therapy of carcinoma of the bladder', *J. Urol.*, 1976, **115**, 150–151. The first published account of clinical use of HpD.

T.J. Dougherty, J.E. Kaufman, A. Goldfarb, K.R. Weishaupt, D. Boyle and A. Mittelman, 'Photoradiation therapy for the treatment of malignant tumours',*Cancer Res.*, 1978, **38**, 2628–2635. Succesful clinical results reported with HpD and a range of tumours.

H. Kato, 'Photodynamic therapy for lung cancer — a review of 19 years experience', *J. Photochem. Photobiol. B. Biol.*, 1998, **42**, 96–99. A useful overview of the work of one laboratory on the clinical use of HpD/Photofrin.

R. Bonnett, R.J. Ridge, P.A. Scourides and M.C. Berenbaum, 'On the nature of 'Haematoporphyrin derivative', *J. Chem. Soc., Perkin Trans. I,* 1981, 3135–3140. A detailed account of the chemistry of HpD Stage I.

T.J. Dougherty, 'Studies on the structure of porphyrins contained in Photofrin II', *Photochem. Photobiol.*, 1987, **46**, 569–573.

D. Kessel, P. Thompson, B. Musselman and C.K. Chang, 'Chemistry of haematoporphyrin-derived photosensitizers', *Photochem. Photobiol.*, 1987, **46**, 563–568. Studies on the nature of HpD using $LiAlH_4$ reduction.

C.J. Byrne, L.V. Marshallsay and A.D. Ward, 'The composition of Photofrin II', *J. Photochem. Photobiol.B Biol.*, 1990, **6**, 13–27. Various studies on the composition of HpD Stage II and of Photofrin. (Originally, the commercial interests called HpD (Stage II) by the name of Photofrin I, and the oligomeric concentrate — which has intellectual property content — by the name of Photofrin II. The Roman numerals seem now to have disappeared).

Chapter 7

SECOND GENERATION PHOTOSENSITISERS

7.1 FIRST GENERATION PHOTOSENSITISERS

HpD and its relatives, including commercial variants, were considered in the last chapter. These compounds comprise the *first generation* of photosensitisers. *Second generation photosensitisers* are those subsequently developed and are generally single substances, not necessarily porphyrins, with improved activity and selectivity. *Third generation photosensitisers* have an additional targetting mechanism, for example by covalent attachment to monoclonal antibodies. These are still at an early stage of development.

It is useful at this point to rehearse the advantages and disadvantages of the first generation compounds (HpD and its relatives).

7.1.1 Advantages

(i) The material is prepared by a simple procedure from readily available starting materials (Figure 6.1).

(ii) Its clinical activity has been repeatedly demonstrated, so that there is now extensive clinical experience recorded in the literature.

(iii) It was the first effective preparation to be demonstrated in this way.

(iv) It was the first substance to receive regulatory approval (Canada, 1993) and now has approvals in several other countries. Even if it is superseded, as seems likely, it is clearly of considerable historical importance.

7.1.2 Disadvantages

(i) It is a very complex mixture in terms of positional isomerism, stereoisomerism, oligomeric composition, and the nature of the inter-porphyrin linkages (section 6.2). The composition of such a mixture is difficult to reproduce; and relating the clinical response to molecular structure is virtually impossible.

(ii) Although HpD is active in cancer PDT, the activity is modest.

(iii) It is not selective enough. Sensitisation of normal tissue (especially skin) remains for some weeks.

(iv) HpD has an etio type spectrum (Panel 8.1) in which the band at lowest energy (Band I, at about λ_{max} 630 nm) has the *weakest* absorption. This is where irradiation is carried out, because tissue transmittance is highest in the red (section 7.2.7). Substances with low ε_{max} in this region require higher photosensitiser doses or higher light doses (or a combination of these) if an adequate level of excited photosensitiser (and therefore singlet oxygen) is to be generated.

7.2 DESIGN CRITERIA FOR SECOND GENERATION PHOTOSENSITISERS

During the development of the present second generation photosensitisers a number of design criteria have become apparent.

7.2.1 Dark Toxicity

It is clearly desirable that the photosensitiser has zero or very low cytotoxicity in the absence of light. This means that tumour damage can be controlled by the light dose for a given drug dose. It has occasionally been suggested that the tumour-targetting properties of some porphyrins could be used to transport antimetabolites (such as nitrogen mustards, covalently attached) to the tumour site, but this idea does not appear to have been effectively developed.

7.2.2 Composition

The photosensitiser should be of constant composition, and preferably a single substance which does not have chiral centres (unlike haematoporphyrin (**2**), which has two). If the presence of one or more chiral centres is allowed, then the optical isomers may be expected to locate differently in cells, and the photodynamic effects will be different: it may be anticipated that regulatory authorities will ask for test results on such stereoisomers. The need for a single substance is, in fact, an overriding requirement, for the simple reason that there are far more variables in photodynamic therapy than there are in conventional pharmacology, as shown in Table 7.1. Each component may respond differently to each variable, and logical analysis of biological results from a photosensitiser which is a complex mixture becomes very difficult indeed.

7.2.3 Synthesis

The synthesis (chapters 8–11) should be as straightforward and as high-yielding as possible.

7.2.4 Solution Behaviour

What is needed is a substance which has some selectivity for the tumour, but which is rapidly cleared from the body after phototherapy, so that general photosensitisation is minimised. It has emerged that photosensitisers with these features are often amphiphilic, that is, they have a blend of lipophilic and hydrophilic properties (Panel 7.1). Since the macrocyclic nuclei (porphyrins, phthalocyanines etc.) on which many of the photosensitisers are based, are themselves hydrophobic, this means that hydrophilic substituents are needed to achieve the correct amphiphilic balance. Hydroxy, sulphonic acid, and quaternary ammonium groups have been most used for this purpose. The effect of variation of hydrophilicity may be illustrated with three examples.

TABLE 7.1 ADDITIONAL VARIABLES IN PDT COMPARED WITH CONVENTIONAL DRUG ADMINISTRATION.

Parameter	Variables
1. Drug administration	Chemical structure Dose (mg kg^{-1}) Method of administration Adjuncts
2. **Drug-light interval**	**Duration** (hours)
3. **Visible light administration**	**Wavelength** (nm) **Total energy** (J cm^{-2}) **Fluence rate** (mW cm^{-2}) **Continuous/fractionated**

7.2.4.1 Tetraphenylporphyrin sulphonic acids

Sulphonation of *meso*-tetraphenylporphyrin (**27**) gives a mixture of mono, di, tri and tetra sulphonic acids, substitution occurring predominantly at the *para* position for steric and electronic reasons. There are two regioisomers of the disubstituted product.

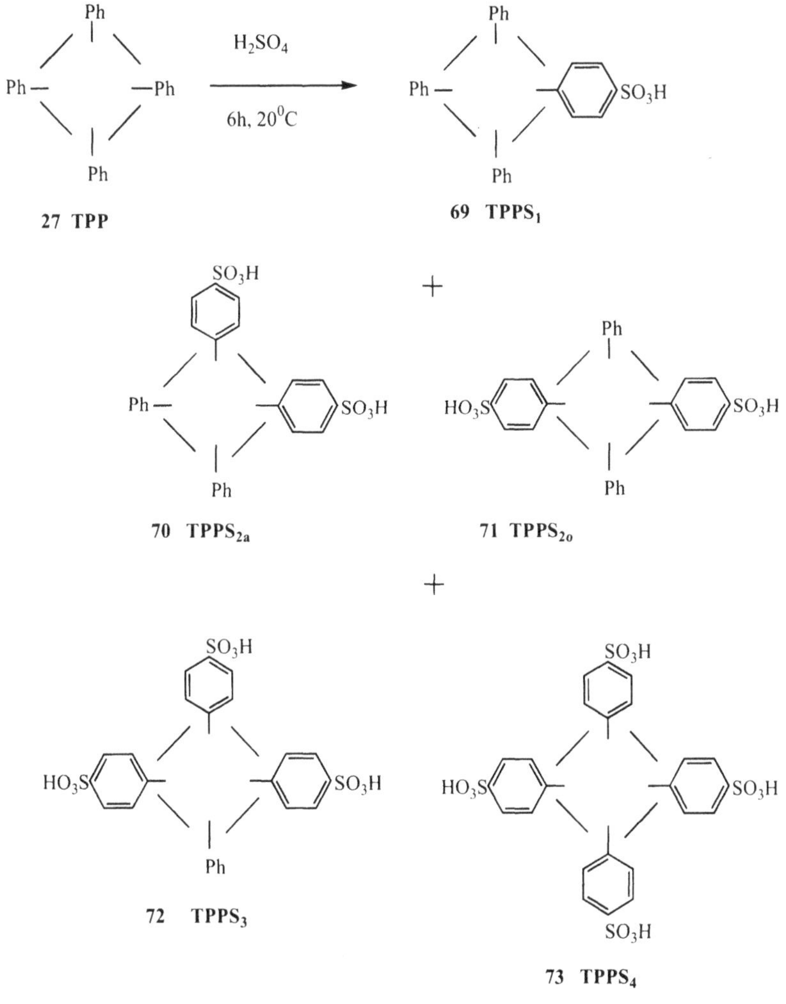

H_2SO_4, 6h, 20^0C

27 TPP → **69 TPPS$_1$** + **70 TPPS$_{2a}$** + **71 TPPS$_{2o}$** + **72 TPPS$_3$** + **73 TPPS$_4$**

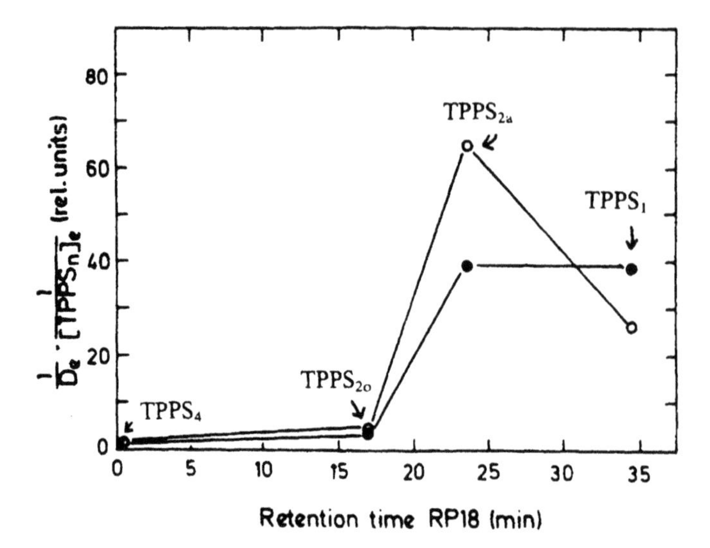

FIGURE 7.1 PHOTOINACTIVATION OF HUMAN CERVIX CARCINOMA CELLS (NHIK 3025) BY VARIOUS SULPHONATED TETRAPHENYL PORPHYRINS (**69**, **70**, **71**, **73**). THE ORDINATE IS THE EFFICIENCY OF CELL INACTIVATION MEASURED AS $1/D_c$ (THAT IS, 1/LIGHT DOSE WHICH INACTIVATES 63% OF THE CELLS) DIVIDED BY THE INITIAL EXTRACELLULAR CONCENTRATION OF THE PHOTOSENSITISER. THE ABSCISSA IS THE RETENTION TIME ON A *REVERSE-PHASE* HPLC COLUMN (SO MORE POLAR MATERIALS ELUTE FIRST). OPEN CIRCLES REFER TO 18 HOURS INCUBATION WITH $TPPS_N$; FILLED CIRCLES REFER TO THAT TREATMENT FOLLOWED BY 1 HOUR IN SENSITISER-FREE MEDIUM. (REPRODUCED FROM K. BERG, J.C. BOMMER, J.W. WINKELMAN AND J. MOAN, *PHOTOCHEM. PHOTOBIOL.*, 1990, **52**, 775–781 WITH PERMISSION).

These compounds, which are normally handled as the sodium salts, are conveniently referred to as $TPPS_1$ (**69**), $TPPS_{2a}$ (a for adjacent, **70**), $TPPS_{2o}$ (o for opposite, **71**), $TPPS_3$ (**72**) and $TPPS_4$ (**73**), and they can, with some difficulty, be separated by reverse phase chromatography. Winkelman had shown in the 1960s that $TPPS_4$ localised in tumour tissue. Hydrophilicity, estimated by the HPLC method (Panel 7.1) increases as expected with increasing substitution, ($TPPS_{2a}$ turning out to be more hydrophobic than $TPPS_{2o}$). In the photodynamic inactivation of human cervix carcinoma cell line NHIK 3025 in cell culture, $TPPS_{2a}$ was shown to be the most effective of these compounds (Figure 7.1).

7.2.4.2 Sulphonated Metallophthalocyanines

Sulphonation of aluminium and gallium phthalocyanines gives a mixture of mono, di, tri, and tetra sulphonic acids, which are increasingly hydrophilic along this sequence. Early experiments were made with this mixture, but the components can be separated according to the number of sulphonic acid groups by HPLC (Figure 7.2).

Although the disulphonated material is most active in cell culture, it is still necessary to consider several regioisomeric possibilities, since sulphonation can occur not only on opposite or adjacent benzenoid rings, but at positions 2 or 6 (the *endo* positions) and 4 or 5 (the *exo* positions) on each ring. The possibilities for $AlPcS_{2a}$ are shown on structure **74.**

Panel 7.1 Hydrophilicity, Lipophilicity, Amphiphilicity

In very basic terms, a *hydrophilic* substance is one that has an affinity for water; a *hydrophobic* substance is one that "hates water" and forms a separate phase in the presence of water. In practical terms this means that a hydrophilic substance will have an appreciable solubility in water, whereas the solubility of a hydrophobic substance in water will be low. A *hydrophilic substituent*, such as OH, will enhance the water solubility of a hydrophobic substance. Thus ethane (hydrophobic) has a low solubility in water, but ethanol is miscible with water in all proportions.

Similarly *lipophilic* implies an affinity for fatty substances (fats, oils, hydrocarbons, chlorinated hydrocarbons): *lipophobic* implies a lack of such affinity. The thermodynamic preference for a lipophilic substance is to interact with itself (ie form a separate phase; "like dissolves like") rather than break up the hydrogen-bonded structure of water (which it is ill-equipped to do).

Amphiphilic implies an affinity both for water and for lipid. Ethanol is such a substance: it is miscible with hexane and with water *separately* in all proportions. Reduce the carbon chain by one and this no longer applies: methanol is miscible with water, of course, but, in the right proportions, forms two phases with hexane.

Although this is what the word amphiphilic actually means, it is usually used with an additional implied meaning related to the structural arrangement of the hydrophobic and hydrophilic parts - that they are at opposite ends of the molecule. The term *amphipathic* is also used here, although usage is somewhat confused. Thus detergents with ionic or neutral polar functions and a long hydrophobic (usually hydrocarbon) chain show amphiphilic properties. They may, or may not, be freely soluble in water and in hydrocarbon, but in a water-hydrocarbon mixture they spontaneously arrange themselves at the interface with the polar end in the aqueous phase, and the hydrophobic end in the hydrocarbon phase (Panel 7.2). Many of the PDT photosensitisers have one or more solubilising groups (SO_3H, OH, CO_2H, $-NR_3Cl$) to modulate the hydrophobic character of the macrocyclic nucleus, and micelle formation may also be expected to be important with such molecules.

Hydrophilicity can be assessed either by measuring the partition coefficient between phosphate buffer pH 7.4 and octanol, where higher values of the partition coefficient $P = [\text{Sens}]_{oct}/[\text{Sens}]_{aq}$ indicate increased hydrophobicity; or by reference to the retention time on a reverse phase HPLC system, where longer retention times indicate increased hydrophobicity (eg. Figure 7.2).

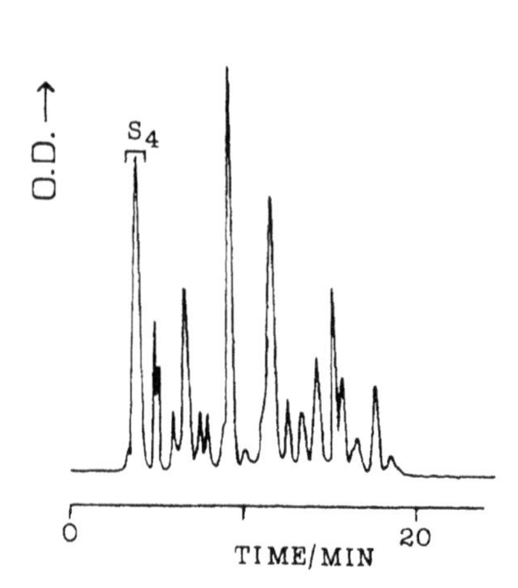

FIGURE 7.2 (A) HPLC ANALYSIS OF $AlPcS_N$ MIXTURE FORMED ON SULPHONATING ClAlPc WITH OLEUM (100°C, 25 MIN). REVERSE PHASE C_{18} COLUMN ELUTED WITH METHANOL-PHOSPHATE BUFFER (pH 5.0) GRADIENT, STARTING WITH 100% BUFFER. (B) HPLC OF $AlPcS_N$ MIXTURE PREPARED BY $AlCl_3$ CONDENSATION OF PHTHALIC ACID AND 4-SULPHOPHTHALIC ACID (1:1) WITH UREA. (REPRINTED FROM *J. PHOTOCHEM. PHOTOBIOL. B BIOL.*, **9**, M. AMBROZ, A. BEEBY, A.J. MACROBERT, M.S.C. SIMPSON, R.K. SVENSEN AND D. PHILLIPS, PREPARATIVE, ANALYTICAL AND FLUORESCENCE SPECTROSCOPIC STUDIES OF SULPHONATED ALUMINIUM PHTHALOCYANINE PHOTOSENSITISERS, 87–95, COPYRIGHT 1991, WITH PERMISSION FROM ELSEVIER SCIENCE).

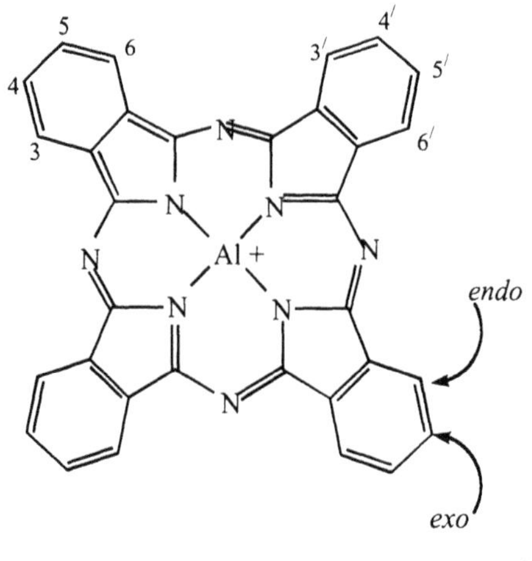

Two SO_3H to be substituted

endo

exo

exo-exo: 4,4′, 4,5′ ; 5, 4′

endo-endo: 3,3′, 3,6′ ; 6,3′

endo-exo: 3,4′, 3,5′ ; 6,4′, 6,5′

74 Regioisomers of $AlPcS_{2a}$

In spite of the mixtures which inevitably result, the product of sulphonation of ClAlPc (essentially a mixture of di and tri sulphonic acids, 'Photosens') has been employed extensively by Stradnadko in clinical trials in Moscow.

Studies have been made on the sulphonated gallium phthalocyanines. The ability of $ClGaPcS_2$ photosensitisers to kill Chinese hamster lung fibroblasts in culture increases with hydrophobicity (increasing retention time on reverse phase HPLC) as follows:

	Substitution	Retention Time	Photoinactivation
$ClGaPcS_{2o}$	4,4"	40	~ 0
	4,5"	42	X
$ClGaPcS_{2a}$	4,5'	46	XX
	4,4'	49	XXX

7.2.4.3 3-(1-Alkyloxyethyl)-3-devinylpyropheophorbide a

The lipophilicity of the 3-(1-alkyloxyethyl)-3-devinyl pyropheophorbide *a* system (**75**) can be varied by changing the carbon chain length of the alkyloxy substituent. As the chain length varies from C_1 to C_{12} the log *P* value (ie $\log_{10}$ $[\text{substrate}]_{\text{octanol}}$ $/[\text{substrate}]_{\text{pH7.4 buffer}}$) changes smoothly from 3.0 to 8.6. When *in vivo* biological activity (reduction of the rate of tumour growth in a tumour implant in the mouse) was determined along this series, maximum activity was observed at R = C_6 (Figure 7.3). This shows clearly the advantage of a system which is amphiphilic, being

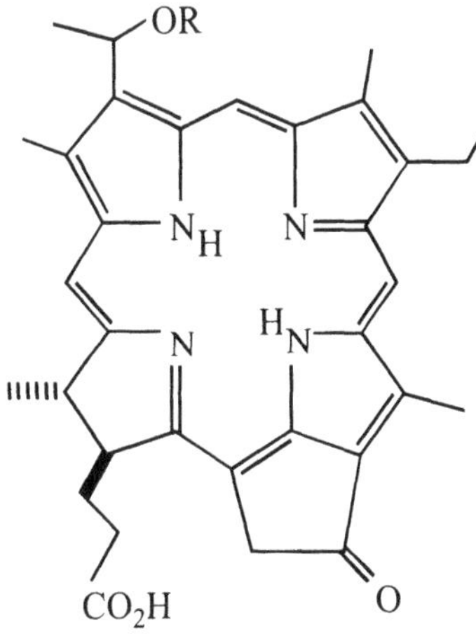

75 3-(1-Alkyloxyethyl)-3-devinylpyropheophorbide *a*

neither too hydrophilic (as in the methoxy derivative, log *P* = 3.05, which was virtually inactive biologically) nor too lipophilic (as in the dodecyloxy derivative, log *P* = 8.6). The hexyloxy derivative was five times more potent than the dodecyloxy derivative, even though the concentration of photosensitiser in the tumour was higher in the latter case (0.5 nmol g^{-1}) than in the former (0.16 nmol g^{-1}). Thus biological activity is not related to *total* photosensitiser concentration in a simple way: it depends on where the drug is located in the cell, which in turn will depend on structure.

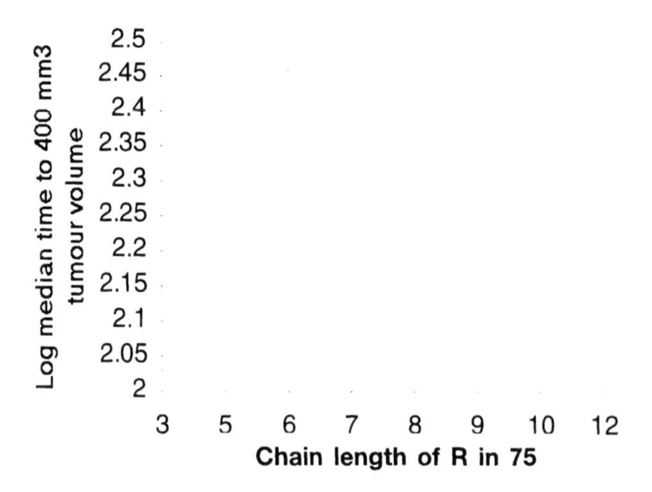

FIGURE 7.3 VARIATION OF THE BIOLOGICAL ACTIVITY OF 3-(1-ALKYLOXYETHYL)-3-DEVINYLPYROPHEOPHORBIDE *A* IN REDUCING THE RATE OF TUMOUR GROWTH IN AN ANIMAL MODEL UNDER STANDARD CONDITIONS (0·4 OR 0·5 μMOL KG^{-1}, I.V. IN TWEEN 80 / BUFFER, DRUG-LIGHT INTERVAL 24H, IRRADIATED AT 665 NM, WITH A LIGHT DOSE OF 135 J CM^{-2} AT A FLUENCE RATE OF 75 MW CM^{-2}) WITH THE CHAIN LENGTH OF THE ALKOXY GROUP. HYDROPHOBICITY AND LOG *P* INCREASE AS CHAIN LENGTH INCREASES. ERROR BARS OMITTED FOR CLARITY. (BASED ON PRIMARY DATA IN B.W. HENDERSON, D.A. BELLNIER, W.R. GRECO, A. SHARMA, R.K. PANDEY, L.A. VAUGHAN, K.R. WEISHAUPT AND T.J. DOUGHERTY, *CANCER RES.*, 1997, **57**, 4000–4007).

As a result of these experiments, 3-(1-hexyloxyethyl)-3-devinylpyropheophorbide *a* (**75**, R = n-C_6H_{13}) has been selected for clinical evaluation by QLT PhotoTherapeutics Inc (Vancouver) (see Panel 14.5).

7.2.4.4 Spacing of hydrophilic substituents on the hydrophobic framework

In rapidly dividing cells, including tumours, the protein tubulin is present in higher amounts than usual since assembly and depolymerisation of microtubules, of which tubulin is a subunit, increases during mitosis. Winkelman, Arad and Kimel have developed the proposition that tubulin is a major receptor for photosensitiser interaction, and that a critical distance of 12Å is needed between the oxygen atoms in solubilising substituents (eg SO_3^-, CO_2^-, OH) in order to facilitate the fit into the tubulin receptor site. Although this hypothesis is, at the very least, a considerable oversimplification (some photosensitisers, such as zinc phthalocyanine, do not have such substituents and are still quite effective PDT agents) it is nonetheless an interesting approach. Figure 7.4 illustrates the situation for isomers of 5,10,15,20-tetrakis(hydroxyphenyl)porphyrin (**76** shows the *meta* isomer, m-THPP), $TPPS_{2a}$ (**70**), $TPPS_{2o}$ (**71**), regioisomers of phthalocyanine disulphonic acid (PcS_2; **74** shows the aluminium(III) complex of PcS_{2a}), HpD (ether and ester linkages) and bis(8-anilinonaphthalene-1-sulphonate) (bis-ANS), a known tubulin binding agent.

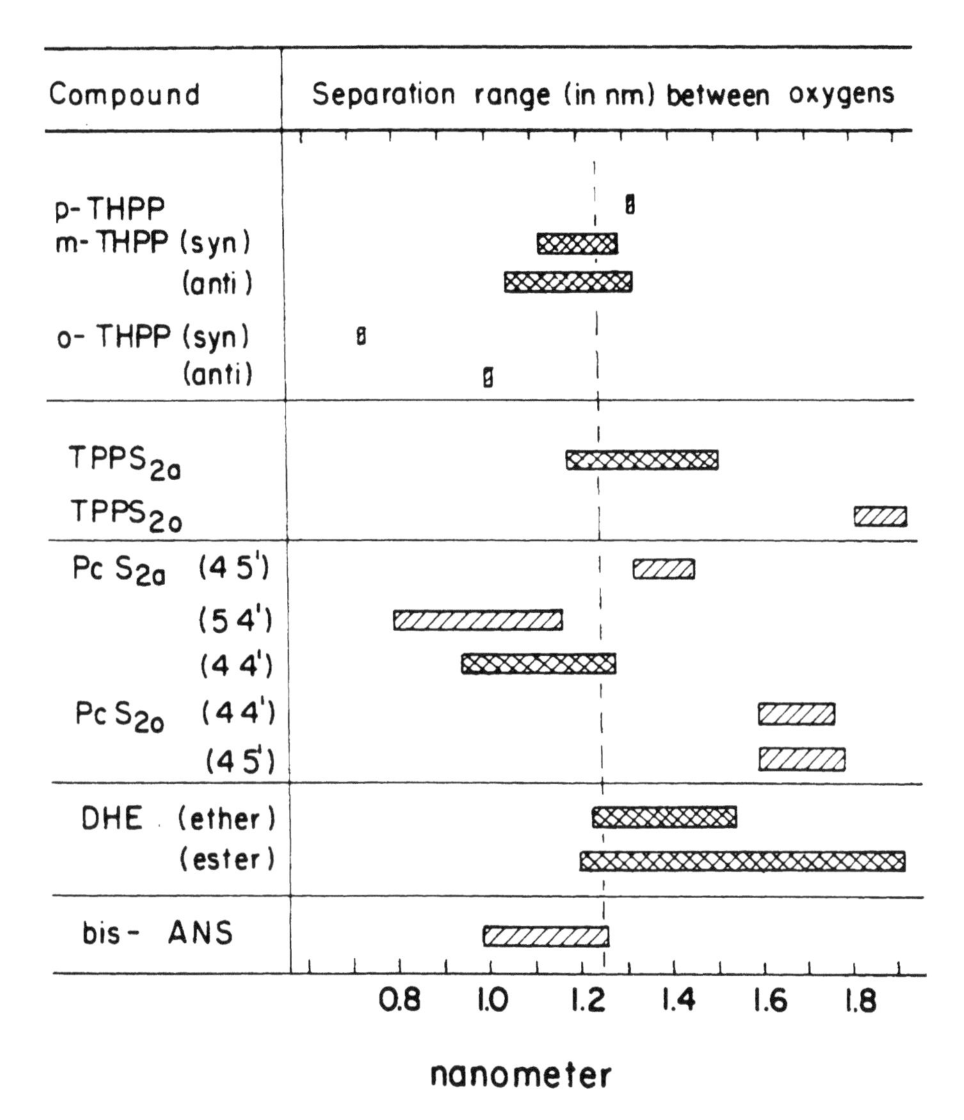

FIGURE 7.4 BIOLOGICAL ACTIVITY OF VARIOUS PHOTOSENSITISERS IN RELATION TO THE SEPARATION RANGES OF THE OXYGEN SUBSTITUENTS. (REPRINTED FROM *J. PHOTOCHEM. PHOTOBIOL., B. BIOL.*, **18**, J.W. WINKELMAN, D. ARAD AND S. KIMEL, STEREOCHEMICAL FACTORS IN THE TRANSPORT AND BINDING OF PHOTOSENSITIZERS IN BIOLOGICAL SYSTEMS AND IN PHOTODYNAMIC THERAPY, 181–189, COPYRIGHT 1993, WITH PERMISSION FROM ELSEVIER SCIENCE).

7.2.5 Delivery Systems

Since the photosensitisers are solids, for intravenous injection they require a delivery system of some sort. Some compounds (eg δ-ALA, $TPPS_4$) are sufficiently water soluble that they can be injected in aqueous buffer. In other cases, some polarity may have been designed into the structure under section 7.2.4, and although these products may not be freely soluble in water, they can be handled in aqueous mixtures. Thus m-THPC (77) can be administered in a solution in ethanol: polyethyleneglycol

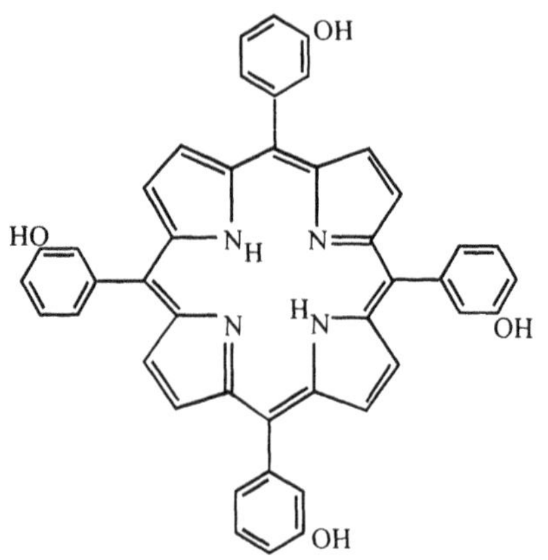

76 m-THPP
5,10,15,20-Tetrakis
(*m*-hydroxyphenyl)porphyrin

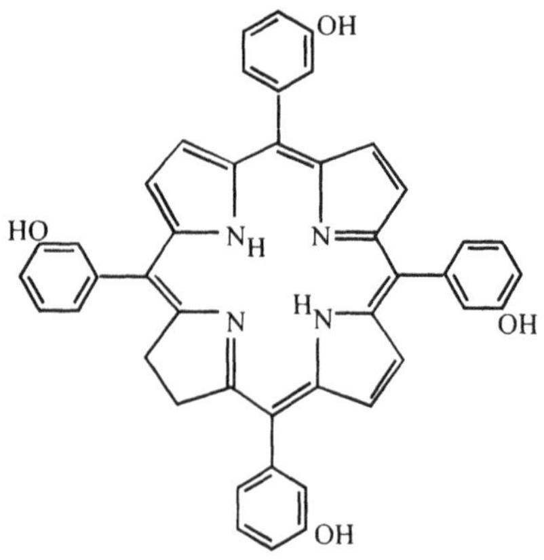

77 m-THPC
5,10,15,20-Tetrakis
(*m*-hydroxyphenyl)chlorin

400: water (2 : 3 : 5, v/v), while with the hexyloxypyropheophorbide *a* (**75**, R = n-C_6H_{13}) an aqueous detergent solution (Tween 80, **82** in Panel 7.2) is employed.

Some researchers have preferred to develop photosensitisers which are highly hydrophobic, and to employ an additional component to transport the photosensitiser. Zinc(II) phthalocyanine (**30**, as Zn(II) complex) is an example here. Such compounds may be delivered using transporter molecules which give the appearance of solubilisation by favourable non-covalent interactions with the photosensitiser molecule. Examples of such systems include oil-based emulsions, inclusion systems such as cyclodextrins, detergents, liposomes and lipoproteins. These are considered in Panel 7.2. A comparison of ethanol solution with various specialised delivery systems (eg oil-based emulsion, γ-cyclodextrin, Panel 7.2) for the hydrophobic sensitiser tin etiopurpurin (**78**) has shown that, in cell culture, sensitiser uptake is greatest when the specialised delivery systems are employed. Delivery systems using microparticulates have been described. Thus biodegradable cyanoacrylate nanospheres (150–250 nm diameter) have been used to encapsulate phthalocyanines and naphthalocyanines.

The incorporation of adjunct components into the photosensitiser inoculum has been found to modulate (usually potentiate) biological activity. Reducing agents appear to increase the effectiveness of PDT. Ascorbic acid appears to be effective in this way, and other bioreductive drugs (nitroimidazoles) show a potentiating effect with $AlPcS_2$ *in vivo*. Glucose (but not galactose) is reported to increase photosensitiser uptake by tumour tissue. When the common antioxidant BHA (3(2)-t-butyl-4-hydroxyanisole) is applied to cells immediately after HpD-PDT the photocytotoxic effect is reported to be 3–4 times greater than when the antioxidant is added before irradiation. Fluoride ion actually protects Chinese hamster cells grown in culture from photodamage by ClAlPc: it has been supposed that $HOAlPc.H_2O$ is the actual photosensitiser and that this is converted into the fluoride complex, which is less active.

Panel 7.2 Delivery Systems

Various delivery systems have been used to solubilise drugs, including photosensitising drugs. In the majority of cases the substrate is not in true homogeneous solution, but is distributed in micelles or vesicles. The stability of such preparations with time is an important consideration: precipitation of the substrate is clearly undesirable.

Detergent-like molecules

Here the aim is to incorporate the hydrophobic substance within a micelle or vesicle (a spherical body, spontaneously formed, the wall of which has a bilayer structure: also called a liposome, see below), thus:-

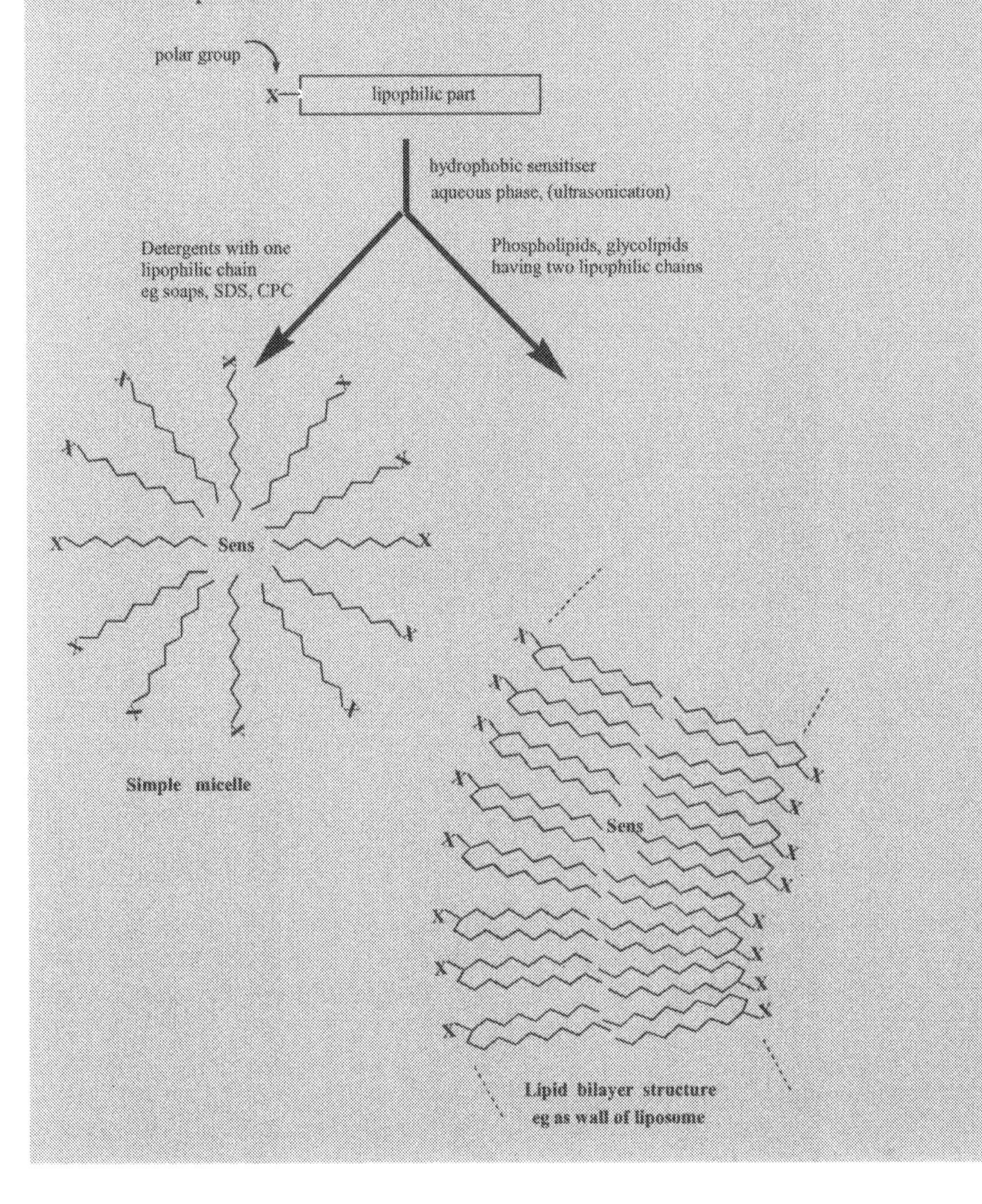

The polar group may be cationic (eg cetyl pyridinium chloride, CPC), anionic (sodium stearate, sodium dodecylsulphate SDS), zwitterionic (phosphatidylcholine) or neutral (polyether/polyalcohol derivatives, as in **82** and **84** below).

The concentration of amphiphile required for micelle formation is termed the *critical micelle concentration.* Micelles may typically have a diameter of <200Å, but the bilayer structure usually formed by amphiphiles with two fatty chains can be much larger (up to 1 mm). Agitation (usually ultrasonication) causes the bilayer structure to rearrange its thermodynamically unstable edges and form bilayer vesicles (also called liposomes) enclosing a sphere of the aqueous phase (see below). The structures (micelles, bilayer sheets, vesicles) are remarkably stable because of the hydrophobic interactions of the hydrocarbon chains with one another, and the hydrophilic interaction of the polar end groups with the aqueous medium. Nonetheless the situation is not a static one: thus, individual amphiphilic molecules can move about laterally in the sheet or vesicle with little energy demand (membrane fluidity); and if the structure should be broken, it readily reseals itself. *Ionic species and polar molecules (apart from water) do not penetrate the bilayer membrane easily, but hydrophobic molecules can get through.*

For the delivery of the more hydrophobic photosensitisers *in vitro* or *in vivo* neutral detergents appear to be favoured over simple ionic detergents. Examples include Tween 80 (polyoxyethylene sorbitan monooleate **82**) and Cremophor EL. The latter is prepared by treating castor oil (which contains glycerides of ricinoleic acid (**83**)) with ethylene oxide, and can be represented by structure (**84**).

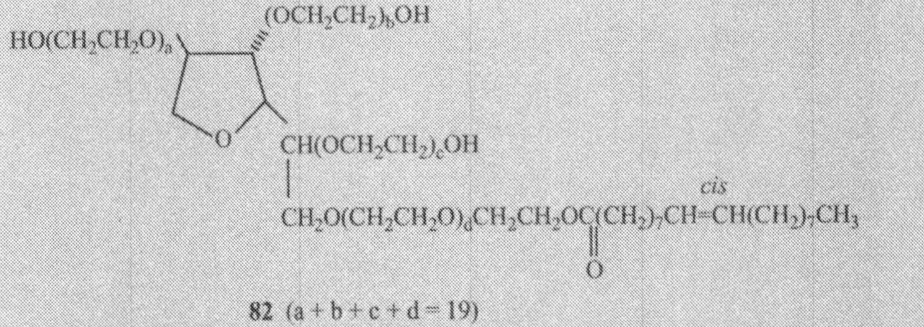

$HO(CH_2CH_2O)_a$ $(OCH_2CH_2)_bOH$
$CH(OCH_2CH_2)_cOH$
cis
$CH_2O(CH_2CH_2O)_dCH_2CH_2OC(=O)(CH_2)_7CH{=}CH(CH_2)_7CH_3$

82 (a + b + c + d = 19)

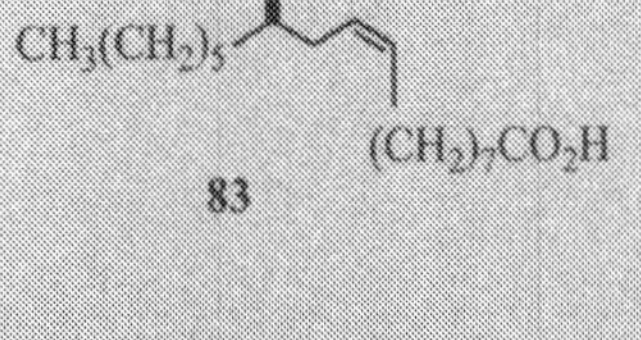

OH
$CH_3(CH_2)_5$
$(CH_2)_7CO_2H$

83

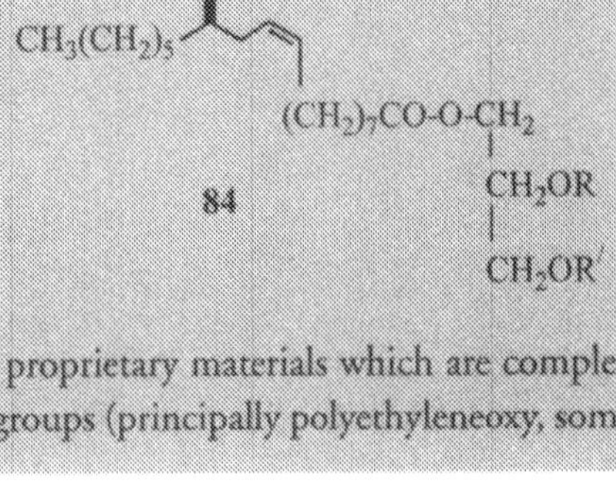

$(OCH_2CH_2)_nOH$
$CH_3(CH_2)_5$
$(CH_2)_7CO\text{-}O\text{-}CH_2$
CH_2OR
CH_2OR'

84

Both Tween 80 and Cremophor EL are proprietary materials which are complex mixtures, and which contain neutral polar groups (principally polyethyleneoxy, some alkanol) together with a hydrophobic feature such as a fatty acid chain. They have the property of forming stable emulsions of hydrophobic substances in aqueous media.

Liposomes are vesicles with a bilayer (or multilayer) structure which have diameters in the range 20 nm – 1 μm. They are generally prepared from synthetic phospholipids by ultrasonication: careful attention to experimental detail is required if reproducible results are to be obtained. Commonly used examples include dipalmitoylphosphatidylcholine (**85**, DPPC), palmitoyloleylphosphatidylcholine (POPC), and dioleylphosphatidylcholine (OOPS). In this case the polar group is zwitterionic, but overall the system is formally neutral. Liposomes have been used, for example, in the administration of phthalocyanines and naphthalocyanines.

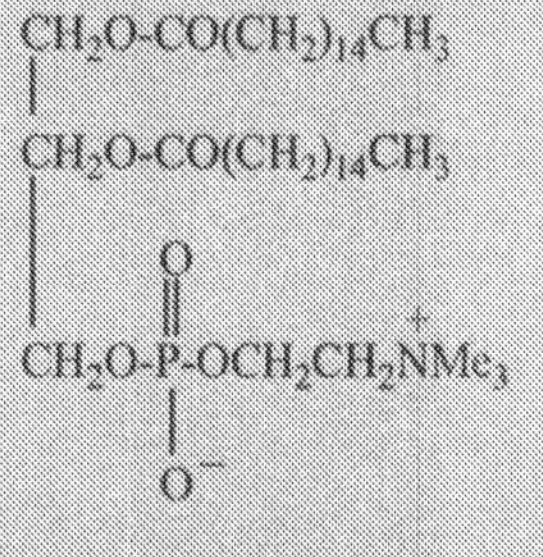

85 Dipalmitoylphosphatidylcholine (DPPC)

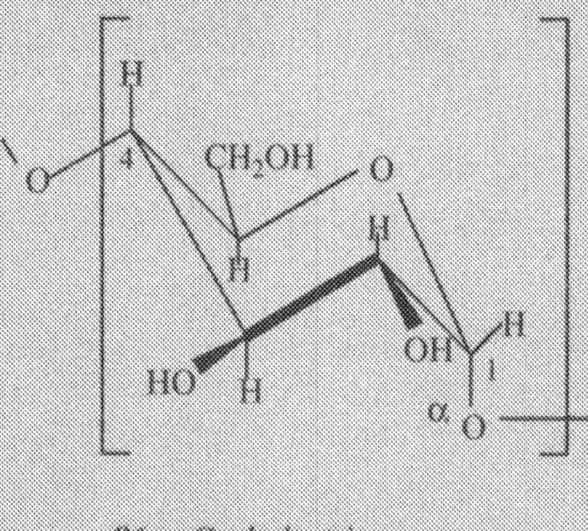

86 Cyclodextrins
n = 6, α; n = 7, β; n = 8, γ

The solubility of oxygen in the lipid phase of cellular constituents is significantly higher (about tenfold) than that in water. Table 7.2 shows oxygen concentrations in some common organic solvents in equilibrium with air to illustrate this point. Consequently, the lipid part of cellular membranes (and micelles and liposomes) concentrates both hydrophobic sensitiser *and* oxygen with respect to bulk aqueous phase. Photooxygenation carried out in the presence of such organised systems (eg in detergent) may be expected to differ from what is found in solution (see section 12.4.1.1 for an example).

TABLE 7.2 CONCENTRATION OF OXYGEN IN VARIOUS SOLVENTS AT 25°C IN AIR.

	10^{-3} mol l^{-1}	mg l^{-1}
Perfluoroheptane	5.2	167
Hexane	3.1	99
Ethanol, methanol	2.1	67
Benzene	1.9	61
Octan-1-ol	1.5	48
Water	0.27	8.6

(S.L. Murov, I. Carmichael and G.L. Hug, *Handbook of Photochemistry*, (2nd edition), Dekker, New York, 1993, Table 12–3)

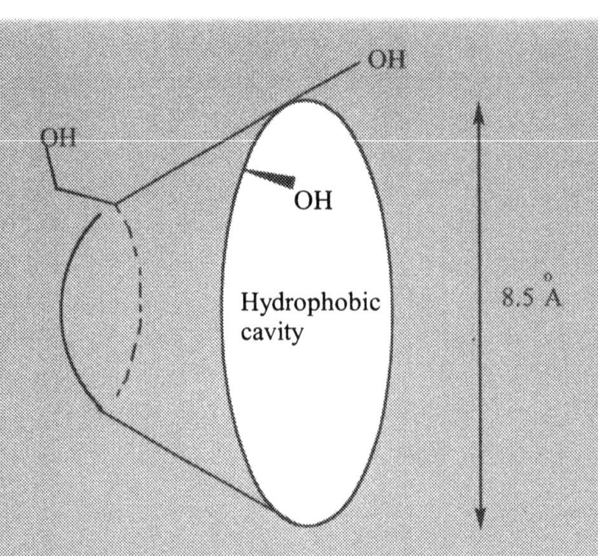

FIGURE 7.6 DIAGRAM OF GROSS STRUCTURE OF γ-CYCLODEXTRIN. THE HYDROXYL SUBSTITUENTS ON ONE GLUCOSYL UNIT ARE SHOWN.

Inclusion complexes

The cyclodextrins (cycloamyloses) are a series of cyclic oligomers of α-glucose joined by 1,4-linkages (**86**) which are produced by the action of an enzyme from *Aerobacillus macerans* on starch. The inside of the ring of glucose units in (**86**) is hydrophobic, while the outside is hydrophilic (Figure 7.6). The annular diameter in γ-cyclodextrin (**86**, n = 8) is about 8.5Å, and is large enough to form an inclusion complex with all or part of a hydrophobic photosensitiser molecule and so transport it in an aqueous medium. Such complexes of tin(IV) etiopurpurin (**78**) caused a greater tumoricidal effect than that observed when Cremophor or DPPC liposomes were used as delivery agents.

Lipoprotein carriers

The *natural* carriers of photosensitising drugs in the bloodstream are the proteins of the serum. Hydrophilic drugs are associated with the larger proteins (albumins, globulins) while the hydrophobic drugs are transported predominantly by the lipoproteins. The hydrophobic zinc(II) phthalocyanine, injected in liposomes or in Cremophor into rabbits, is taken up initially mainly by lipoprotein.

Lipoproteins from blood serum can be separated by centrifugation into very low density (VLDL), low density (LDL) and high density (HDL) lipoprotein. These have been isolated and themselves examined as delivery systems for photosensitisers. In comparative experiments on the effect of haematoporphyrin (**2**, Hp) on MS-2 fibrosarcoma implants in mice, a Hp-LDL preparation (a mixture, not a covalently bound system) showed significantly greater selectivity for tumours than did free Hp or Hp-HDL or Hp-VLDL preparations. Analogous experiments with zinc(II) phthalocyanine showed that tumour levels following ZnPc-LDL administration are twice those observed with liposomes (ZnPc-DPPC) as the delivery agent.

The problem with all these special delivery systems and adjuncts is that they add to the complexity of the inoculum. If it is at all possible, it is much better for the chemist to build such benefit as they may confer into the covalent structure of the photosensitiser.

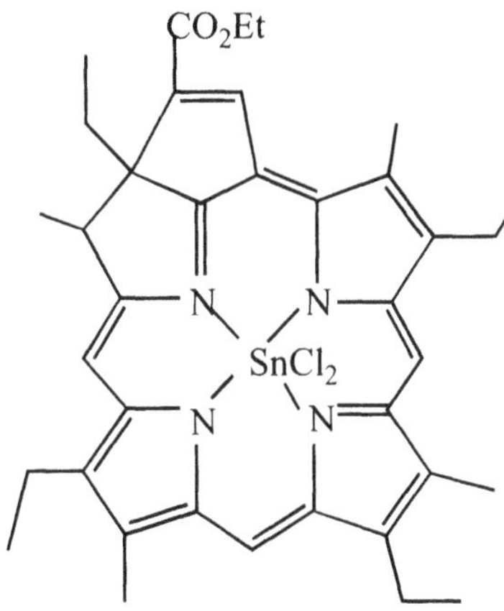

78 Tin etiopurpurin (SnEt2)

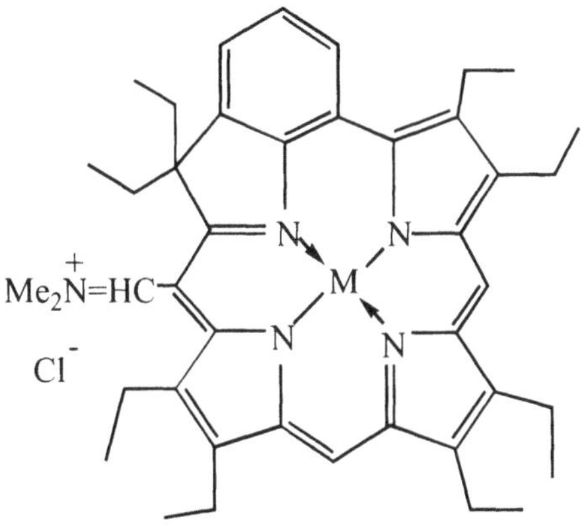

79 Octaethylbenzochlorin iminium salt

7.2.6 Photophysical Properties

For use in the diagnostic mode, where photosensitiser fluorescence is being employed to delineate the tumour, it is the quantum yield of fluorescence (ϕ_f) that is important. Photosensitisers which have high ϕ_f values cannot have high ϕ_T (and therefore ϕ_Δ) values because $\phi_f + \phi_T \leq 1$, according to the Stark-Einstein Law of Photochemical Equivalence (section 1.1). Consequently it makes sense to develop diagnostic compounds which are different from those used in therapy. So far, with the emphasis on therapy, relatively little has been done on this. But the high ϕ_f values of rare earth complexes of porphyrins make these compounds potential candidates; and delayed laser-induced fluorescence is developing as a promising imaging technique using porphyrins as the fluorescent agents.

For photodynamic therapy the requirements are (i) that E_T for the photosensitiser must be greater than 94 kJ mol^{-1} (the energy of singlet oxygen above its ground state: section 3.1), (ii) the triplet must be generated in satisfactory quantum yield, and (iii) it must have a sufficient lifetime so that singlet oxygen is formed efficiently. This presupposes that the singlet oxygen mechanism is the predominant one, which is currently the accepted view: it needs to be recognised, however, that this is still a matter for experiment and discussion (section 4.4.4). If singlet oxygen is a key intermediate, then it is the quantum yield of singlet oxygen formation, ϕ_Δ, that is important. Table 2.1 presents ϕ_Δ values for a range of different structural types, while ϕ_T and ϕ_Δ values for some porphyrin compounds are given in Table 4.3. To be a useful photodynamic sensitiser it appears that a ϕ_Δ value >0.3 is desirable.

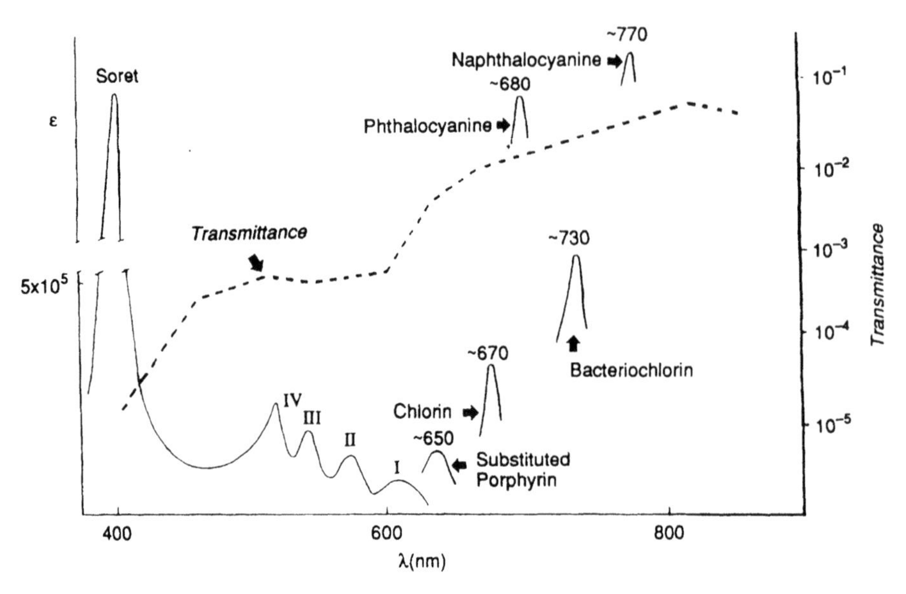

FIGURE 7.5 THE ABSORBANCE OF PORPHYRIN COMPOUNDS IN RELATION TO TISSUE TRANSMITTANCE (SCHEMATIC). BAND I ONLY IS SHOWN, EXCEPT FOR THE PORPHYRIN ABSORPTION SPECTRUM ON THE LEFT. THE TRANSMITTANCE CURVE REFERS TO A SAMPLE OF HUMAN SCROTAL SAC, 7MM THICK (S. WAN, J.A. PARRISH, R.R. ANDERSON AND M. MADDEN, *PHOTOCHEM. PHOTOBIOL.*, 1981, **34**, 679–681). THE BROAD FEATURE AT 500-600 NM IS ATTRIBUTED TO HAEMOGLOBIN. (FROM R. BONNETT, *CHEM. SOC. REV.*, 1995, **24**, 19–33, WITH PERMISSION).

There have been occasional reports of effective photosensitisers which have low or zero ϕ_Δ values, and which presumably therefore operate by a Type I (or some other) mechanism. The copper(II) octaethylbenzochlorin iminium salt (**79**, M = Cu(II)) is a case in point. It is a sensitiser *in vivo* and *in vitro*, but τ_T is <20 ns. Singlet oxygen formation is not important, but oxygen nevertheless is required. Since superoxide dismutase and catalase inhibit activity *in vitro*, an electron transfer mechanism is implicated (section 4.4.1). However, the metal-free compound (**79**, M = 2H) has a higher biological activity, and bioactivity increases further still with the zinc(II) complex (**79**, M = Zn(II)), presumably due to the capacity of these two compounds to act as singlet oxygen photosensitisers.

7.2.7 Red Absorption

The transmission of visible light through human tissue varies to some extent with the tissue sample, but principally depends on the wavelength of the light. At the blue end of the visible the absorption (eg by haemoglobin) and scattering are greatest, but both factors are much lessened in the red region. The transmittance of a slab of human tissue 0.7 mm thick is about 10^{-1} – 10^{-2} at 800 nm, but only 10^{-5} at 400 nm (Figure 7.5). Consequently if we want light to be picked up efficiently by a photosensitiser *in vivo*, the photosensitiser needs to have a strong absorption band in the red (say 650–800 nm). Figure 7.5 shows the relation between tissue transmittance, the absorption spectrum of a typical porphyrin (including HpD), and

Panel 7.3 Synthesis of the [26]Porphyrin dication System (80)

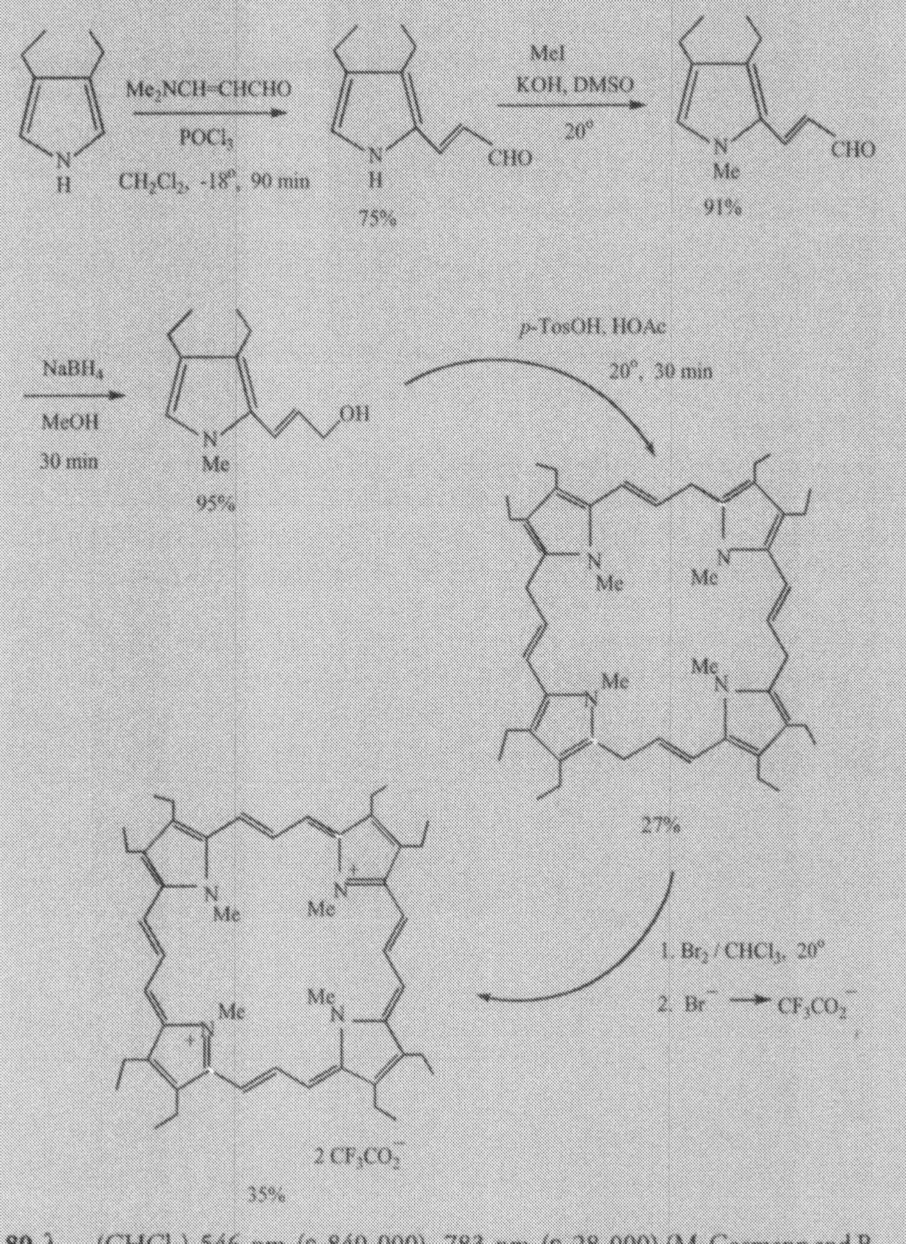

$Me_2NCH{=}CHCHO$, $POCl_3$, CH_2Cl_2, -18°, 90 min — 75%

MeI, KOH, DMSO, 20° — 91%

$NaBH_4$, MeOH, 30 min — 95%

p-TosOH, HOAc, 20°, 30 min — 27%

1. Br_2 / $CHCl_3$, 20°
2. $Br^- \rightarrow CF_3CO_2^-$

$2\ CF_3CO_2^-$ — 35%

80 λ_{max} ($CHCl_3$) 546 nm (ε 840 000), 783 nm (ε 28 000) (M. Gosmann and B. Franck, *Angew. Chem. Intl. Ed. Engl.*, 1986, **25**, 1100-1101).

Band I of a number of selected photosensitiser types. There are huge differences. Band I of a typical porphyrin is at about 630 nm with $\varepsilon \sim 5000$; whereas for a disiloxysilicon(IV) naphthalocyanine Band I has $\lambda_{max} \sim 775$ nm ($\varepsilon \sim 500\,000$).

However, in seeking to improve the sensitiser, this pushing into the red and infrared cannot be carried too far, for three reasons. Firstly, absorption by water, the main constituent of living tissue, begins to cut in, and transmittance falls. Secondly, to get the λ_{max} of the photosensitiser to move to lower energy, conjugation is increased, and with increased electron availability the redox potential is lowered: the photosensitiser system itself becomes more sensitive to oxidation (see photobleaching, Chapter 12). Thirdly, since λ_{max} for Band I represents approximately the energy of S_1, and since the energy of T_1 is lower than this (for porphyrins by about 40–50 kJ mol^{-1}, Figure 4.3), a point will be reached (for Band I at about 830 nm = 12050 cm^{-1} = 144 kJ mol^{-1}) where the *triplet* energy of the photosensitiser is below 94 kJ mol^{-1} (1270 nm) or thereabouts, and singlet oxygen can no longer be efficiently generated.

An instructive example where the triplet energy is lowered in this way is to be found in a group of expanded porphyrins prepared by Franck in Münster. These compounds are named using a numeral to denote the number of p electrons in the essential aromatic system (porphyrin would be [18] porphyrin in this nomenclature). The [26] porphyrin dication (**80**) has Band I at λ_{max} 783 nm (ε 28 000), and has a low but appreciable ϕ_Δ value (0.19). However when the aromatic macrocycle is expanded by a further 4 double bonds ([34] porphyrin dication, **81**) Band I has shifted to λ_{max} 997 nm (ε 24 000) and $\phi_\Delta = 0$. Presumably, therefore, the first excited triplet energy value for **81** must be less than 94 kJ mol^{-1}.

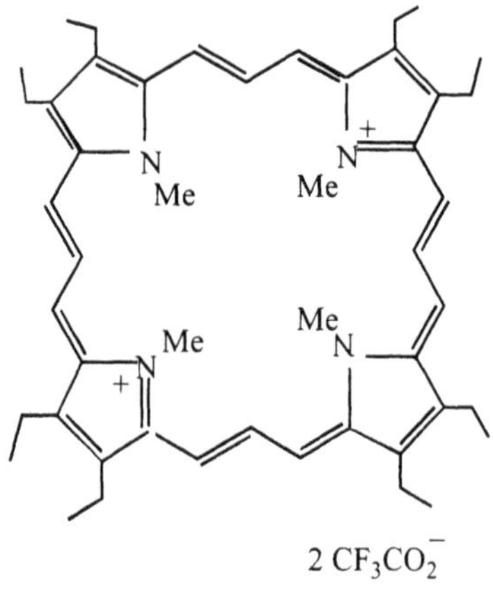

80 [26]Porphyrin derivative

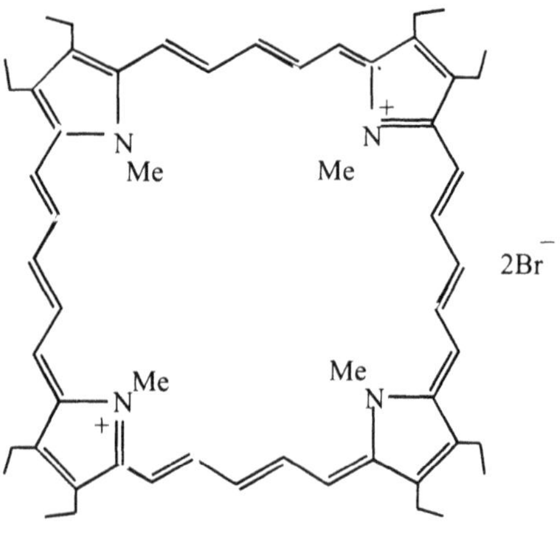

81 [34]Porphyrin derivative

The synthesis of **80** is shown in Panel 7.3: compound **81** is prepared in an analogous manner.

Thus in the design of PDT photosensitisers two sorts of physical characteristics are important — the photophysical and the solution/partition. It is not difficult to find substances which have suitable absorption in the red, and suitable ϕ_Δ values. A large range of macrocyclic pigments satisfy these photophysical requirements (see Table 4.3). From the viewpoint of physical chemistry the principal hurdle here, as with other drugs, is in getting the solution chemistry right, that is in designing in the features which give the proper balance of hydrophilicity and lipophilicity so as to confer an effective and safe therapeutic result.

BIBLIOGRAPHY

J.W. Winkelman, D. Arad and S. Kimel, 'Stereochemical factors in the transport and binding of photosensitisers in biological systems and in photodynamic therapy'. *J. Photochem. Photobiol., B. Biol.*, 1993, **18**, 181–189. An attempt to relate photosensitiser activity to the separation of hydrogen bonding functions within it, based on the proposition that tubulin is the major acceptor site. See section 7.2.4.4.

M. Kreimer-Birnbaum, 'Modified porphyrins, chlorins, phthalocyanines and purpurins: second generation photosensitizers for photodynamic therapy', *Seminars in Hematology*, 1989, **26**, 157–173.

D. Phillips, 'Chemical mechanisms in photodynamic therapy with phthalocyanines', *Prog. Reaction Kinetics*, 1997, **22**, 175–300. Includes structures for isomers of $AlPcS_2$ (see section 7.2.4.2).

G. Schermann, R. Schmidt, A. Völcker, H-D. Brauer, H. Mertes and B. Franck, 'Photophysical properties of [26] porphyrin'. *Photochem. Photobiol.*, 1990, **52**, 741–744. Includes a note of comparison with [34] porphyrin (**81**).

Chapter 8

PORPHYRIN PHOTOSENSITISERS

"Examination of the formula for aetioporphyrin $C_{37}H_{36}N_4$*, shows that the number of hydrogen atoms is small when compared with the number of carbon atoms in the molecule, whence it may be deduced that the four pyrrole nuclei must be united to each other in such a way as to leave only a few positions free for hydrogen atoms to occupy.

W. Küster (*Z. physiol. Chem.*, 1912, **82**, 463) has suggested a constitutional formula for haemin which takes this point into consideration, but as his formula contains a sixteen-membered ring, it will hardly recommend itself to the ordinary chemist unless no simpler structure can be suggested which will give equally satisfactory results."

A.W. Stewart, *Ann. Rep. Prog. Chem.*, 1913, **10**, 157.

(illustrating the attitude of incredulity adopted towards the Küster structure for the porphyrin nucleus, which subsequently turned out to be correct).

*The correct molecular formula is $C_{32}H_{38}N_4$.

8.1 INTRODUCTION

Because of the importance of haematoporphyrin (Chapter 6) the porphyrins have a special position amongst the sensitisers considered for PDT applications. It will emerge that the hydroporphyrins, and especially the chlorins (Chapter 9) appear at present to be the most promising second generation compounds. Nevertheless some porphyrins with extended chromophoric systems (eg *meso*-alkynyl porphyrins, section 8.4.2) and endogenously produced protoporphyrin (section 8.5) deserve special attention.

Porphyrin nomenclature has been considered in Panel 4.2. Various key structural features of porphyrins are summarised in Panel 8.1.

8.2 PORPHYRIN SOURCES

There are in essence three sources of porphyrins which are important in the present context:

(i) porphyrins derived from haemoglobin
(ii) porphyrins prepared by total synthesis in the laboratory
(iii) porphyrins generated by manipulation of the biosynthetic pathway to protohaem (Panel 4.1) in some way.

We shall now examine each of these approaches in more detail.

Panel 8.1 Porphyrin Structure and Properties

1. Shape and size. The porphyrin system is a planar tablet about 9Å across, with a central cavity about 4Å diameter, thus:

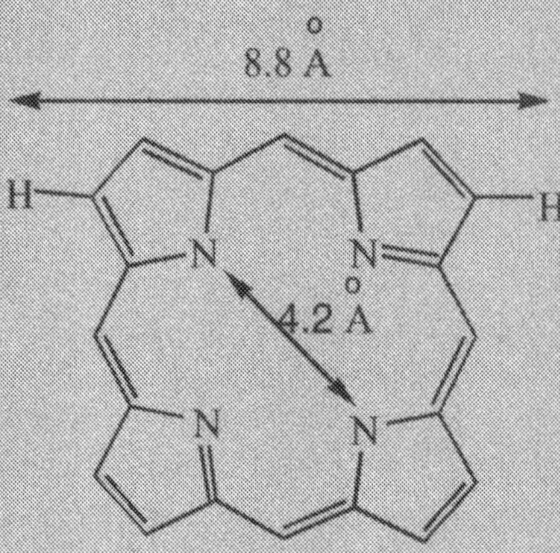

Porphyrin itself is planar, and is not much soluble in organic solvents and insoluble in water (ie it is hydrophobic — Panel 7.1). Substitution at the periphery generally leads to increased solubility in organic solvents. (Octamethylporphyrin is an exception: it is a rigid centrosymmetric molecule and virtually insoluble in organic solvents. Octaethylporphyrin is frequently employed as a model compound because the increased side chain flexibility reduces crystal forces and increases solubility in organic solvents).

Increased peripheral substitution causes overcrowding and leads to the deformation of the macrocycle from planarity. Deformation occurs with less loss of aromaticity than might be expected because of an orbital overlap 'swings and roundabouts' effect, thus:

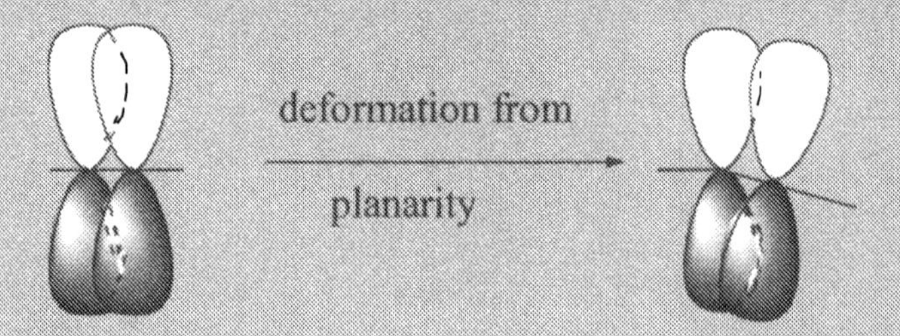

2. Metal complexes. The central hole can accommodate a wide variety of metal ions: complexes are known with nearly every metal, and some metalloids (eg silicon), in the Periodic Table. For PDT applications metals which have empty (eg Al(III)) or complete (eg Zn(II)) *d* shells are most important because the excited states of metal complexes with incomplete *d* shells (first row transition metals) have short lifetimes. Metallation of a porphyrin can usually be effected using a metal salt in a buffering solvent system (eg copper(II) is readily inserted using $Cu(OAc)_2$, NaOAc, HOAc,

$CHCl_3$ under reflux for a few minutes). Some metals require special conditions (eg chromium(II) with chromium hexacarbonyl in *n*-decane). The completion of the reaction is assessed by visible spectroscopy and by TLC.

The rate of demetallation varies over a wide range. Magnesium(II) complexes are so readily demetallated that they are not generally suitable as PDT sensitisers: on the other hand, aluminium(III) complexes are very difficult to demetallate. A practical scale of ease of demetallation refers to the acid conditions required to effect demetallation, exemplified as follows:-

Acid	*Porphyrin complex demetallated*
H_2O	Li(I), Na(I), K(I), Hg(II), Ca(II), Hg(II)
HOAc	Mg(II)
Dilute HCl (~2*M*)	Zn(II), Cd(II), Fe(II), Sn(II)
Conc H_2SO_4	Cu(II), Co(II), Ni(II), Fe(III)
Resist demetallation	Al(III), VO(IV), Sn(IV)

3. Aromaticity

The porphyrin system is aromatic. The aromaticity is due to a cyclic 18π electron system (obeying Huckel's 4n+2 rule) which persists in the chlorin, bacteriochlorin and isobacteriochlorin systems. When picked out from the delocalised system, the cyclic conjugated path is generally written as in A:

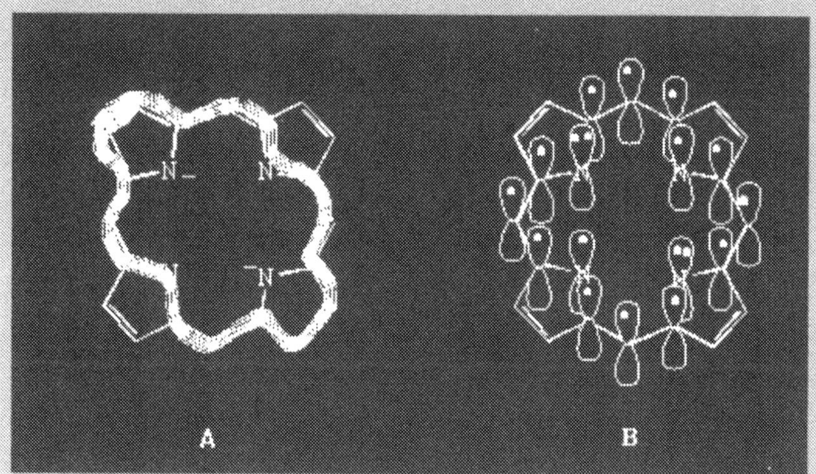

Some authors prefer to consider the 18π electron systems as associated with the inner 16 membered ring as displayed in B. However, this seems unlikely, since if the 16 membered cycle in B truly represents the stable aromatic system, then it should be possible to make stable β-octahydro derivatives of porphyrins and metalloporphyrins, and it is not. (β-Tetrahydroporphyrins ie bacteriochlorins and isobacteriochloris are well known, but even β-hexahydro derivatives have not often been described, and there are no well characterised β-octahydro compounds).

The resonance stabilisation energy of porphyrin has not been measured, but is estimated to be about 1200 kJ mol^{-1}. The system is rugged, and in the absence of

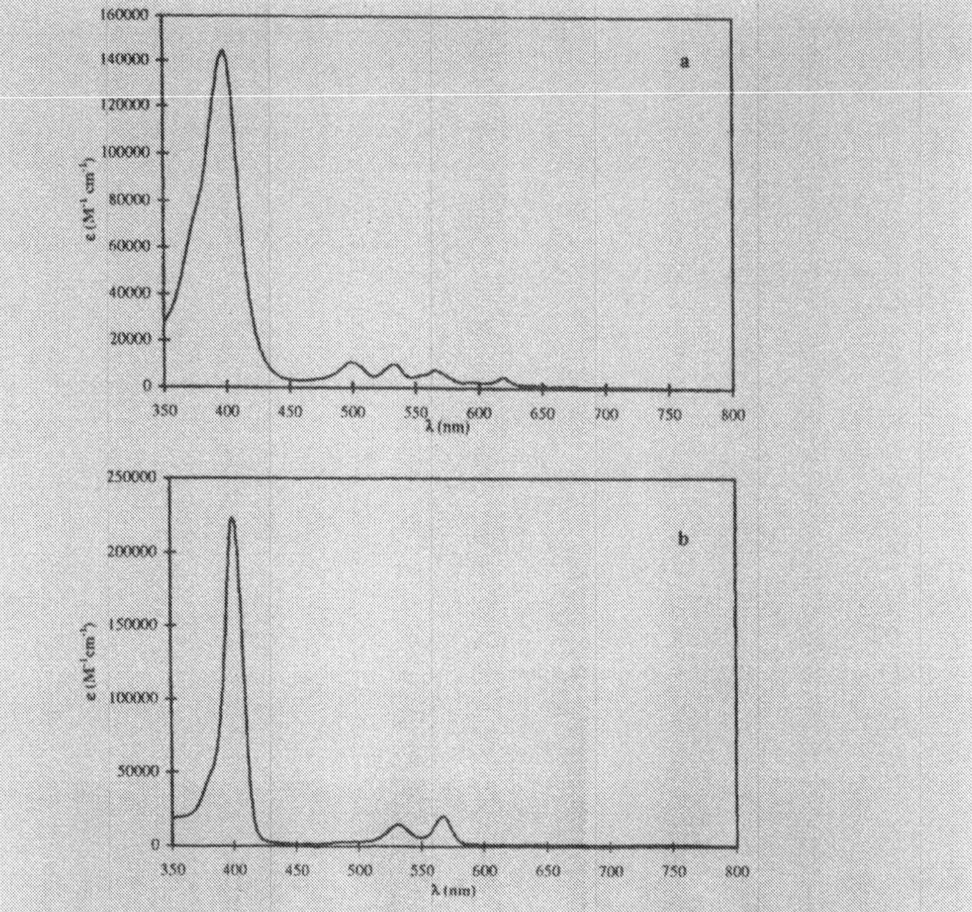

FIGURE 8.1 ELECTRONIC SPECTRA: (A) OF OCTAETHYLPORPHYRIN IN $CHCl_3$; AND (B) ZINC(II) OCTAETHYLPORPHYRIN IN $CHCl_3$.

reactive substituents shows considerable thermal stability in the absence of oxygen: for example octamethylporphyrin sublimes slowly at *ca.* 400°C *in vacuo*. The ^1H NMR spectrum shows striking evidence of an induced ring current (*meso*-proton *ca.* δ 10, NH *ca.* δ −2). Electrophilic aromatic substitution (eg nitration, halogenation) occurs at free *meso*- and β-positions. Free radical substitution is also known (eg benzoyloxylation with $(PhCOO)_2$ occurs at *meso* and benzylic positions).

4. Electronic spectra

The spectra of porphyrins are very characteristic. At about 400 nm there is an intense band (ε ~ 200 000) called the Soret or B band, while in the region of 500-600 nm there are usually four distinct bands, called Q bands. These are denoted by Roman numerals from the low energy side. In the spectra of the divalent metal complexes, and the dianion and dication, the Q bands are reduced to two, and are termed α and β from the low energy side. Figure 8.1 shows the spectra of octaethylporphyrin and zinc(II) octaethylporphyrin to illustrate these points. The B / Q nomenclature was introduced by Platt, who made important early contributions to our understanding of porphyrin spectra (J.R. Platt in *Radiation Biology*, (A. Hollaender, ed.), Vol III, chapter 2. McGraw-Hill, New York, 1956).

In the 1930s the physical chemists associated with Hans Fischer drew up a series of correlations between structure and the relative intensities of the Q bands of the free bases. These correlations were subsequently extended and are as follows:-

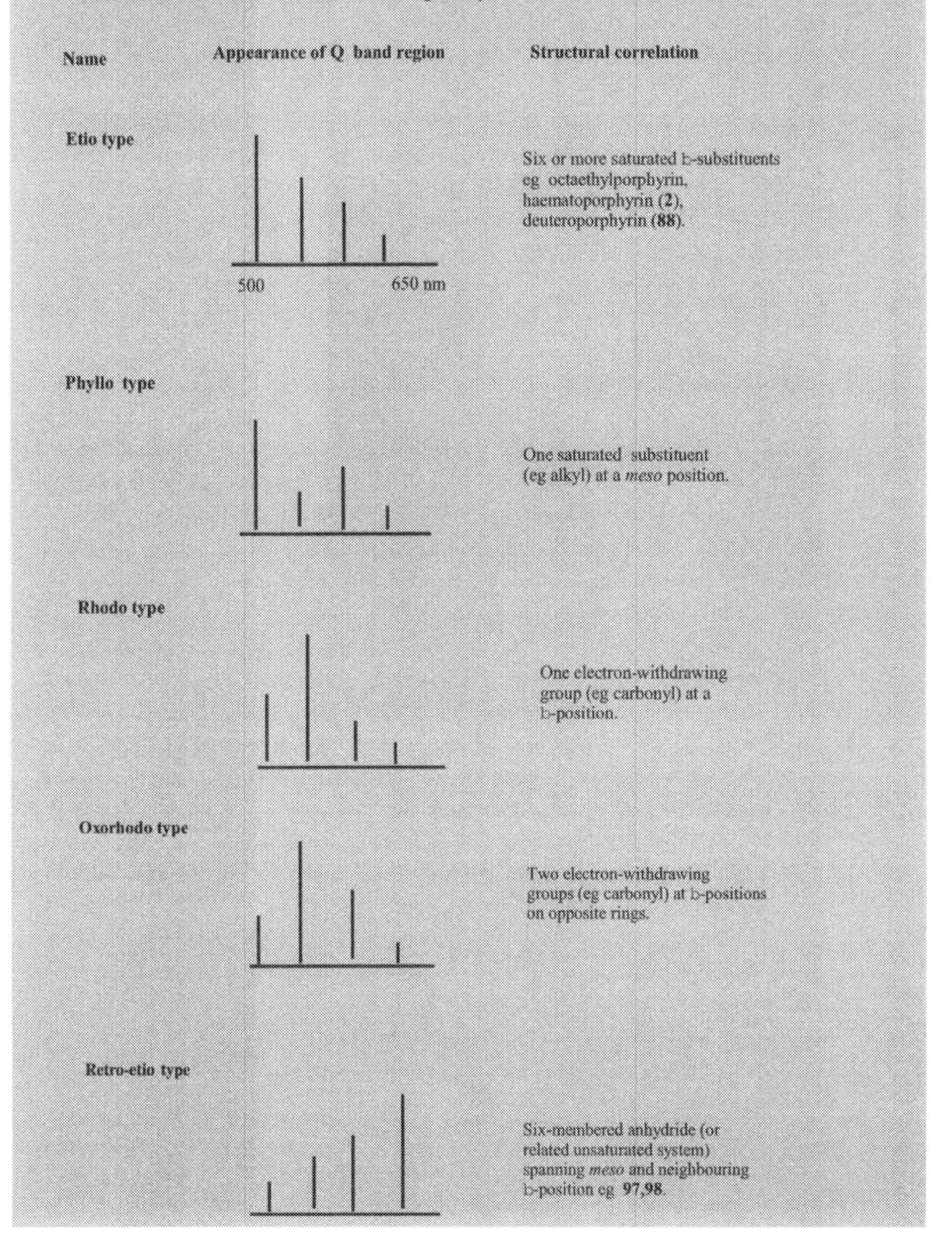

These correlations are empirical: although there are numerous exceptions, historically they have had considerable use. Today they are mainly used as a shorthand way of referring to the appearance of the Q-band region.

The theoretical interpretation of porphyrin spectra has attracted considerable interest. The absorption bands have high molecular extinction coefficients (B ~ 200 000; Q ~ 10 000 l mol^{-1} cm^{-1}) and are regarded as being due to $\pi \rightarrow \pi^*$ transitions. In principle $n \rightarrow \pi^*$ transitions involving non-bonding electrons on nitrogen might also contribute, but such absorption is forbidden (section 2.3.2.2) and would be expected to be weak.

The most successful interpretation of porphyrin spectra is due to Gouterman - the four orbital model. If we start by looking at a simple divalent metalloporphyrin, the macrocycle of which has D_{4h} symmetry, it has been found experimentally that the energy gap between the α- and β-bands is fairly constant (about 1200 cm^{-1}) even though the energies of these bands vary. This suggests that the two bands are not due to different electronic transitions, but are associated with vibrational substructure of one electronic excitation, such that the α-band is regarded as the 0–0 transition and the β-band is regarded as the 0-1 transition. $\pi \rightarrow \pi^*$ transitions in a metalloporphyrin of D_{4h} symmetry are required to be of E_u symmetry ie they consist of two equivalent dipole transitions in the x and y directions. (For the purposes of this argument it is convenient to rotate the porphyrin system anti clockwise by 45° with respect to that normally drawn here, so that the x axis is horizontal, the y vertical as in Figures 8.2 and 8.3). The π molecular orbitals for such a metalloporphyrin have symmetry a_{1u}, a_{2u}, b_{1u}, b_{2u} and e_g. Of these, the HOMOs are a_{2u} and a_{1u}: transitions occur between these and the LUMOs, which have e_g symmetry, as shown in Figure 8.2.

These are the four orbitals of Gouterman's "four orbital" model. Hückel calculations show that the a_{2u} orbital has a higher energy than the a_{1u} orbital, and hence the $a_{2u} \rightarrow e_g$ transition has been assigned to the Q band, and the $a_{1u} \rightarrow e_g$ transition to the B band. These assignments do not account for the intensity differences, however. It is therefore assumed that the upper HOMO's (a_{2u}, a_{1u}) are, *pace* the Hückel calculations, accidentally degenerate: configuration interaction then occurs between the configurations resulting from the transition $a_{2u} \rightarrow e_g$, $a_{1u} \rightarrow e_g$ with the result that the band of lower energy becomes forbidden ie the Q band is the less intense.

Generation of the free base from the metalloporphyrin reduces the symmetry to D_{2h} since the imino protons, while involved in tautomerism on a longer timescale, are located on opposite nitrogen atoms during the time of excitation (~ 10^{-15} s). The x and y axes are no longer equivalent, and consequently the Q band is split into two electronic transitions, Q_x and Q_y. Since each component is associated with vibrational substructure (the 0-1 bands), four bands now appear in the visible (Figure 8.3).

Diprotonation of the free base porphyrin restores D_{4h} symmetry: the visible spectrum returns to the two-banded sort found in metal complexes and the dianion, and the B band sharpens. Splitting of the B band might be expected in the free base.

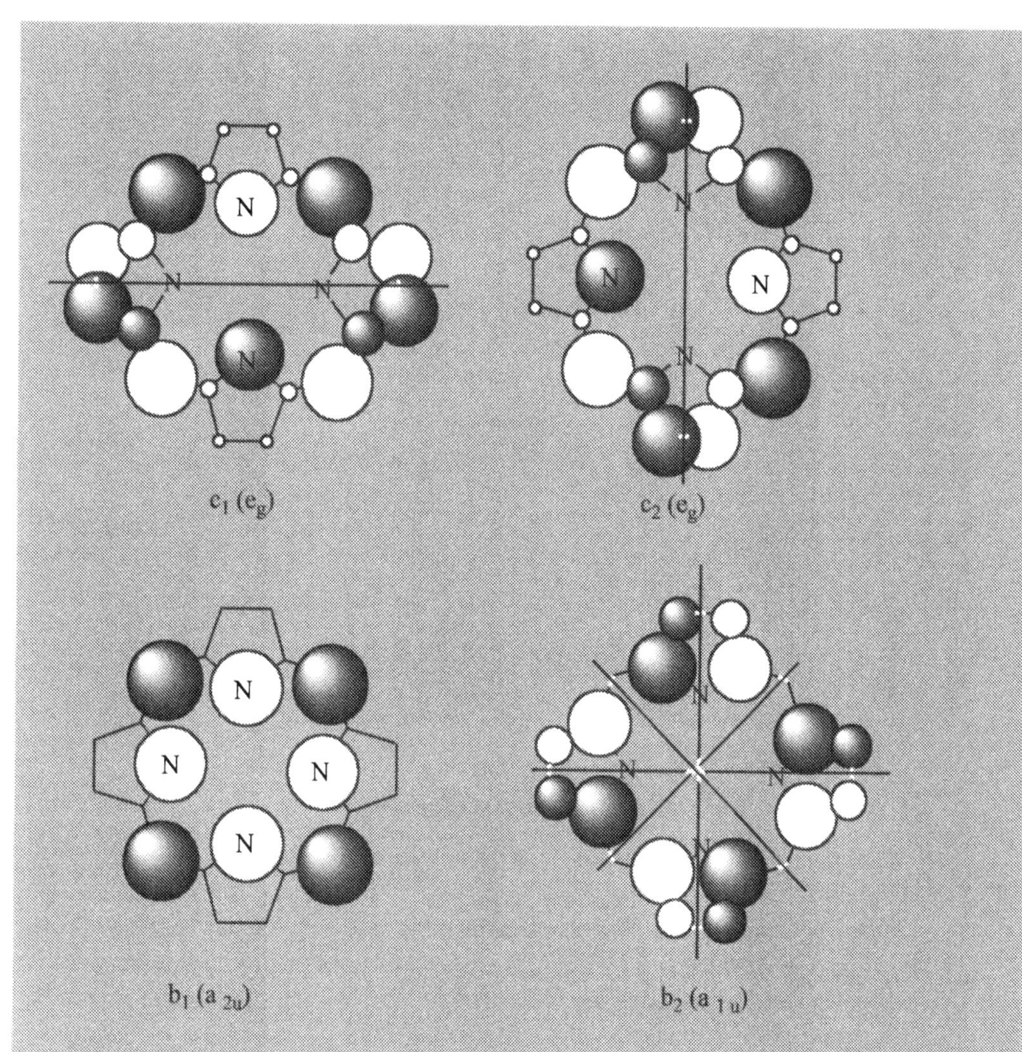

FIGURE 8.2 MOLECULAR ORBITALS FOR A D_{4h} PORPHYRIN SYSTEM. THE ATOMIC ORBITAL COEFFICIENTS ARE PROPORTIONAL TO THE SIZE OF THE CIRCLES; SHADING OR LACK OF IT INDICATES SIGN.

This is not found, although the band is usually broader than in simple divalent metal complexes (see Figure 8.1 for an example of this effect). Splitting of the B band is observed in tetrabenzoporphyrins, however. For further detail, see M. Gouterman, *J. Mol. Spectroscopy*, 1961, 6, 138–163 and M. Gouterman, G.H. Wagnière and L.C. Snyder, *idem.*, 1963, 11, 108–127.

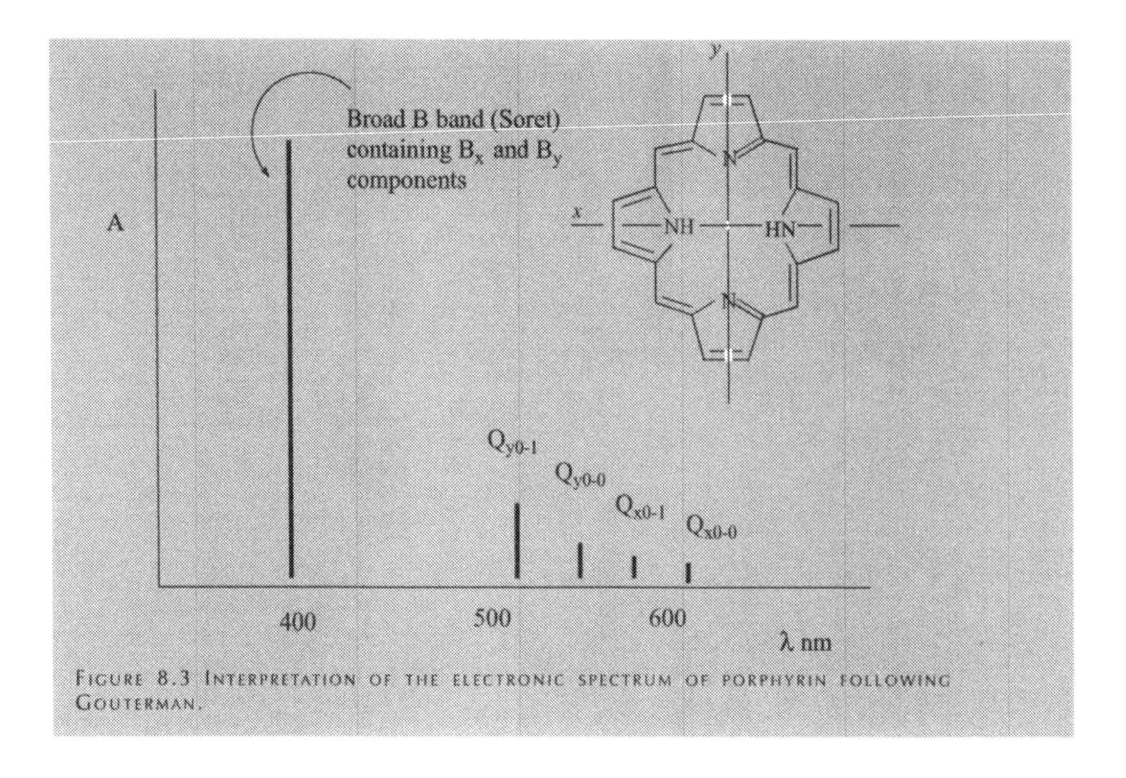

FIGURE 8.3 INTERPRETATION OF THE ELECTRONIC SPECTRUM OF PORPHYRIN FOLLOWING GOUTERMAN.

8.3 PORPHYRINS DERIVED FROM HAEMOGLOBIN

Blood from the slaughterhouse is the starting material here, and from it, after treatment with heparin to prevent coagulation, haemolysis and removal of the erythrocyte ghosts by centrifugation, haemoglobin can be isolated and crystallised. Alternatively by suitable treatment with acetic acid and sodium chloride the chloroiron(III) protoporphyrin (**87**), otherwise known as protohaemin (or simply haemin), can be crystallised. Both these substances are available from chemical suppliers, but they are quite simple, and inexpensive, to prepare in a chemical laboratory. Demetallation of protohaemin gives protoporphyrin (**61**). Since iron(II) (the larger cation) is easier to remove than is iron(III), the demetallation with acid is carried out in the presence of iron(II) sulphate as a reductant.

87 Protohaemin

2 Haematoporphyrin

The preparation of haematoporphyrin (**2**) from protohaemin (**87**) involves treatment with concentrated HBr/HOAc, which demetallates the complex with concomitant Markovnikov addition of hydrogen bromide to the two vinyl groups. Subsequent work-up causes hydrolysis of the benzylic bromide functions to give haematoporphyrin (**2**), which usually comes on the market as the dihydrochloride. As explained earlier (see section 6.1) commercial samples are generally impure.

Treatment of protohaemin (**87**) with molten resorcinol causes protiodevinylation to produce deuterohaemin, which on demetallation gives deuteroporphyrin (**88**).

(**87**) —— 1. Resorcinol,190^0, 15 min; 2. H$^+$, demetallation ——→

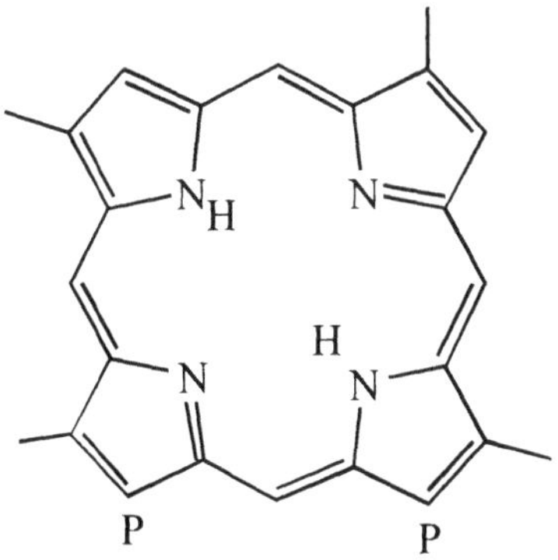

(**88**) Deuteroporphyrin

Protodevinylation is not a commonly encounted aromatic desubstitution, but has been known in the porphyrin series since 1928, when it was reported by Schumm. Iron is not essential for the reaction, which also proceeds with free base vinyl porphyrins: experiments with a model system (vinylheptaethylporphyrin) have allowed the characterisation of an intermediate which supports the view that the reaction proceeds by C-alkylation of the resorcinol, as illustrated in Figure 8.4 .

Deuteroporphyrin, protoporphyrin and haematoporphyrin have been the most common starting materials for the synthesis of PDT photosensitisers originating from haemoglobin. Commonly the dimethyl esters have been used in chemical manipulation, since these esters are conveniently soluble in organic solvents. Thus, reaction of protoporphyrin dimethyl ester with Eschenmoser's reagent occurs at both vinyl groups, and subsequent quaternisation with methyl iodide gives the 3, 8-bis quaternary salts:

$$-CH{=}CH_2 \xrightarrow[CH_2Cl_2]{CH_2{=}\overset{+}{N}Me_2\ \overset{-}{I}} -CH{=}CHCH_2NMe_2 \xrightarrow{MeI} -CH{=}CHCH_2\overset{+}{N}Me_3\ \overset{-}{I}$$

Deuteroporphyrin dimethyl ester undergoes analogous electrophilic reaction at the two free β-positions, and the procedure is a useful route to cationic sensitisers.

As mentioned earlier (section 6.1), haematoporphyrin (**2**) has the disadvantage from the pharmaceutical point of view of having two chiral centres. The two diastereoisomers can be separated by HPLC using reverse phase chromatography. Enantiomers have been generated by enantioselective reduction of 3,8-

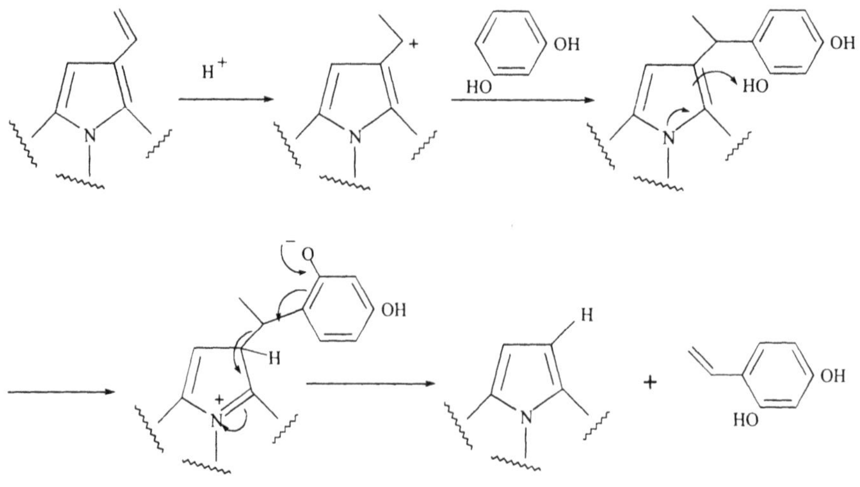

FIGURE 8.4 PROPOSED MECHANISM FOR PROTIODEVINYLATION IN RESORCINOL (SCHUMM REACTION).

diacetyldeuteroporphyrin dimethyl ester (**89**) with the methyloxazaborolidine (**90**) as homochiral catalyst and borane-dimethyl sulphide as the stoichiometric reductant.

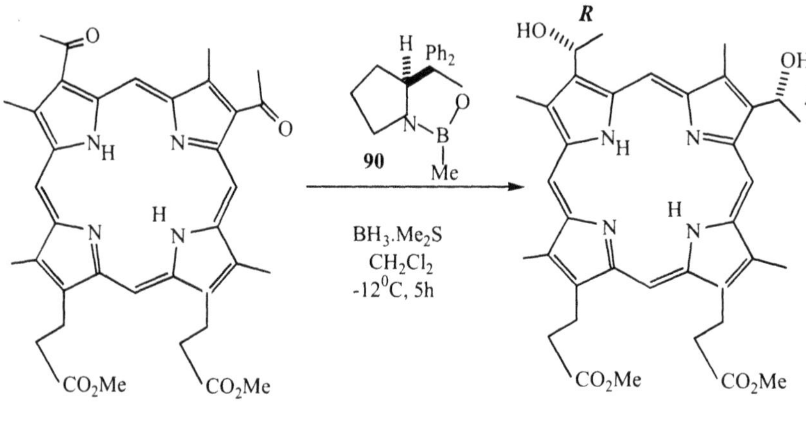

$BH_3.Me_2S$
CH_2Cl_2
-12^0C, 5h

(**89**) 3,8-Diacetyldeuteroporphyrin dimethyl ester

(**91**)
Total yield 86%, containing 69% of the enantiomer shown

The (*R, R*) enantiomer (**91**) can be separated from the small amount of its enantiomer (*S, S* 2.9%, 92% ee), and from more substantial amounts of the diastereoisomer (*RS, SR*), by HPLC of the dibenzoates on a homochiral HPLC column. While scaling up such a process is feasible, it is expensive, and does not fit happily with production economics.

Haematoporphyrin (and protoporphyrin) have been functionalised at the C-3/C-8 substituents and at the propionic acid groups. Of a series of bis(alkyl ethers), bioassay indicates that the bis-pentyl ether (**92**) is the most active. Thioether and amino derivatives have also been evaluated, as well as compounds with various carboxyamide functions at C-17 and C-18. Although biological activity is observed with these compounds, many of them, besides being diastereoisomeric mixtures, retain the inherently unstable bis-benzylic chemical reactivity: none of them appears to have been taken further.

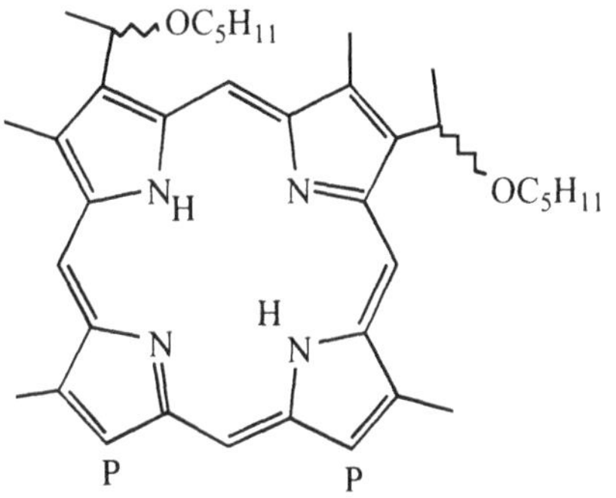

OC_5H_{11} OC_5H_{11} NH N H N N P P

92

However, protoporphyrin dimethyl ester is the starting material for a promising compound of the chlorin series, the so-called benzoporphyrin derivative (section 9.4.2).

8.4 SYNTHETIC PORPHYRINS

8.4.1 General Approaches

The chemical synthesis of porphyrins is a well-developed subject. The majority of synthetic approaches fall into one of the various categories indicated schematically in Panel 8.2, which seeks to divide the field according to the manner in which the four rings are assembled. For systems with a substitution pattern with low symmetry, such as the natural porphyrins, a stepwise synthesis is necessary (Classes A or B): for compounds with a symmetrical (D_{4h}) substitution pattern the Class C reactions (4×1) are appropriate. Since the latter type of synthesis involves fewer steps than reactions in Classes A and B, it is advantageous on economic grounds if a commercial photosensitiser can be made by a Class C reaction: even though the cyclisation step may be low-yielding, it is a one-pot reaction from which the product crystallises directly in many cases.

We shall now look at examples selected from each of these three classes, but two general points are helpful before we do this. Firstly, the starting materials are pyrroles, and the *meso* bridges are put in place either via the pyrromethene, formed by the electrophilic attack of a (protonated) pyrrole aldehyde on an α-free pyrrole (Piloty reaction):

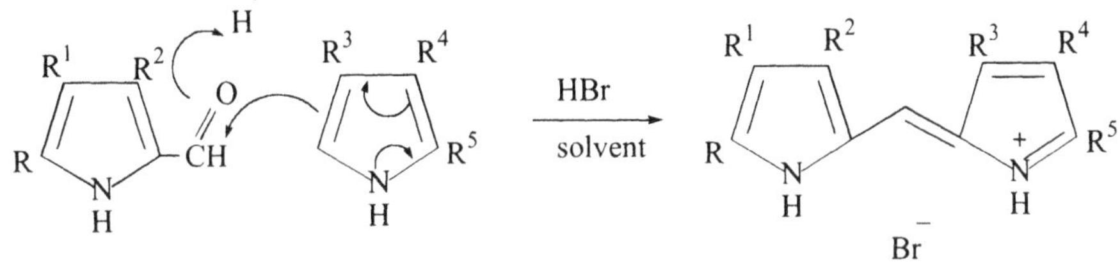

H^+ R^1 R^2 O CH R N H R^3 R^4 R^5 N H HBr solvent R^1 R^2 R N H R^3 R^4 R^5 N^+ H Br^-

Pyrromethene hydrobromide
(orange-yellow)

Panel 8.2 A Classification of Porphyrin Syntheses

The type of synthesis adopted depends on the symmetry of the substitution pattern. The syntheses can be formally classified as shown schematically below. A similar classification (using different building blocks as appropriate) can be used with other macrocycles (such as benzoporphyrins and phthalocyanines).

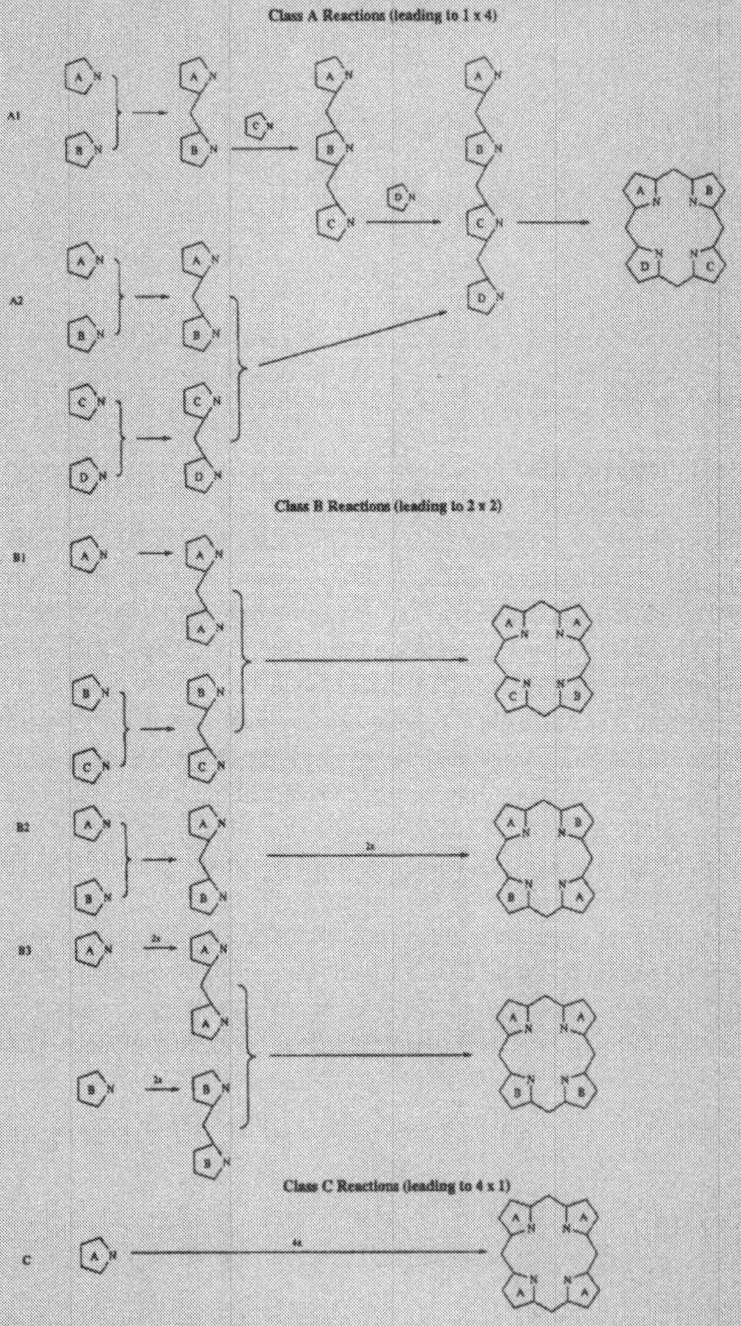

Class A Reactions (leading to 1 x 4)

A1

A2

Class B Reactions (leading to 2 x 2)

B1

B2

B3

Class C Reactions (leading to 4 x 1)

C

(Reproduced with permission from R. Bonnett and K.A. McManus, *J. Chem. Soc., Perkin Trans . 1*, 1996, 2461–2466) or via the dipyrrylmethane, formed by nucleophilic displacement on a substituted α-methyl function, thus:

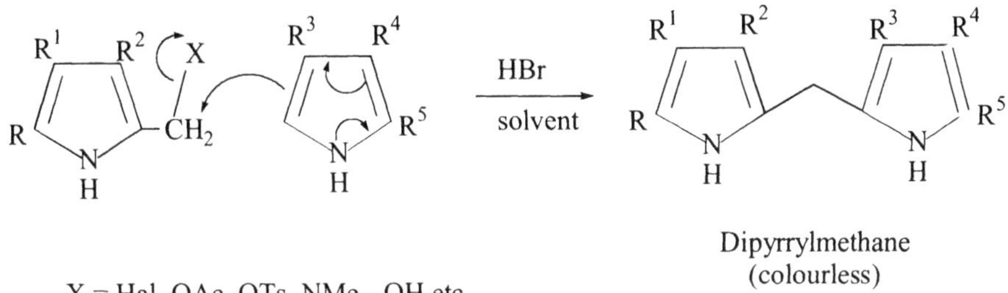

R^1 R^2 X R^3 R^4
R CH_2 R^5
N H N H

HBr
solvent

R^1 R^2 R^3 R^4
R R^5
N H N H

Dipyrrylmethane
(colourless)

X = Hal, OAc, OTs, NMe_2, OH etc

Secondly, the macrocycle is initially formed at a reduced level, with extra hydrogens at the *meso* positions: these hydrogens are readily removed by oxidation with aromatisation, sometimes involving oxygen of the air, and sometimes involving an added oxidant (often a high potential quinone such as DDQ).

Class A

A1. "3 + 1" Approach

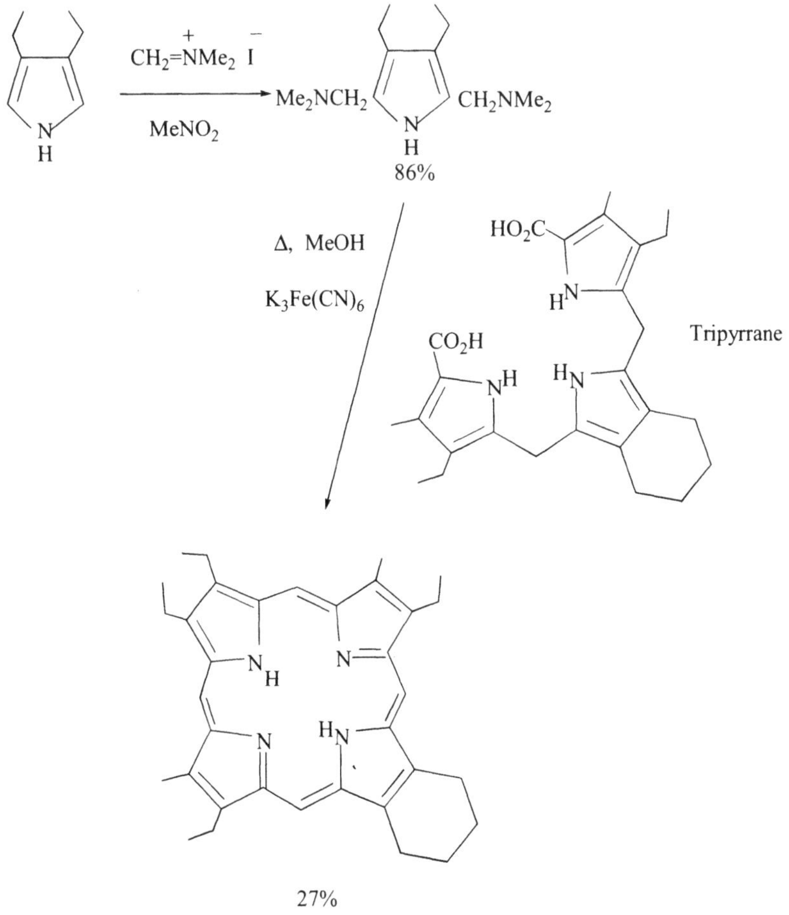

N H

$CH_2=\overset{+}{N}Me_2\ \overset{-}{I}$

$MeNO_2$

Me_2NCH_2 N H CH_2NMe_2

86%

Δ, MeOH

$K_3Fe(CN)_6$

HO_2C HN

CO_2H NH HN

Tripyrrane

NH N N HN

27%

(L.T. Nguyen, M.O. Senge and K.M. Smith, *J. Org. Chem.*, 1996, **61**, 998-1003).

A2. Johnson's Biladiene-ac Synthesis

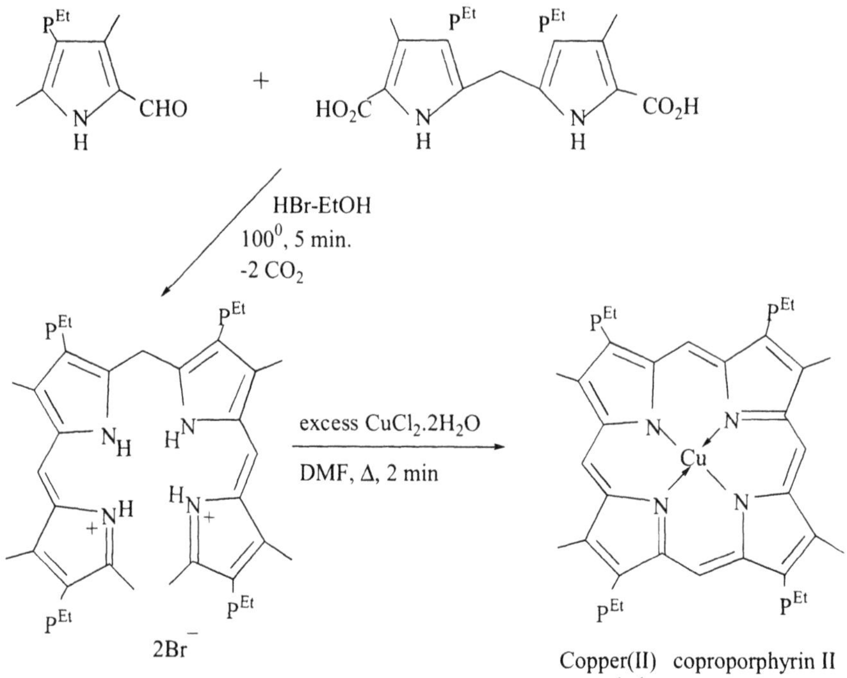
PEt
CHO
N
H
+
HO2C
CO2H
HBr-EtOH
100⁰, 5 min.
-2 CO2
NH
HN
excess CuCl2.2H2O
DMF, Δ, 2 min
Cu
2Br⁻

Biladiene-ac dihydrobromide
48%

Copper(II) coproporphyrin II
tetraethyl ester
39%

(R. Grigg, A.W. Johnson, R. Kenyon, V.B. Math and K. Richardson, *J. Chem. Soc. C*, 1969, 176-182).

The dealkylation in the last step of this reaction is remarkable, but is not uncommon in the porphyrin series, the driving force being conjugation / aromatisation:

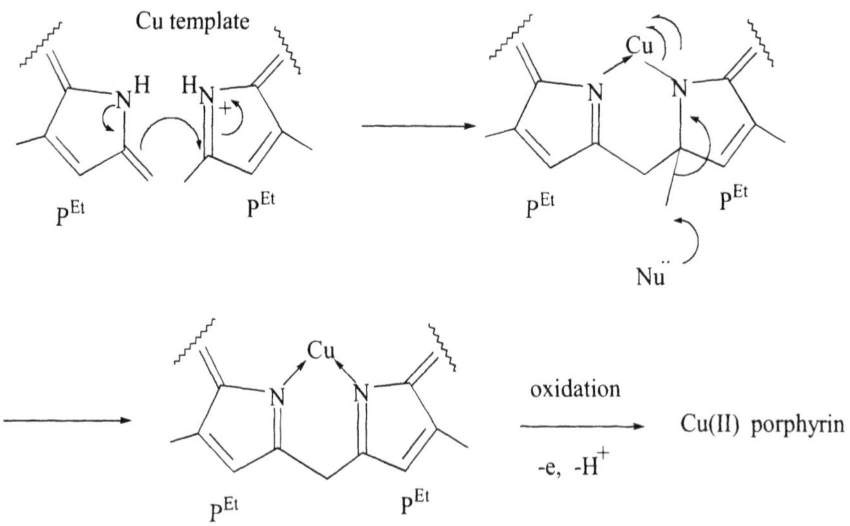
Cu template
PEt
Cu
Nu
oxidation
-e, -H+
Cu(II) porphyrin

A2. Woodward's Chlorophyll Synthesis

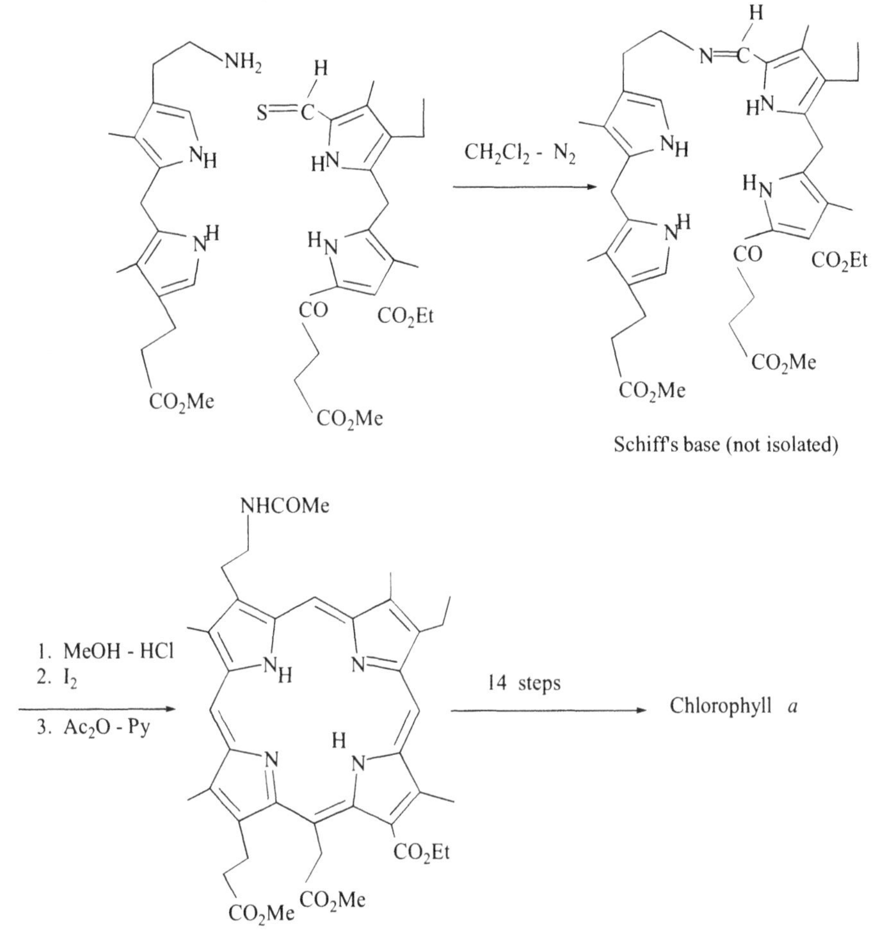

NH_2 H S=C NH HN NH HN CO CO_2Et CO_2Me CO_2Me

CH_2Cl_2 - N_2

H N=C HN NH HN NH CO CO_2Et CO_2Me CO_2Me

Schiff's base (not isolated)

NHCOMe

1. MeOH - HCl
2. I_2
3. Ac_2O - Py

NH N H N N CO_2Et CO_2Me CO_2Me

14 steps

Chlorophyll *a*

(R.B. Woodward *et al., Tetrahedron,* 1990, **46,** 7599-7659).

Class B

B1 MacDonald Synthesis

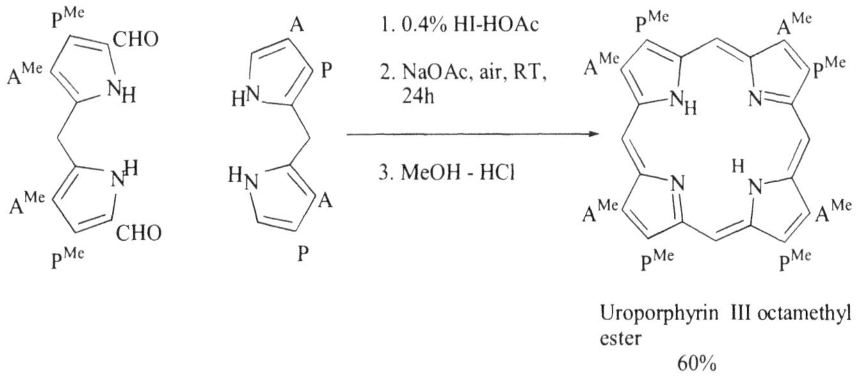

P^{Me} CHO A^{Me} NH NH A^{Me} CHO P^{Me}

A P HN HN A P

1. 0.4% HI-HOAc
2. NaOAc, air, RT, 24h
3. MeOH - HCl

P^{Me} A^{Me} A^{Me} P^{Me} NH N H N N A^{Me} A^{Me} P^{Me} P^{Me}

Uroporphyrin III octamethyl ester
60%

(E.J. Tarlton, S.F. MacDonald and E. Baltazzi, *J. Am. Chem. Soc.,* 1960, **82,** 4389-4395).

Class C

β-Substituted Porphyrins

Ph Ph NO$_2$ → Barton-Zard reaction, $CNCH_2CO_2Et$, DBU,THF, RT, 14h → (pyrrole, Ph, Ph, CO_2Et, N–H) 71% → 1. $LiAlH_4$, THF, 0^0C, 1h; 2. *p*-TsOH, *p*-chloranil, 24h, RT → (octaphenylporphyrin) 30%

(N. Ono *et al., J. Chem. Soc., Perkin Trans. 1,* 1998, 1595–1601).

meso-Substituted Porphyrins (Rothemund-Adler reaction)

(pyrrole) + PhCHO → $C_2H_5CO_2H$, Δ, air, 30min → [intermediate] → **27** 5,10,15,20-Tetraphenylporphyrin (20%) (corresponding chlorin removed with DDQ); → Linear polymer

(A.D. Adler *et al., J. Org. Chem.,* 1967, **32,** 476).

The last reaction is probably the most common porphyrin synthesis carried out in the laboratory. The general route was discovered by Rothemund in the 1930s, but the experimental method was much improved by Adler and his colleagues. It is the most convenient of these syntheses, provided that the product crystallises out of the propionic acid on cooling. Often a small proportion (<5%) of chlorin is crystallised with the porphyrin. It is removed in a separate step by treatment with DDQ.

The reaction is sometimes carried out with two different reactant aldehydes, so as to generate crossed products: under these conditions mixtures are produced, and they need to be separated, usually chromatographically, and identified.

Two modifications of this synthesis are available:

(i) Lindsey's Mild Method

In this variant the intermediate porphyrinogen is formed in dilute solution ($\approx 10^{-2}$ M) in a separate step at room temperature or below, and then dehydrogenated with a high potential quinone.

$PhCH_2O$–C_6H_4–CHO

dry CH_2Cl_2
$BF_3.Et_2O$, N_2

Ar

DDQ

OCH_2Ph

$PhCH_2O$

8%

(J.S. Lindsey *et al.*, *J. Org. Chem.*, 1987, **52,** 827-836).

The yield in this example is increased to 24% when triethyl orthoacetate is included as a water scavenger. With benzaldehyde itself, tetraphenylporphyrin is isolated (after chromatography) in 45% yield. This method is generally better than the Rothemund-Adler method for compounds with reactive substituents, and for *meso*-tetraalkylporphyrins.

(ii) Bonar-Law's Micellar Assembly Method

Here micelles (Panel 7.2) based on sodium dodecyl sulphate (0.5 M) are employed with the intention of assembling the reactants (pyrroles, benzaldehydes) in the lipophilic interior for reaction to give the porphyrinogen. The anionic surfactant is presumed to stabilise the cationic intermediate. The medium is aqueous, the catalyst being hydrochloric acid. Dehydrogenation is carried out with *p*-chloranil.

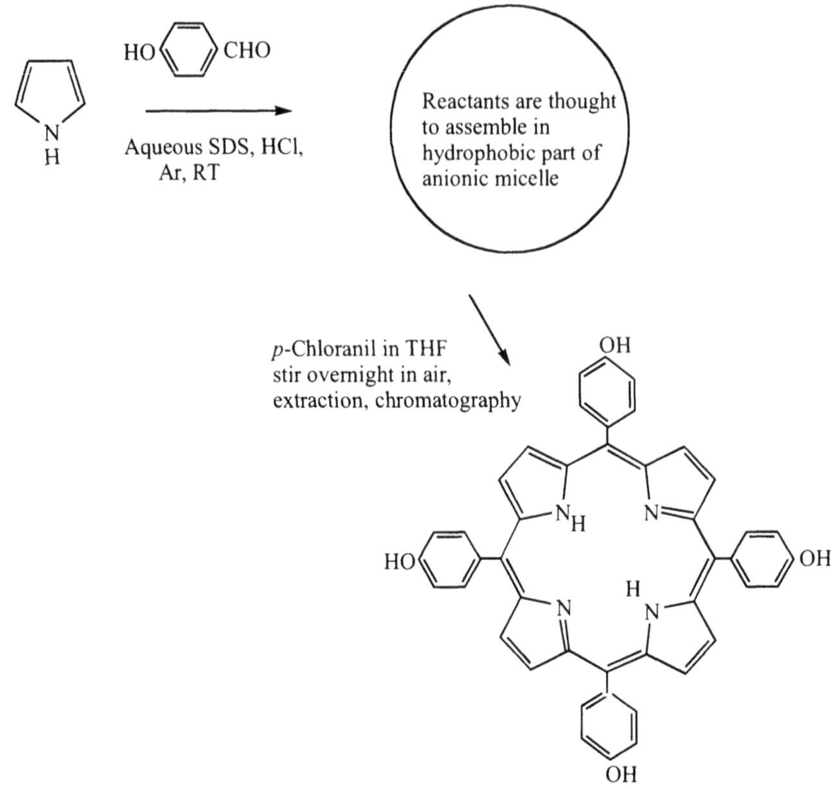

Aqueous SDS, HCl, Ar, RT

Reactants are thought to assemble in hydrophobic part of anionic micelle

p-Chloranil in THF
stir overnight in air,
extraction, chromatography

38%
94, p-THPP

(R.P. Bonar-Law, *J. Org. Chem.*, 1996, **61,** 3623-3634).

This procedure appears to be particularly useful for porphyrins with sensitive polar functions, where the Lindsey method is less good, due to solubility problems. Removing all traces of the detergent can be a problem. In both of these methods, the final treatment with a high potential quinone ensures that any chlorin impurities are dehydrogenated.

8.4.2 Specific Examples

(i) Tetraphenylporphyrin sulphonic acids (see section 7.2.4.1)

Treatment of 5,10,15,20-tetraphenylporphyrin (**27**) with fuming sulphuric acid causes sulphonation at the *p*-positions of the benzene rings. Usually these compounds are isolated as the sodium salts. Under vigorous conditions the tetrasulphonic acid (**74**) is formed, and this was the subject of early PDT studies by Winkelman (1962): while effective as a photosensitiser for PDT, the tetrasulphonic acid seems to have some neurotoxicity in the rat.

Under less forcing conditions, a mixture of mono ($TPPS_1$, **69**), di ($TPPS_{2a}$, **70**, $TPPS_{2o}$, **71**) and tri ($TPPS_3$, **72**) sulphonic acids is formed, which requires chromatographic separation. These compounds have quite distinct intracellular localisations (section 13.2.2) and photocytotoxicities (Figure 7.1) in cell cultures.

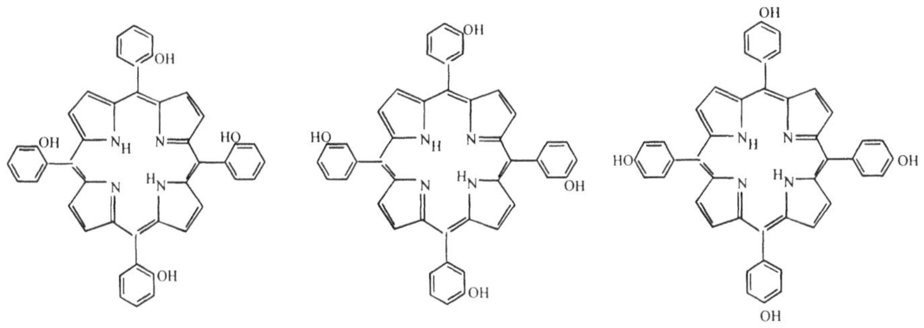

93 o-THPP **76** m-THPP **94** p-THPP

FIGURE 8.5 STRUCTURES AND ACRONYMS OF *MESO*-TETRA(HYDROXYPHENYL)PORPHYRINS.

(ii) Tetra(hydroxyphenyl)porphyrins

Three of these are available. They are known by the acronyms o-THPP, m-THPP and p-THPP (Figure 8.5).

They are conveniently prepared by the Rothemund-Adler route, usually with protection of the phenolic function, eg

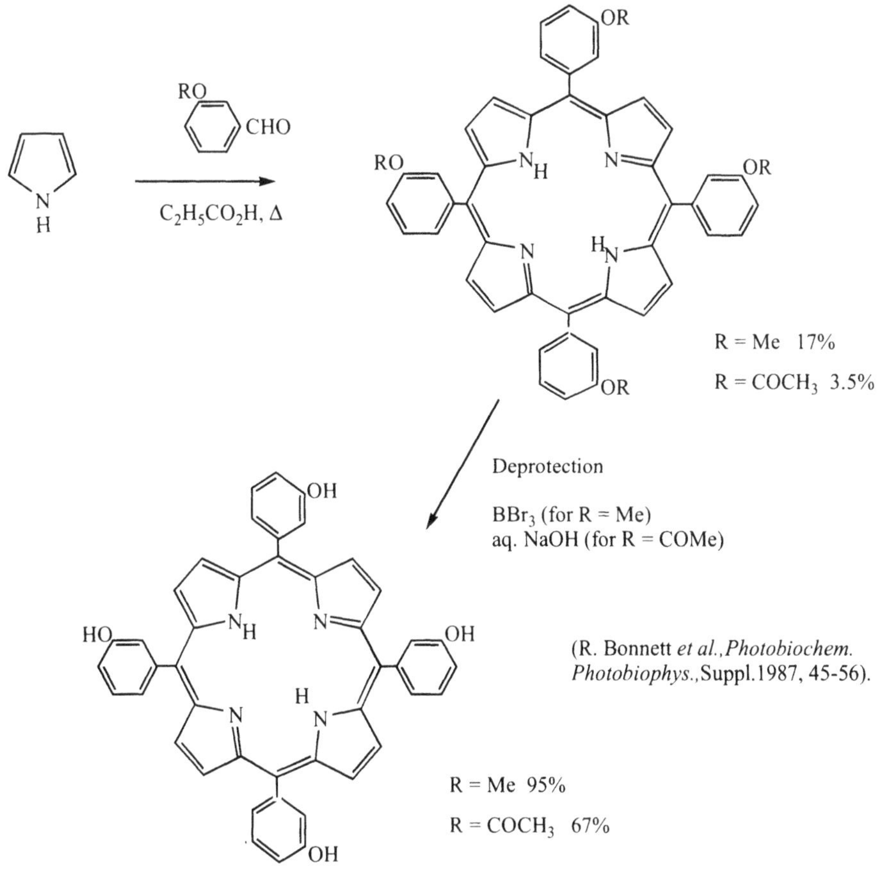

$C_2H_5CO_2H$, Δ

R = Me 17%

R = $COCH_3$ 3.5%

Deprotection

BBr_3 (for R = Me)
aq. NaOH (for R = COMe)

(R. Bonnett *et al.*, *Photobiochem. Photobiophys.*, Suppl. 1987, 45-56).

R = Me 95%

R = $COCH_3$ 67%

76 m-THPP

TABLE 8.1 *IN VIVO* BIOASSAY (DEPTH OF PHOTONECROSIS IN A TUMOUR IMPLANT IN THE MOUSE) OF SOME TETRA-SUBSTITUTED *MESO*-TETRAPHENYLPORPHYRINS. STANDARD CONDITIONS: DRUG DOSE: 50 μ MOL KG^{-1}, I.P., DLI = 24 HOURS, LIGHT FLUENCE: 10 J CM^{-2}.

Substituent in each phenyl ring in TPP (**27**)	Wavelength of irradiation λ (nm)	Depth of tumour necrosis (mm) ±SE (number of experiments)
[HpD (67μM kg^{-1})	625	2.33 ± 0.4 (6)]
p-NO_2	647	0.05 ± 0.06 (8)
o-OH	648	4.25 ± 0.50 (7)
m-OH	648	4.63 ± 0.64 (6)
p-OH	656	5.06 ± 0.48 (4)
o-OMe	650	0.13 ± 0.06 (6)
p-OAc	650	0.09 ± 0.04 (8)

(R. Bonnett, S. Ioannou, R.D.White, U-J. Winfield and M.C. Berenbaum,*Photobiochem. Photobiophys.* Suppl., 1987, 45–56).

The hydroxy functions make the system more amphiphilic in a general sense (Panel 7.1) and also cause small hyperchromic and bathochromic effects on Band I. In experiments measuring the depth of photonecrosis in tumour implants in mice, these compounds are all potent photosensitisers, being up to about 25 times as effective as HpD. Unlike the other isomers, o-THPP shows severe skin photosensitisation, and hence experiments with this substitution pattern were discontinued at an early stage.

The activity of the *meso*-tetra(hydroxyphenyl)porphyrins emerged from *in vivo* bioassay of a range of substituted *meso*-tetraphenylporphyrins. Substituents were varied to provide a range of hydrogen bonding capability to change solubility, polarity, and interactions with receptor sites. Table 8.1 shows the comparisons using an animal model under standard experimental conditions, and a drug-light delay of 24 hours.

(iii) meso-Alkynylporphyrins

The periphery of the porphyrin ring soon becomes overcrowded on polysubstitution: for example the phenyl groups in *meso*-tetraphenylporphyrin are twisted out of plane (dihedral angle about 60° in the crystal) in spite of the effect of conjugation. On the other hand, the alkynyl function has excellent capacity for conjugation, but being rod-like, has low space demand. This accounts for the current activity with alkynyl porphyrins and phthalocyanines for electroactive applications.

meso-Tetra(arylethynyl)porphyrins can be synthesised by the Lindsey route, thus:

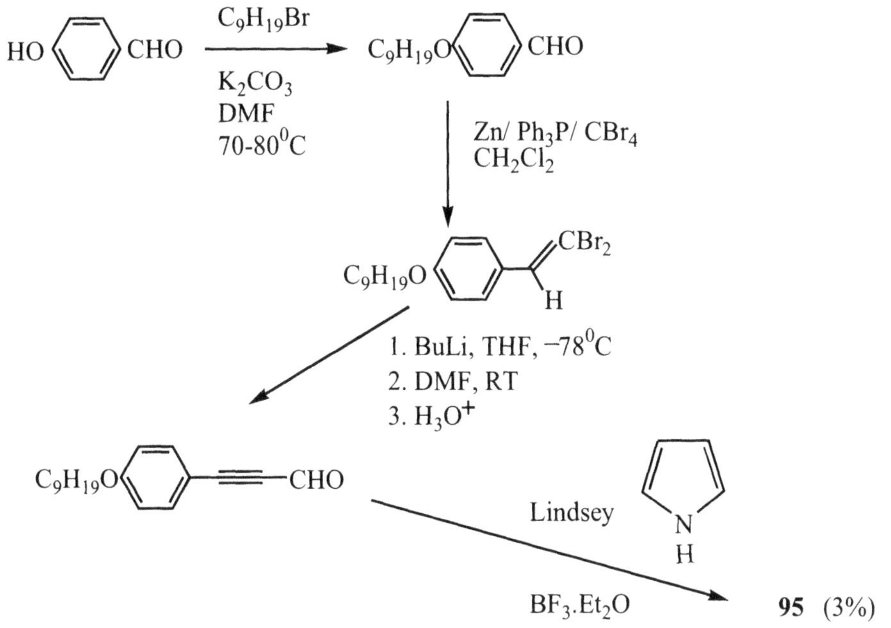

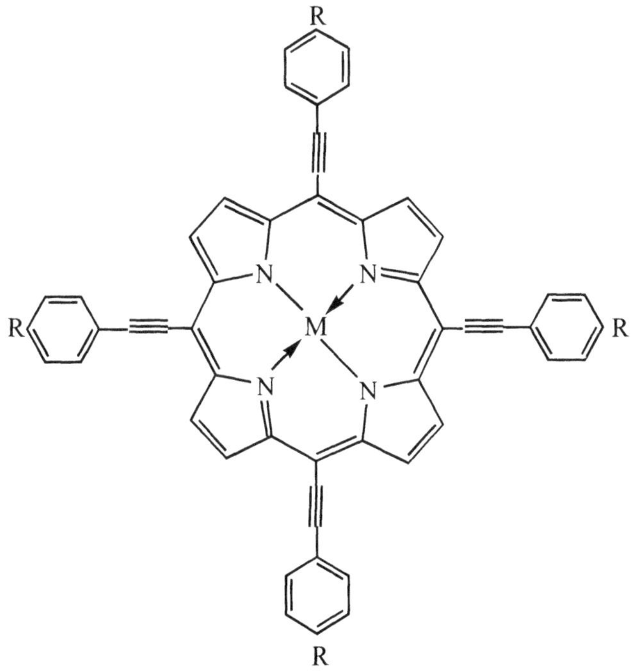

HO–C6H4–CHO → ($C_9H_{19}Br$, K_2CO_3, DMF, 70-80^{0}C) → $C_9H_{19}O$–C6H4–CHO

→ (Zn/ Ph_3P/ CBr_4, CH_2Cl_2) → $C_9H_{19}O$–C6H4–CH=CBr_2 (H)

→ (1. BuLi, THF, −78^{0}C; 2. DMF, RT; 3. H_3O^+) → $C_9H_{19}O$–C6H4–C≡C–CHO

→ (Lindsey, pyrrole N–H, $BF_3.Et_2O$) → **95** (3%)

R, R, R, R, N, N, N, N, M

95 R = OC_9H_{19}, M = 2H
96 R = C_8H_{17}, M = Zn

(J.A. Lacey *et al., Photochem. Photobiol.,* 1998, **67,** 97–100).

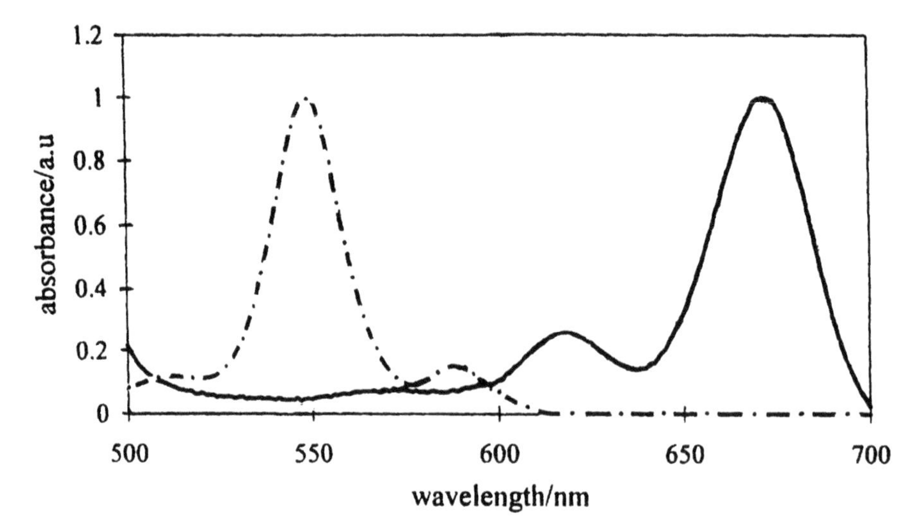

FIGURE 8.6 COMPARISON OF THE Q-BAND REGIONS IN THE ELECTRONIC SPECTRA OF ZINC(II) *MESO*-TETRAPHENYLPORPHYRIN (**27**, AS ZINC COMPLEX) AND ZINC(II) *MESO*-TETRA(*P*-OCTYLPHENYLETHYNYL)PORPHYRIN (**96**). SOLVENT: BENZENE (ADAPTED WITH PERMISSION FROM J.A. LACEY, D. PHILLIPS, L.R. MILGROM, G. YAHIOGLU AND R.D. REES, *PHOTOCHEM. PHOTOBIOL.*, 1998, **67**, 97–100).

The visible bands in (**95**) are bathochromically shifted to such an extent that this porphyrin is green in solution, with the Soret band at λ_{max} 472nm (ε = 297 000). The Q band region is essentially two-banded [λ_{max} 653 nm (89 000) and 744nm (56 000)]. The shift is illustrated for the zinc(II) complex (**96**) in Figure 8.6 by a striking conparison with zinc(II) tetraphenylporphyrin. The zinc complex (**96**) has ϕ_Δ = 0.97. It appears that, with suitable hydrophilic substitution (such as –OH, $-CO_2H$), this chromophore might well provide a useful PDT photosensitiser.

(iv) retro-Etio systems

Porphyrins bearing a six-membered anhydride, imide, or isoimide ring fused across a *meso*-position and a neighbouring β-position also appear greenish and not purple in solution. This is because, of the four bands, Band I is the most intense (retro-etio spectrum: Panel 8.1.4). This behaviour was first commented on by Stern and Wenderlein in 1936 who recognised the spectrum of 15-carboxyrhodoporphyrin anhydride methyl ester (**97**) as of a novel type. They proposed to call it "anhydride type", but this has not found favour because other porphyrins with unsaturated fused rings (eg the mesoverdin methyl ester **98**) also show Band I as the most intense of the Q bands.

Mesoverdin (**98**, but as free acid) might be worth consideration as a PDT sensitiser, since it is available from mesoporphyrin dimethyl ester (**99**) (Figure 8.7). However, while the chemistry looks simple, the intramolecular Friedel-Crafts reaction gives regioisomers (13,15 and 15,17), which have to be separated by fractional crystallisation and chromatography.

SO_3
H_2SO_4
aromatic electrophilic substitution at C-15

99 Mesoporphyrin dimethyl ester

100 Mesorhodin methyl ester (separate from 15,17-isomer) 36%

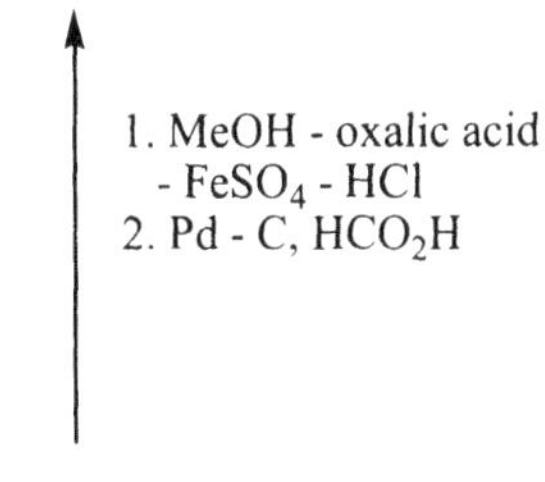

Δ, HOAc (O_2)

98 (56%)

87 Protohaemin

$KMnO_4$, acetone

97

FIGURE 8.7 CHEMICAL ROUTE FROM A READILY AVAILABLE PORPHYRIN, MESOPORPHYRIN DIMETHYL ESTER **99**, TO PORPHYRINS **97** AND **98**, IN WHICH BAND I IS RED SHIFTED AND INTENSIFIED.

97 15-Carboxyrhodoporphyrin anhydride methyl ester
λ_{max} nm(ε), dioxan
520 (4 250)
554 (5 280)
591 (12 900)
652.5 (14 050)

98 Mesoverdin methylester (13,15 isomer shown)
λ_{max} nm(ε), $CHCl_3$
642 (5 190)
704 (10 500)

DDQ

$CHCl_3$
RT, 2h

101 **102** (45%)

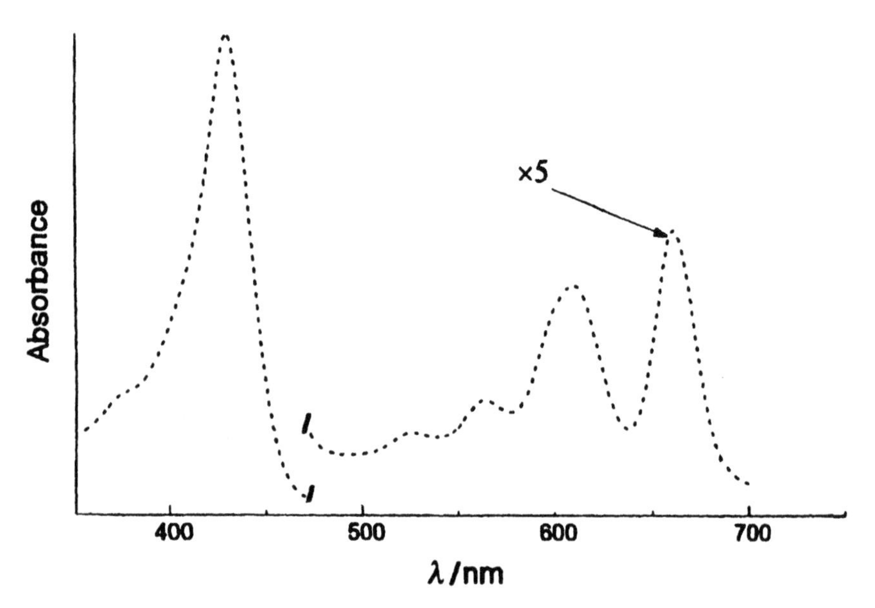

FIGURE 8.8 THE ELECTRONIC SPECTRUM OF THE PORPHYRIN **102** (PURPUROPORPHYRIN 18 METHYL ESTER) IN CHLOROFORM SHOWING AN EXAMPLE OF A RETRO-ETIO SPECTRUM. (REPRODUCED WITH PERMISSION FROM A.F. MIRONOV, A.V. EFREMOV, O.A. EFREMOVA, R. BONNETT AND G. MARTINEZ, *J. CHEM. SOC., PERKIN TRANS. 1*, 1998, 3601–3608).

However, such anhydrides and related fused ring compounds are more commonly prepared from chlorophyll *a* or bacteriochlorophyll (section 9.5). Thus the chlorophyll-derived purpurin 18 methyl ester (**101,** Figure 9.3) on dehydrogenation with DDQ in chloroform gives the corresponding porphyrin (**102**), the visible spectrum of which is shown in Figure 8.8. Compounds of this sort are now being examined as PDT sensitisers.

v. Dimeric Systems

In Panel 6.1 the chemistry of some model dimers derived from porphyrins was outlined. These have some relevance in attempts to understand the complex chemistry of HpD, but have not generally been seen as practicable photosensitisers for PDT.

However the heterodimer (**103**), synthesised from two unsymmetrically substituted tetraphenylporphyrin derivatives, and having the polar functions (OMe) localised on one of the macrocycles, has attracted interest because it is reported to have excellent selectivity (~100 with respect to normal tissue) for fibrosarcoma implants in mice, and causes a significant delay in tumour growth after illumination.

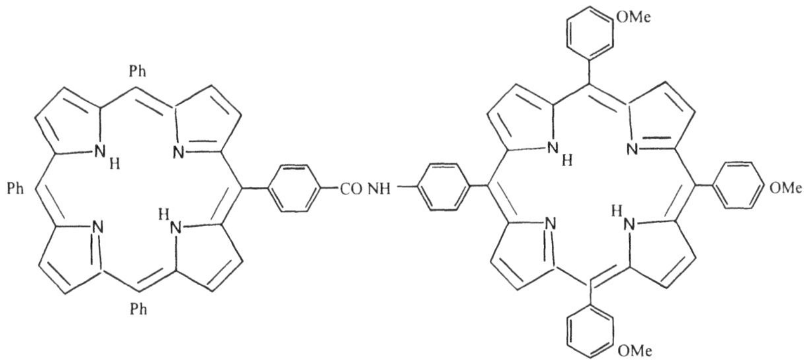

103

8.5 ENDOGENOUS PORPHYRIN: δ-ALA AS A PRO-DRUG

Currently the most exciting aspect of porphyrin sensitisation is the generation of protoporphyrin endogenously, that is from biosynthetic precursors by making use of the body's enzymatic pathway to make the porphyrin naturally.

The biosynthetic pathway has been outlined in Panel 4.1. The pathway is normally controlled by the concentration of protohaem which acts by a negative feed-back mechanism on the activity of the enzyme ALA synthase, thus controlling the levels of δ-aminolaevulinic acid (**104**). If this control is bypassed by adding δ-ALA to the normal living system, the biosynthetic production line inevitably takes over. The intermediate porphyrinogens are formed in excess, and these may oxidise adventitiously to porphyrins; and excess of protoporphyrin, which is on the direct biosynthetic line (Panel 4.1) is produced. It appears that protoporphyrin (rather than coproporphyrin or uroporphyrin) is the key photosensitiser in this form of therapy.

When tested using an *in vivo* bioassay with a tumour implant (Chapter 13), protoporphyrin shows little activity, possibly because it does not reach the tumour effectively. Exogenous protoporphyrin that does reach the tumour site localises in the plasma membrane. But when generated within the tumour using δ-ALA as a pro-drug, it goes to the mitochondria and proves to be very photoactive indeed. It has potential both as a photodiagnostic agent (fluorescence) and as a phototherapeutic agent.

δ-Aminolaevulinic acid (**104**) is commercially available as its hydrochloride.

$$NH_2CH_2COCH_2CH_2CO_2H$$

104

It is a sensitive substance, but can be kept satisfactorily in the freezer when not in use. It is administered topically, iv., or orally. Topical application, as a cream or lotion, works because δ-ALA is able to penetrate the abnormal keratin of the tumour epidermis (and also of wound tissue) more readily than it penetrates the normal epidermis. Kennedy and his colleagues (1990, 1992) reported that over a three year period, 300 basal cell carcinomas were given a single treatment of topical δ-ALA, followed by irradiation (600–800 nm, filtered projector lamp source), with a complete response rate of about 79% at three months.

This report led to intense activity on the use of this method in the PDT of superficial tumours, particularly basal cell carcinoma, squamous cell carcinoma, Bowen's disease and actinic keratoses (section 14.3.1 and Panel 14.1). Results worldwide have been somewhat uneven, but generally complete response figures are good. (See Peng *et. al.* (1997) in Bibliography). The procedure suffers from two limitations: the protoporphyrin tends to be rapidly photobleached when generated *in situ* in this way; and the procedure is suitable only for superficial neoplasms.

Because of the dependence on a specific enzyme sequence, there is little that the chemist can do to enhance the effectiveness of the endogenous protoporphyrin. Until we can do some heavy genetic engineering on *Euglena* (or catch a Triffid) it does not seem possible for humans to tap into chlorophyll-generating enzymes. Minor things can be done. Penetration can be improved by using a carrier such as DMSO or EDTA in the cream formulation, or mechanically by curettage. Using alkyl esters (eg the hexyl ester **105**) enhances protoporphyrin concentrations in cell culture, presumably because of increased membrane penetration. And the use of an iron chelator such as 1,2-diethyl-3-hydroxy-4-pyridone (**106**) also increases porphyrin levels, presumably by limiting [Fe^{++}] required in the ferrochelatase reaction.

$NH_2CH_2COCH_2CH_2CO_2C_6H_{13}$

105

O
OH
N
Et
Et

106

BIBLIOGRAPHY

General

K.M. Kadish, K.M. Smith and R. Guilard (eds.) *The Porphyrin Handbook*, Vols 1–10, Academic Press, San Diego, 2000. Multi-authored project of broad coverage: large and up-to-date but expensive. Synthesis is in volumes 1 and 2, PDT in volume 6.

D. Dolphin (ed.) *The Porphyrins*, Vols I–VII, Academic Press, London and New York, 1978-1979. A multi-authored work with broad coverage, including synthesis (volumes I and II), electronic spectroscopy (volume III), NMR (volume IV), and biochemical aspects (volumes VI and VII).

J.E. Falk *Porphyrins and Metalloporphyrins*, Elsevier, Amsterdam, 1964. Good on laboratory methods.

K.M. Smith (ed.) *Falk's Porphyrins and Metalloporphyrins*, Elsevier, Amsterdam, 1975. Extended updated version of Falk's book, with a more chemical emphasis.

M. Gouterman, 'Spectra of porphyrins', *J. Mol. Spectroscopy*, 1961, **6**, 138–163. First of a series of papers on the four orbital model for the electronic spectra of porphyrins.

E.D. Sternberg, D. Dolphin and C. Brückner, 'Porplyrin-based photosensitises for use in photodynamic therapy', *Tetrahedron*, 1998, **54**, 4151–4202.

TPPS

J.W. Winkelman, 'The distribution of tetraphenylporphinesulfonate in the tumor bearing rat', *Cancer Res.*, 1962, **22**, 589–596.

K. Berg, K. Madslien, J.C. Bommer, R. Oftebro, J.W. Winkelman and J. Moan, 'Light induced relocalisation of sulphonated *meso*-tetraphenylporphyrins in NHIK 3025 cells and the effects of dose fractionation', *Photochem. Photobiol.*, 1991, **53**, 203–210.

THPP

M.C. Berenbaum, S.L. Akande, R. Bonnett, M. Kaur, S. Ioannou, R.D. White and U-J. Winfield, '*meso*-Tetra(hydroxyphenyl)porphyrins, a new class of potent tumour photosensitisers with favourable selectivity', *Br. J. Cancer*, 1986, **54**, 717–725.

meso-Phenylalkynyl porphyrins

J.A. Lacey, D. Phillips, L.R. Milgrom, G. Yahioglu and R.D. Rees, 'Photophysical studies of some 5,10,15,20-tetraarylethynylporphyrinatozinc(II) complexes as potential lead compounds for PDT', *Photochem. Photobiol.*, 1998, **67**, 97-100.

retro-Etio systems

A. Stern and H. Wenderlein, 'On the light absorption of porphyrins. V.', *Z. physikal. Chem. A*, 1936, **176**, 81–124. Early discussion of the spectra of green porphyrins ("the green anhydride").

A.N. Kozyrev, G. Zheng, E. Lazarou, T.J. Dougherty, K.M. Smith and R.K. Pandey, 'Syntheses of emeraldin and purpurin 18 analogues as target-specific photosensitizers for PDT', *Tetrahedron Letters*, 1997, **38**, 3335–3338.

Dimers

M.A.F. Faustino, M.G.P.M.S. Neves, M.G.H. Vicente, J.A.S. Cavaleiro, M. Neumann, H-D. Brauer and G. Jori, '*meso*-Tetraphenylporphyrin dimer derivative as a potential photosensitiser in PDT', *Photochem. Photobiol.*, 1997, **66**, 405–412. Selectivity of a porphyrin dimer in fibrosarcoma implants.

δ-ALA

J.C. Kennedy and R.H. Pottier, 'Endogenous protoporphyrin, a clinically useful photosensitiser for photodynamic therapy', *J. Photochem. Photobiol. B. Biol.*, 1992, **14**, 275–292. Early report of success with δ-ALA in treating superficial BCC.

J.C. Kennedy, S.L. Marcus and R.H. Pottier, 'Photodynamic therapy and photodiagnosis using endogenous photosensitization induced by δ-aminolevulinic acid: mechanisms and clinical results', *J. Clin. Laser Med. Surg.*, 1996, **14**, 289–304. Includes colour photographs illustrating photodiagnostic mode.

Q. Peng, T. Warloe, K. Berg, J. Moan, M. Kongshaug, K-E. Giercksky and J.M. Nesland, '5-Aminolevulinic acid-based photodynamic therapy', *Cancer*, 1997, **79**, 2282–2308. A review with 225 references: contains useful tables summarising clinical studies.

K. Tabata, S. Ogura and I. Okura, 'Photodynamic efficiency of protoporphyrin IX: comparison of endogenous protoporphyrin IX induced by 5-aminolevulinic acid and exogenous protoporphyrin IX', *Photochem. Photobiol.*, 1997, **66**, 842–846. Localisation of exogenous and endogenous protoporphyrin in tumour cells in culture.

CHAPTER 9

CHLORINS AND BACTERIOCHLORINS

"If a petroleum ether solution of chlorophyll is filtered through a column of an absorbant (I use mainly calcium carbonate which is stamped firmly into a narrow glass tube), then the pigments, according to the adsorption sequence, are resolved from top to bottom in various coloured zones, since the stronger adsorbed pigments displace the weaker adsorbed ones and force them further downwards. This separation becomes practically complete if, after the pigment solution has flowed through, one passes a stream of pure solvent through the absorbent column ..... Such a preparation I call a chromatogram and the corresponding method, the chromatographic method"

Michael Tswett (1906) (as quoted by L. Zeichmeister, 1948).

First description of chromatography, and the separation of chlorophylls *a* and *b*.

9.1 STRUCTURAL DEFINITION

The chlorins and bacteriochlorins are β-dihydroporphyrins and β-tetrahydroporphyrins, respectively. There is only one type of β-dihydroporphyrin, chlorin (**107**) but two types of β-tetrahydroporphyrin (adjacent, isobacteriochlorin **108** and opposite, bacteriochlorin **109**) and one type of β-hexahydroporphyrin (**110**) are known. Unless the possibility is blocked (eg by β, β-disubstitution), all of these hydroporphyrins can be dehydrogenated to the more stable porphyrin. The reaction occurs slowly in solution by autoxidation, but is usually rapid and virtually quantitative in the presence of a high-potential quinone (eg DDQ).

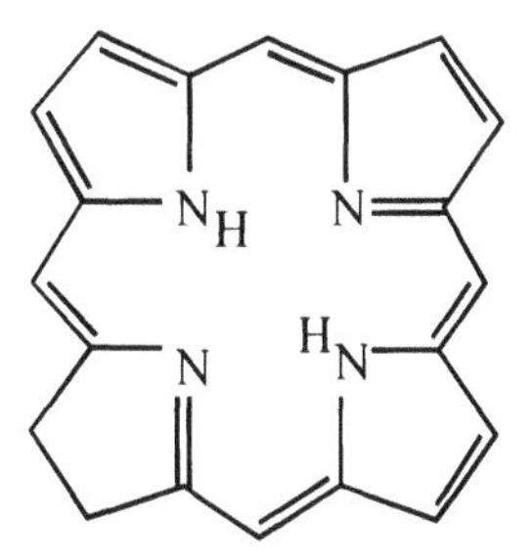

107 Chlorin

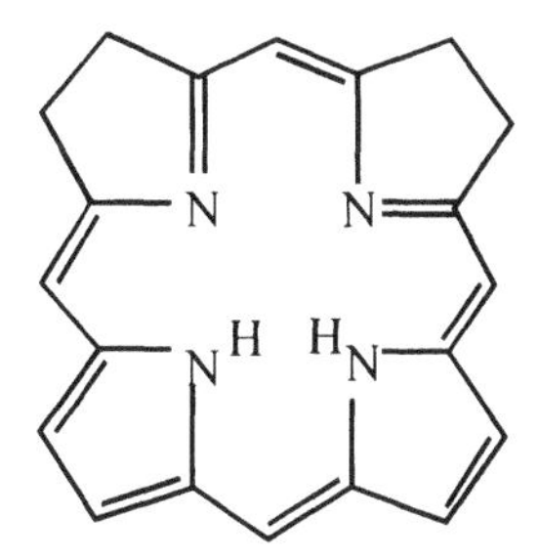

108 Isobacteriochlorin

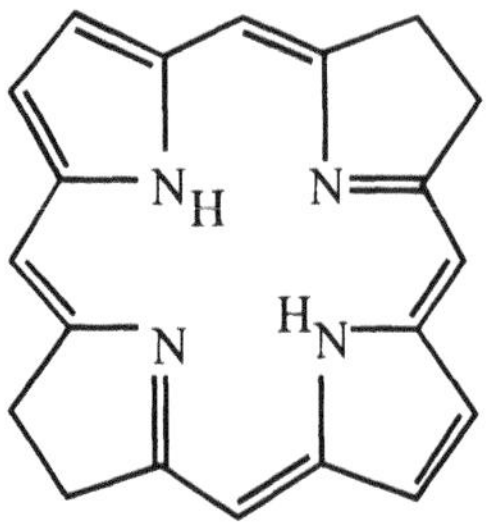

109 Bacteriochlorin

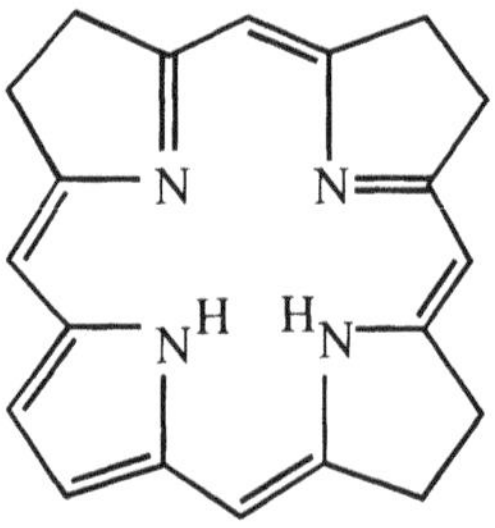

110 *β*- Hexahydroporphyrin

All of these compounds have an 18-π electron aromatic system, though in **110** the macrocyclic π-ring requires interaction through nitrogen (as it does, of course, in pyrrole). Other hydroporphyrins are known with reduction elsewhere eg phlorins, which are *meso*, N-dihydro derivatives; porphyrinogens, which are $(meso)_4$ N_2-hexahydro derivatives (eg protoporphyrinogen in Panel 4.1), but in these compounds macrocyclic conjugation is interrupted, and visible absorption is diminished in intensity or is absent.

These macrocycles have essential functions in living things. The commonest example of a chlorin, for example, is the chromophore of chlorophyll *a* and chlorophyll *b*; bacteriochlorin is the chromophore of the bacterial photosynthetic pigment, bacteriochlorophyll *a*; while the isobacteriochlorin system occurs at an intermediate stage (precorrin 2 and precorrin 3A, both *meso*,N-dihydroisobacteriochlorins) in the biosynthesis of vitamin B_{12} and, as the iron complex sirohaem, is the prosthetic group in the sulphite and nitrite reductases which occur in plants and bacteria.

Although chlorins are often drawn with the reduced ring at ring B, chlorophylls are always drawn with ring D reduced. For consistency in this book, all chlorins are shown with ring D reduced, as at **107.**

9.2 ABSORPTION SPECTRA

The absorption spectra of the chlorins show striking differences from those of the porphyrins. Band I is much more intense in the chlorins and the colour changes from purple to green. Hence, blood is red and grass is green, although the chromophores are both based on porphyrin-type macrocycles. The relative intensity of Band I is further enhanced in the bacteriochlorins, but not in the isobacteriochlorins. Substitution of an electron-withdrawing group (formyl, ketone, or anhydride carbonyl in the naturally-derived compounds, which also carry a carbonyl at C-13) at a *meso*-position adjacent to the reduced ring of a chlorin changes the green colour to a curious puce kharki: this is associated with enhanced absorption in the 550nm region, and a bathochromic shift of Band I, which is especially marked in the case of the 13,15-cyclic anhydride (eg purpurin 18, **116**) where Band I occurs at λ_{max} 702nm. The term *purpurin* is applied to this behaviour.

The three main types of spectra are shown schematically in Panel 9.1. A comparison of the Band I region for a porphyrin, chlorin and bacteriochlorin in a real

Panel 9.1 Schematic representation of spectral types amongst hydroporphyrins

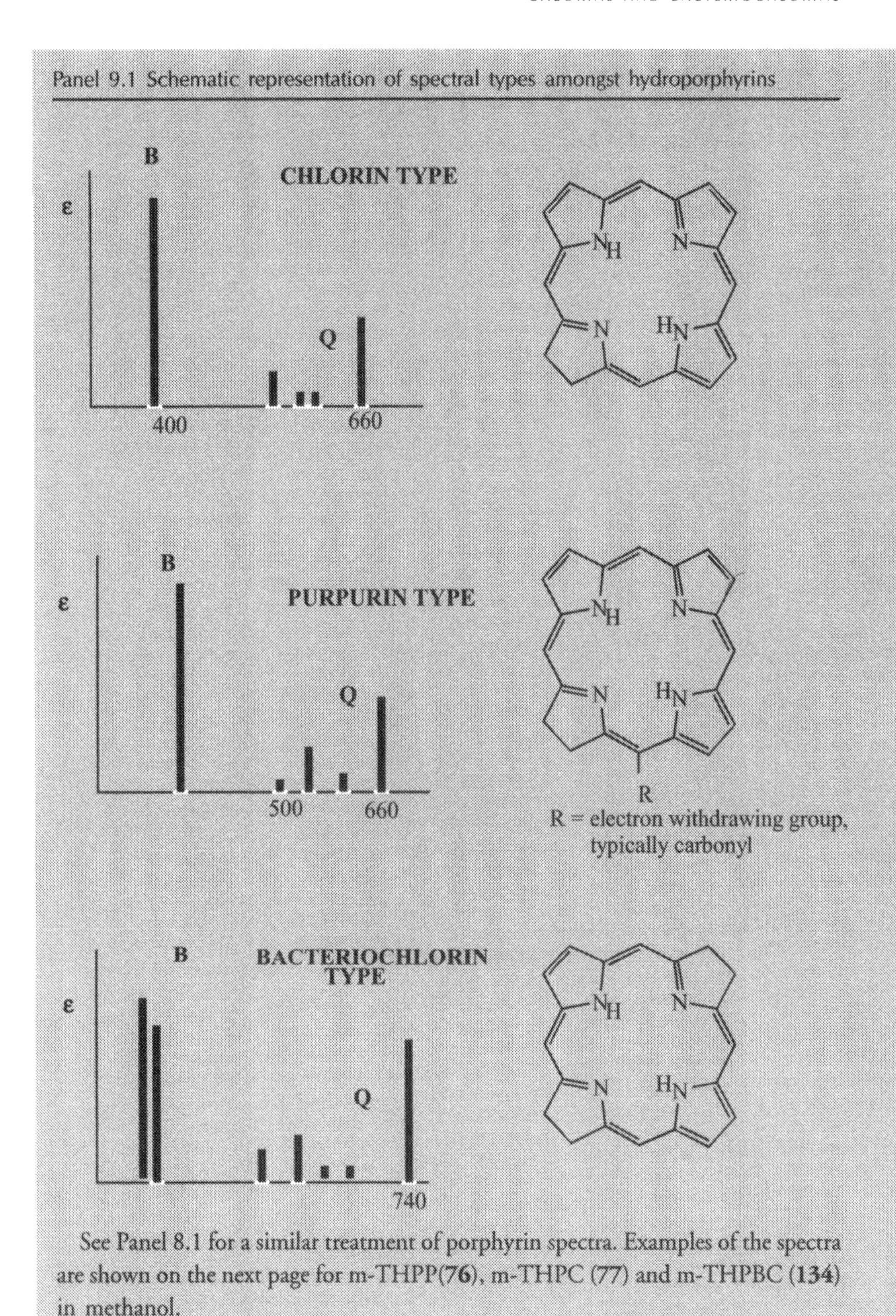

See Panel 8.1 for a similar treatment of porphyrin spectra. Examples of the spectra are shown on the next page for m-THPP(**76**), m-THPC (**77**) and m-THPBC (**134**) in methanol.

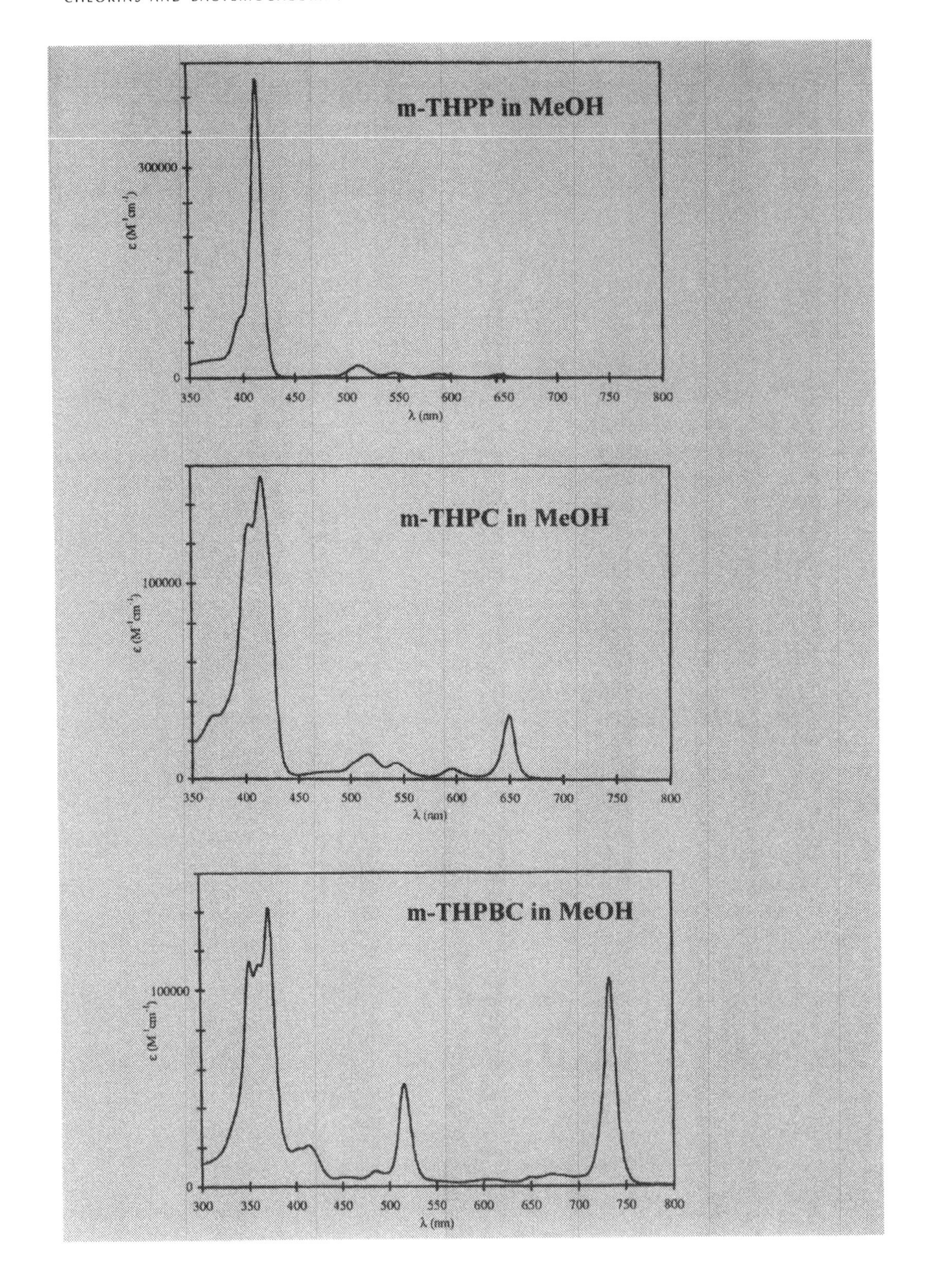
m-THPP in MeOH
300000
0
ε (M⁻¹cm⁻¹)
350 400 450 500 550 600 650 700 750 800
λ (nm)
m-THPC in MeOH
100000
0
ε (M⁻¹cm⁻¹)
350 400 450 500 550 600 650 700 750 800
λ (nm)
m-THPBC in MeOH
100000
0
ε (M⁻¹cm⁻¹)
300 350 400 450 500 550 600 650 700 750 800
λ (nm)

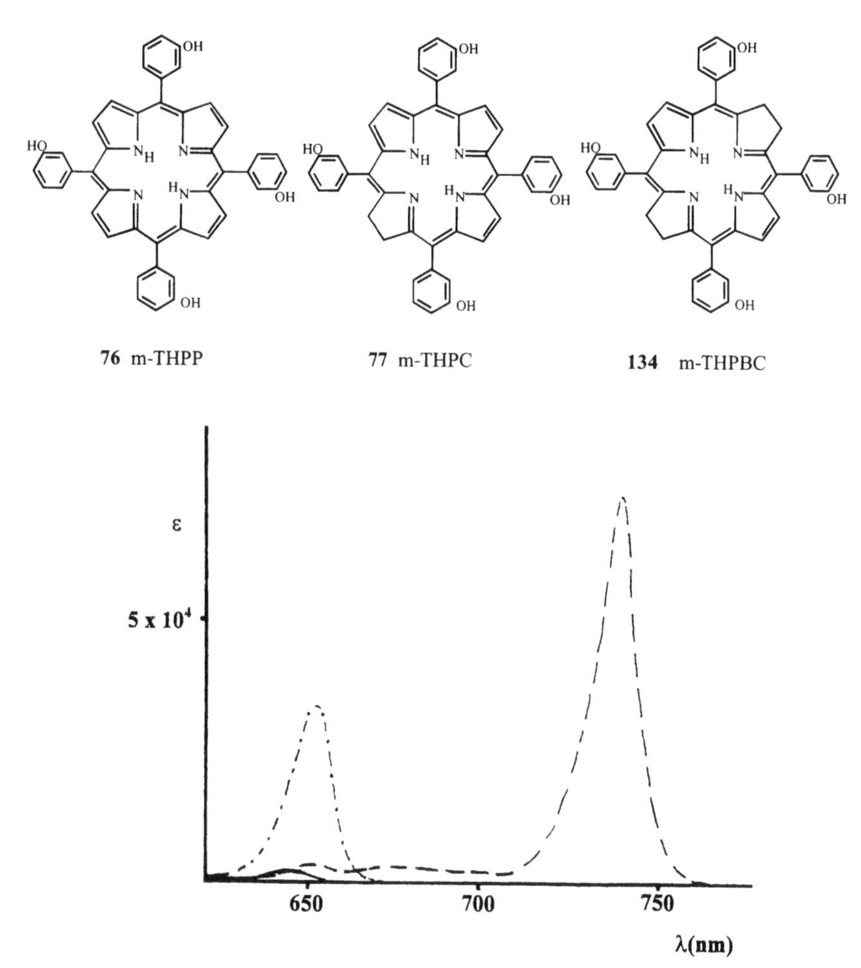

FIGURE 9.1 SPECTRA AT BAND I OF A PORPHYRIN, A CHLORIN, AND A BACTERIOCHLORIN WITH THE SAME SUBSTITUTION PATTERN. — M-THPP (**76**) λ_{MAX} 646NM, — · — · M-THPC (**77**) λ_{MAX} 652NM, ——— M-THPBC (**134**) λ_{MAX} 735NM. SOLVENT: MEOH. (REPRODUCED FROM B.D. DJELAL, PHD THESIS, LONDON 1994, WITH PERMISSION). THE COMPLETE SPECTRA IN THE VISIBLE REGION ARE SHOWN IN PANEL 9.1.

example (the 5,10,15,20-tetra(*m*-hydroxyphenyl) series) is shown in Figure 9.1 to illustrate these spectroscopic differences of Band I energy and intensity of absorption which are of particular relevance in PDT applications.

Panel 9.2 provides an explanation of the bathochromic shift of Band I and increasing ease of oxidation along the sequence porphyrin - chlorin - bacteriochlorin.

Panel 9.2 Molecular orbital interpretation of absorption spectra and redox properties of chlorins and bacteriochlorins with respect to porphyrins

Normally the cutting down of the chromophore of an organic molecule, *eg* by one conjugated double bond, leads to a hypsochromic (blue) shift of the absorption band. Yet when a porphyrin is reduced to a chlorin and then to a bacteriochlorin, the lowest energy transition undergoes a bathochromic (red) shift and intensifies.

This experimental result is contrary to intuition (which is based on much simpler systems), but can be understood in terms of the energies of the HOMO and LUMO (see Panel 8.1 and Figure 8.2). Calcuation shows that the relative orbital energies are as shown in Figure 9.2. In both the chlorin and the bacteriochlorin the component orbital energies of both the e_g (π^*) and the a_{nu} (π) levels become well separated. Although the energy of the lowest e_g (π^*) orbital does not increase much, that of the HOMO (a_{1u} in both chlorin and bacteriochlorin) is successively raised. The overall result is that the energy of the lowest transition (Band I, heavy arrows in Figure 9.2) decreases along the sequence porphyrin-chlorin-bacteriochlorin. At the same time symmetry restrictions are removed or modified, and Band I becomes more intense.

Figure 9.2 also offers an interpretation of redox behaviour. Along the sequence porphyrin-chlorin-bacteriochlorin the energy of the HOMO increases. Thus the removal of an electron becomes progressively easier, that is, the ease of oxidation increases along this sequence.

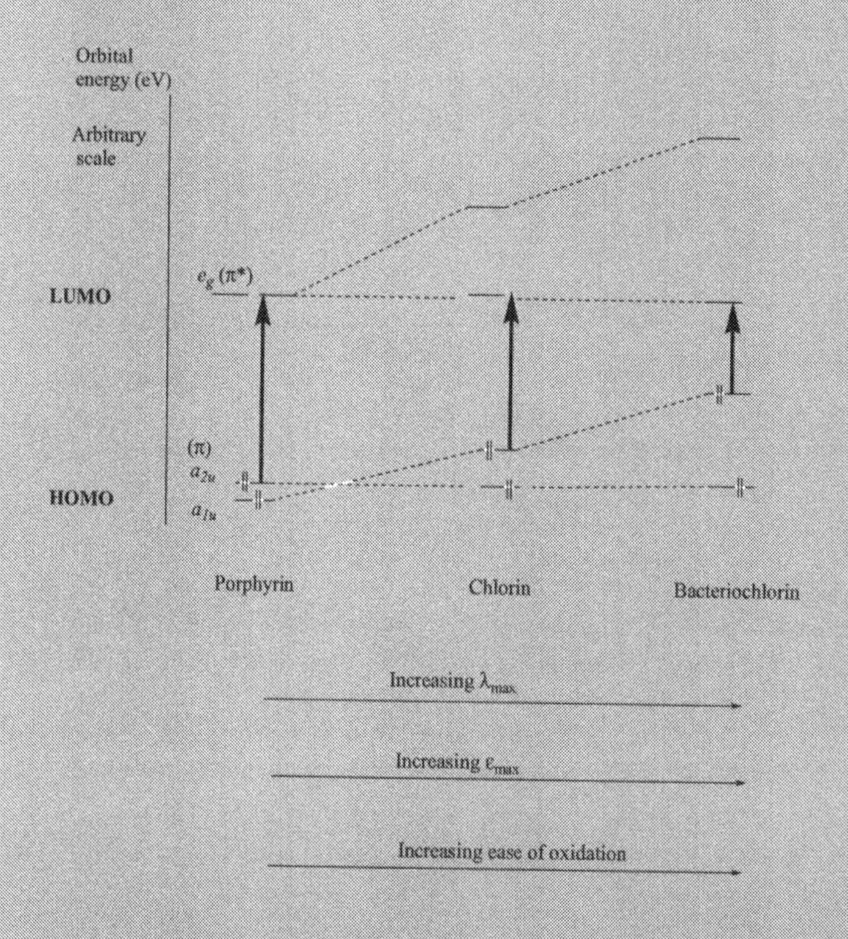

(Based on C.K. Chang, L.K. Hanson, P.F. Richardson, R. Young and J. Fajer, *Proc. Natl. Acad. Sci. USA*, 1981, **78**, 2652-2656).

FIGURE 9.2 SCHEMATIC ENERGY LEVEL DIAGRAM FOR THE HOMOS AND LUMOS OF PORPHYRIN, CHLORIN AND BACTERIOCHLORIN (ZINC(II) COMPLEXES).

There are two sources for these compounds — isolation/modification of natural examples, and total synthesis. Both are important in PDT applications.

9.3 NATURALLY DERIVED CHLORINS AND BACTERIOCHLORINS

The common chlorophylls and bacteriochlorophylls are waxy substances which are not themselves suitable for administration. Chemical modification generates products which are more easily purified and formulated for injection.

In the higher plants, photosynthesis is based on a mixture of chlorophyll *a* (**17**) and chlorophyll *b* (7-formyl-7-demethyl **17**) in a ratio of about 3:1. The separation of these and their transformation products is tedious, and for most applications it is far better to start with a source which essentially contains only chlorophyll *a*. Fortunately, the alga *Spirulina maxima* provides such a source, and it is readily available (health food shops). Chlorophyll *a* has a rich and fascinating chemistry: some of the most relevant chemical modifications are summarised in Figure 9.3.

Two steps may be taken to simplify the handling of these compounds. Firstly, the vinyl group at C-3 may be catalytically reduced to ethyl, giving compounds of the so-called meso series (eg **112**, **117**). The hypsochromic shift associated with this change is generally small. Secondly, the β-ketoester function in ring E, which is a particular site of instability ("allomerisation") because it is subject to autoxidation (involved in **113** → **101**, for example), can be thermally demethoxycarbonylated (as in **17** → **114**), leading to the so-called pyro series in which ring E is a fused cyclopentanone.

Apart from the reduction of the C-3 vinyl group, the chemical transformations in Figure 9.3 affect only the lower part of the molecule. This is the same as found in bacteriochlorophyll *a* (**118**) and hence analogous reactions can be wrought on that system, Band I in all cases being at longer wavelength than in the corresponding chlorophyll *a* derivative. However the bacteriochlorophylls are not so conveniently available, and must be isolated from photosynthetic bacteria, for example *Rhodobacter spheroides* or *Rhodobacter capsulatus,* specially grown in deep culture for the purpose. Yields are about 4 mg per gram of bacterial culture (dry powder).

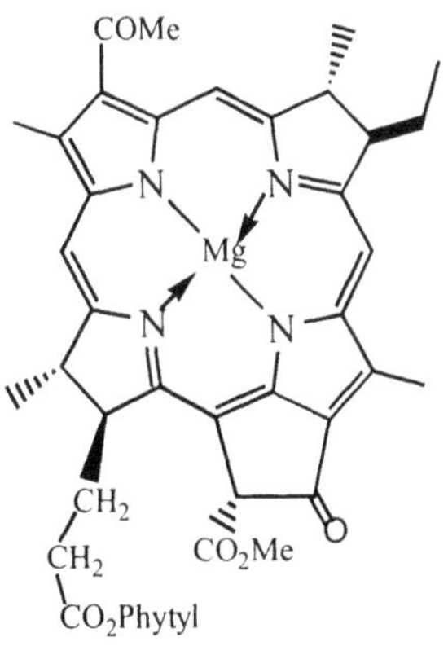

118 Bacteriochlorophyll *a*

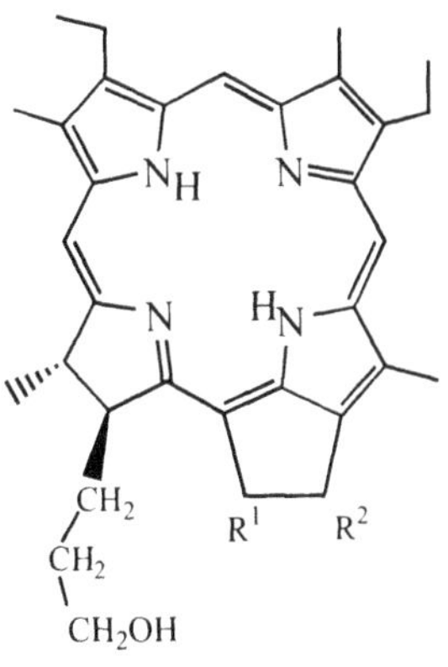

119 $R^1 = R_2^2 = H$
120 $R^1 = H$; $R^2 =$ ⁄ OH
121 $R^1 =$ - - CH_2OH; $R^2 =$ ▬OH

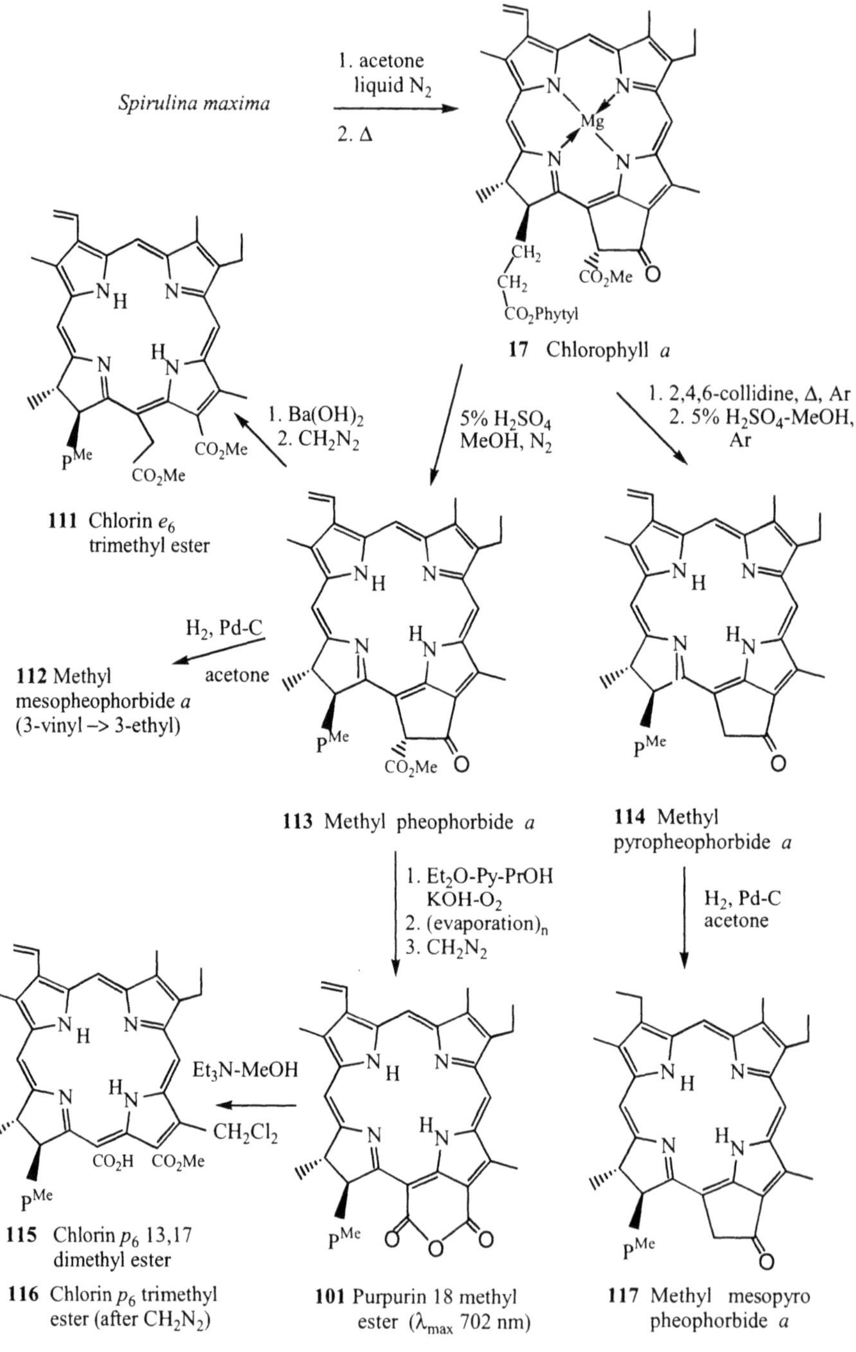

Spirulina maxima
1. acetone
liquid N_2
2. Δ
Mg
CH_2
CH_2
CO_2Phytyl
CO_2Me
17 Chlorophyll *a*
1. $Ba(OH)_2$
2. CH_2N_2
5% H_2SO_4
MeOH, N_2
1. 2,4,6-collidine, Δ, Ar
2. 5% H_2SO_4-MeOH,
Ar
111 Chlorin e_6
trimethyl ester
P^{Me}
CO_2Me
H_2, Pd-C
acetone
112 Methyl
mesopheophorbide *a*
(3-vinyl -> 3-ethyl)
113 Methyl pheophorbide *a*
114 Methyl
pyropheophorbide *a*
1. Et_2O-Py-PrOH
KOH-O_2
2. $(evaporation)_n$
3. CH_2N_2
H_2, Pd-C
acetone
Et_3N-MeOH
CH_2Cl_2
CO_2H CO_2Me
115 Chlorin p_6 13,17
dimethyl ester
116 Chlorin p_6 trimethyl
ester (after CH_2N_2)
101 Purpurin 18 methyl
ester (λ_{max} 702 nm)
117 Methyl mesopyro
pheophorbide *a*

FIGURE 9.3 CHEMICAL MODIFICATIONS OF CHLOROPHYLL *a*. CHLOROPHYLL *b* (**17**, BUT WITH 7-CHO IN PLACE OF 7-ME) UNDERGOES A SIMILAR SERIES OF REACTIONS, BUT IS NOT SUCH A READILY ACCESSIBLE STARTING MATERIAL. (H. FISCHER AND A. STERN, 'DIE CHEMIE DES PYRROLS', AKADEMISCHE VERLAGSGESELLSCHAFT, LEIPZIG, 1940, VOLUME II(II), P.40 AND ELSEWHERE SUMMARISED IN R.K. DINELLO AND C.K. CHANG, THE PORPHYRINS (ED. D. DOLPHIN), ACADEMIC PRESS, NEW YORK, VOL I, PP324–329. K.M. SMITH, D.A. GOFF AND D.J. SIMPSON, *J. AM. CHEM. SOC.*, 1985, **107**, 4946–4954).

Since natural chlorins and bacteriochlorins are available in this way, it is not surprising that PDT studies with them have been very active. The range of the investigations is illustrated by the following examples, selected to try to represent the worldwide activity.

(i) *chlorin e_6* (**111**, as triacid). Gurinovich's group in Minsk has studied the photophysical and photobiological properties of this compound.
(ii) *hexyl ether of pyropheophorbide a* (**75**) has been developed to the clinical level by Dougherty in Buffalo (see section 7.2.4.3).
(iii) *chlorin polyols.* Vallés (Barcelona) has shown that lithium aluminium hydride reduction of methyl mesopyropheophorbide *a* (**117**) gives a mixture of the hydrogenolysis product (**119**) and the diol (**120**), while reduction of methyl mesopheophorbide *a* (**112**) gives the triol (**121**). *In vitro* assay (MTS method, section 13.1.2) shows cell inactivation increases in the series mono-ol <diol <triol.
(iv) *chlorin* p_6 *derivatives* (**115**). Bioassay of a series of chlorin p_6 derivatives indicated that the 3-formyl-3-devinyl derivative (λ_{max} 698 nm) was the most active (Mironov, Moscow). Smith and Kessel (USA) have reported various conjugates of (**115**). For example the C_{13} lysyl derivative, prepared by opening the cyclic anhydride of purpurin 18 ester (**101**) with a large excess of lysine in pyridine, was fast acting, and more selective (no severe cutaneous photosensitivity) in *in vivo* assays.
(v) *bacteriopurpurin derivatives.* Both Smith (Davis) and Mironov (Moscow) have studied analogues of purpurin 18, but in the bacteriochlorophyll series. In this way they have prepared imide (eg **122**) and isoimide analogues. These appear to be surprisingly rugged substances, and to have absorption maxima at > λ_{max} 800 nm. Preliminary screening experiments with tumour implants in mice are promising (0.25 μmol kg^{-1}, 75 mW cm^{-2}, 24h DL I, 804 nm irradiation).

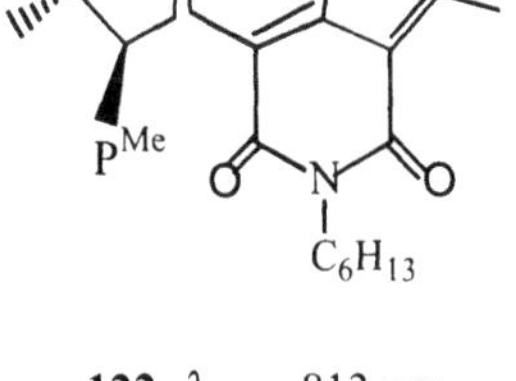

122 λ_{max} 813 nm 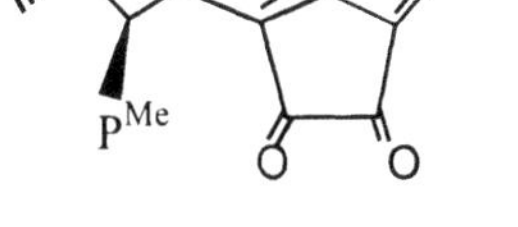 **123** λ_{max} 678 nm 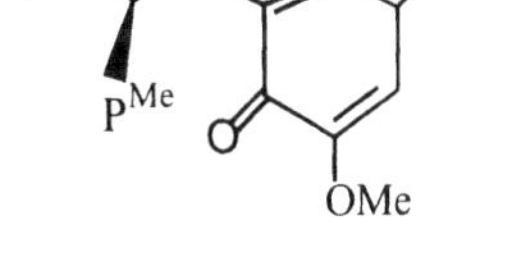 **124** λ_{max} 777 nm

(vi) *carbene insertion into 13²-oxopyropheophorbide a.* Kozyrev *et al.* (Buffalo) have shown that carefully controlled autoxidation of methyl pyropheophorbide *a* (**114**) in LiOH/THF gives the 13^2-oxo derivative (**123**, λ_{max} 678 nm). Treatment with a large excess of CH_2N_2 (RT, 6h) causes carbene insertion to give three products, the main one (29%) being (**124**). This structure recalls that of mesoverdin (**98**, section 8.4.2) and like that substance it shows a considerable bathochromic shift at Band I (in this case to λ_{max} 777 nm). Bioassay results are awaited.

The chlorophyll-derived PDT photosensitiser with the longest track record is monoaspartyl chlorin e_6 (MACE, **125**). This is prepared from chlorin e_6 (**111**, tricarboxylic acid) by DCC coupling with di-t-butyl aspartate. For some years it was thought that coupling occurred preferentially at the 17-propionic acid function for steric reasons, but NMR studies have recently indicated that the major product is formed by reaction at the 15-acetic acid. Some diamide (15,17-diamide, DACE) is also formed, so a chromatographic separation is normally required. Cleavage of the t-butyl esters with trifluoroacetic acid gives (**125**).

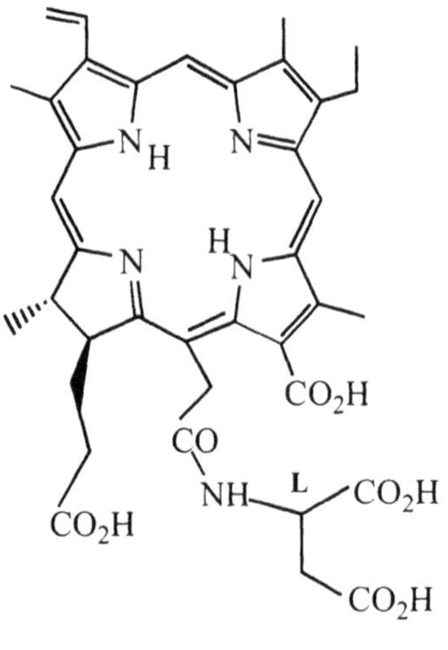

125 MACE

MACE is being developed in Japan, under the acronym NPe6. It has λ_{max} 660 nm (ε 40,000) in PBS (*ie* it is water soluble), and $\phi_\Delta = 0.8$. It has the advantage of a short drug-light interval (3–4h) and any generalised photosensitisation is short-lived. PDT using it has been recorded to be effective against recurrent breast adenocarcinoma, as well as basal cell carcinoma and squamous cell carcinoma (see also section 14.3.4 and Panel 14.4).

9.4 CHLORINS DERIVED FROM PROTOHAEM

In section 8.3 we saw that some *porphyrin* sensitisers (including HpD) were derived in one way or another from the haem of haemoglobin. By appropriate chemical transformations it is also possible to generate *chlorins* from this source.

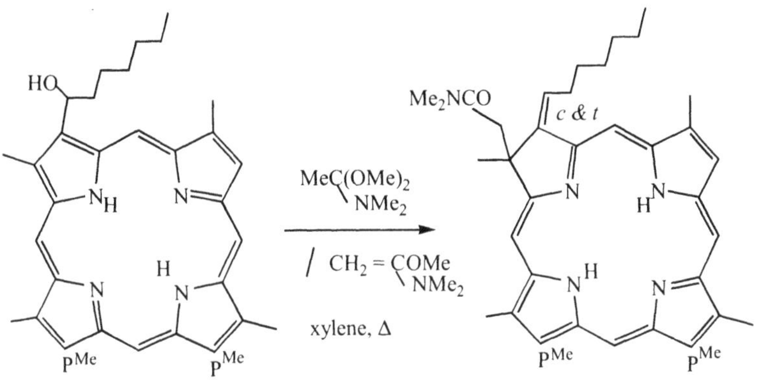

$MeC(OMe)_2NMe_2$ / $CH_2=C(OMe)NMe_2$, xylene, Δ

126 → **127** 76%

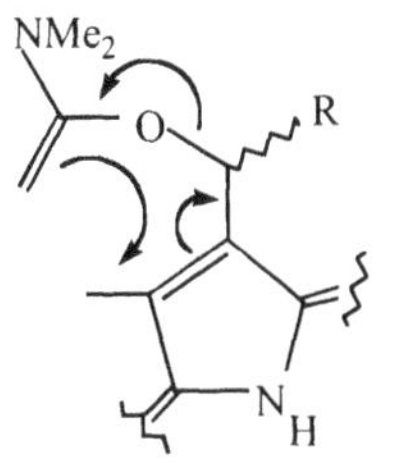

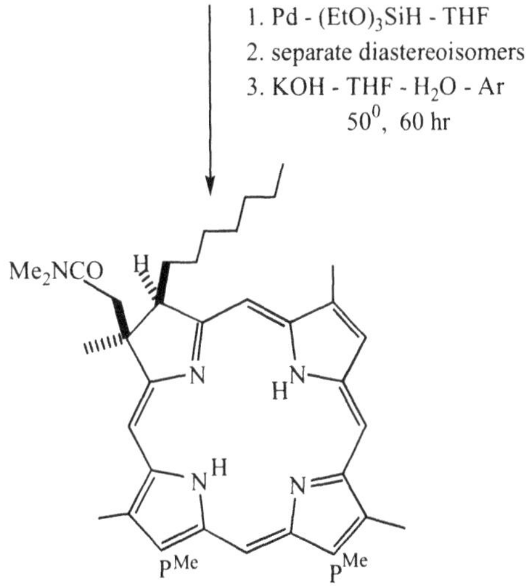

1. Pd - $(EtO)_3SiH$ - THF
2. separate diastereoisomers
3. KOH - THF - H_2O - Ar, 50^0, 60 hr

128 [2*RS*,3*SR*] isomer shown (racemic)
λ_{max} 645 nm (ε 33 700)

FIGURE 9.4 MONTFORT'S ROUTE TO CHLORIN SYSTEMS FROM DEUTEROPORPHYRIN (**88**, DME). (D. KUSCH, A. MEIER AND F-P. MONTFORTS, *LIEBIGS ANN. CHEM.*, 1995, 1027–1032).

9.4.1 From Deuteroporphyrin (88)

Treatment of copper(II) deuteroporphyrin dimethyl ester with heptanoic anhydride under Friedel-Crafts conditions [$(C_6H_{13}CO)_2O$, $SnCl_4$, CH_2Cl_2, -15°] gives a mixture of the 3- and 8-heptanoyldeuteroporphyrin dimethyl esters, which are demetallated (conc. H_2SO_4, RT). Borohydride reduction then gives the mixed 3- and 8-(1-hydroxyheptyl) derivatives (eg **126**, total yield 83%) which are separated chromatographically (Figure 9.4). Thermal Claisen rearrangement of the amideacetal formed with *N*, *N*-dimethylacetamide dimethyl acetal converts (**126**) into the *chlorin* (**127**, cis/trans isomers). Reduction, chromatographic separation of the isomers, and (separate) hydrolysis then gives the racemic *cis*-chlorin (**128**) (and the corresponding

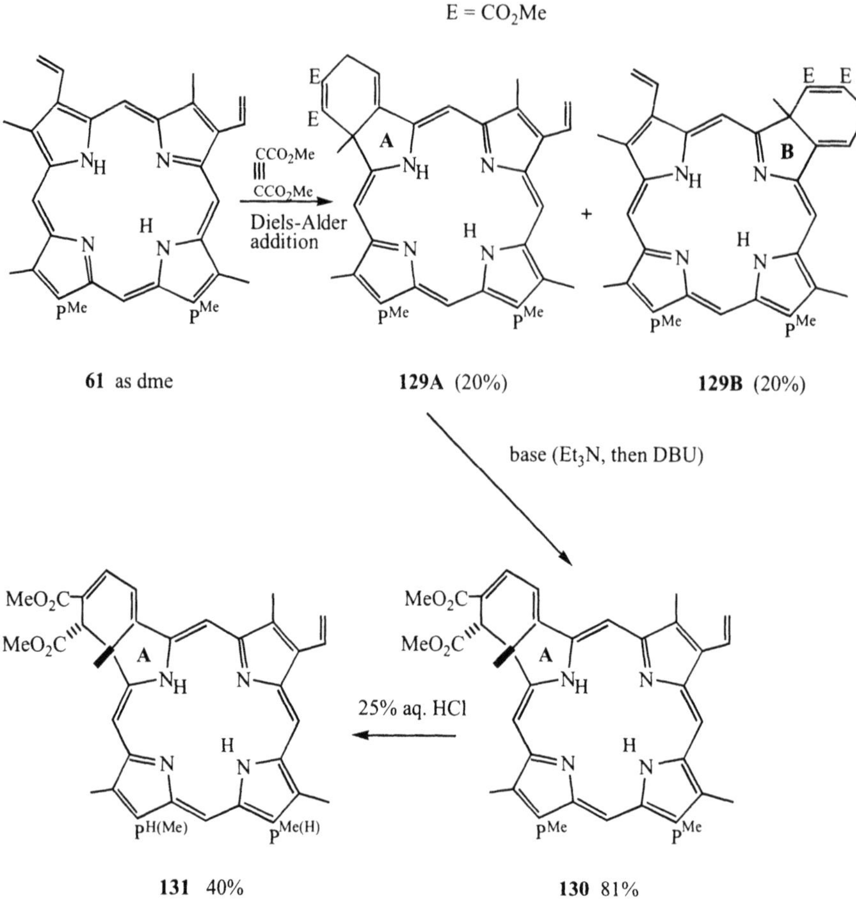

$E = CO_2Me$

$CCO_2Me \equiv CCO_2Me$, Diels-Alder addition

61 as dme — **129A** (20%) + **129B** (20%)

base (Et_3N, then DBU)

130 81% — 25% aq. HCl → **131** 40%

FIGURE 9.5 DOLPHIN'S SYNTHESIS OF BENZOPORPHYRIN DERIVATIVE. ELABORATION OF THE A ADDUCT TO GIVE **131** (BPDMA) IS SHOWN. (E.D. STERNBERG, D. DOLPHIN AND C. BRÜCKNER, *TETRAHEDRON*, 1998, **54**, 4151–4202).

racemic *trans*-chlorin). This possesses a typical chlorin absorption spectrum (λ_{max} ($CHCl_3$) 645 nm, ε 33 700) and a ϕ_Δ value of 0.56. Compounds of this series show photobiological activity *in vitro*, but the need to separate diastereoisomers makes the route rather unattractive. What is pertinent, however, is that here we have an example (**128**) of a *blocked* chlorin system which is protected by β,β-disubstitution against dehydrogenation, a structural feature which will reappear in some other second generation photosensitisers (eg **131**, **140**, **145**, **150** and **152**).

9.4.2 From Protoporphyrin (61)

With the adjacent β,β′-double bond, a vinyl substituent in protoporphyrin (**61**) can provide a 1,3-diene unit for Diels-Alder chemistry. With protoporphrin the reaction can occur once or, under forcing conditions, twice. If it occurs once the product is a chlorin: if it occurs twice, an isobacteriochlorin system (**108**) is formed, but an aromatic ring (18πe) is still preserved.

Powerful dienophiles are needed to generate adducts of this sort. Thus, Diels-Alder addition of acetylene dicarboxylic acid dimethyl ester to protoporphyrin dimethyl ester (**61**, diester) gives a mixture of the two regioisomers (**129A** and **B**) which have chlorin spectra (λ_{max} 666 nm, ε 50 000) (Figure 9.5) and which have to be separated. Base-catalysed isomerisation of (**129A/B**) brings the isolated double bond into conjugation with the main chromophore, and generates the thermodynamically more stable transoid arrangement of the methyl and methoxycarbonyl groups. The next step is partial hydrolysis, which, for steric reasons, occurs preferentially at the propionic ester groups, but does not distinguish between them, so the product is a mixture of mono acids (**131**). It has become known as benzoporphyrin derivative mono acid, ring A (BPDMA), although it is not a benzoporphyrin but a chlorin (λ_{max} 686 nm, ε 34 000 in MeOH).

Although the preparation thus involves a chromatographic separation of regioisomers (which is a costly operation on a large scale) and the molecule (**131**) contains two chiral centres, this substance is under commercial development (proprietary name 'Verteporfin', section 14.3.7 and Panel 14.7). Its photophysical properties are appropriate, since besides the strong chlorin band just mentioned, it has a high singlet oxygen quantum yield (ϕ_Δ = 0.7). Although a monocarboxylic acid, it remains a hydrophobic substance, and is administered in a liposomal preparation. The substance has been under development for the treatment of cutaneous malignancies and psoriasis. In Phase III trails the dose is approximately 0.3 mg kg^{-1} in unilamellar liposomes with a fluence of 50–75 J cm^{-2} using LED panels with a fluence rate of 60–160 mW cm^{-2}. A feature of this photosensitizer is that it does not cause prolonged photosensitivity because it is quickly removed from the system.

Other applications, particularly in opthalmology (age-related macular degeneration) and in immunomodulation (rheumatoid arthritis), are under active study (section 14.5.3).

9.4.3 From Photoprotoporphyrin (132)

When protoporphyrin (**61**) as its dimethyl ester is irradiated in solution in the presence of oxygen, singlet oxygen is formed. The singlet oxygen then acts as a dienophile (section 3.3.3), adding to the vinyl — ββ-double bond system of the protoporphyrin as in the previous section (see also section 12.4.1.1). The mono adducts rearrange spontaneously to the hydroxy aldehydes, which were identified by Inhoffen and his colleagues as shown in Figure 9.6 (see also Figure 12.3 for an extension of this).

Photoprotoporphyrin (**132**) is a chlorin (λ_{max} 671 nm, ε = 57 500) and has one chiral centre. It is not appreciably soluble in water, but when converted into the oxime and conjugated with aspartic acid then, in the form of its tetrasodium salt (**133**), it becomes water-soluble. The chromophore shows λ_{max} 670 nm, similar to that of **132** as expected, but the molar absorption coefficient at 16 000 is lower than expected, possibly due to aggregation. Compound (**133**), which has been designated ATX-S10(Na) by its discoverers, Nakajima and Sakata, shows rapid uptake by tumour transplants in mice. Clinical experiments, for both photodiagnosis and phototherapy applications, have started.

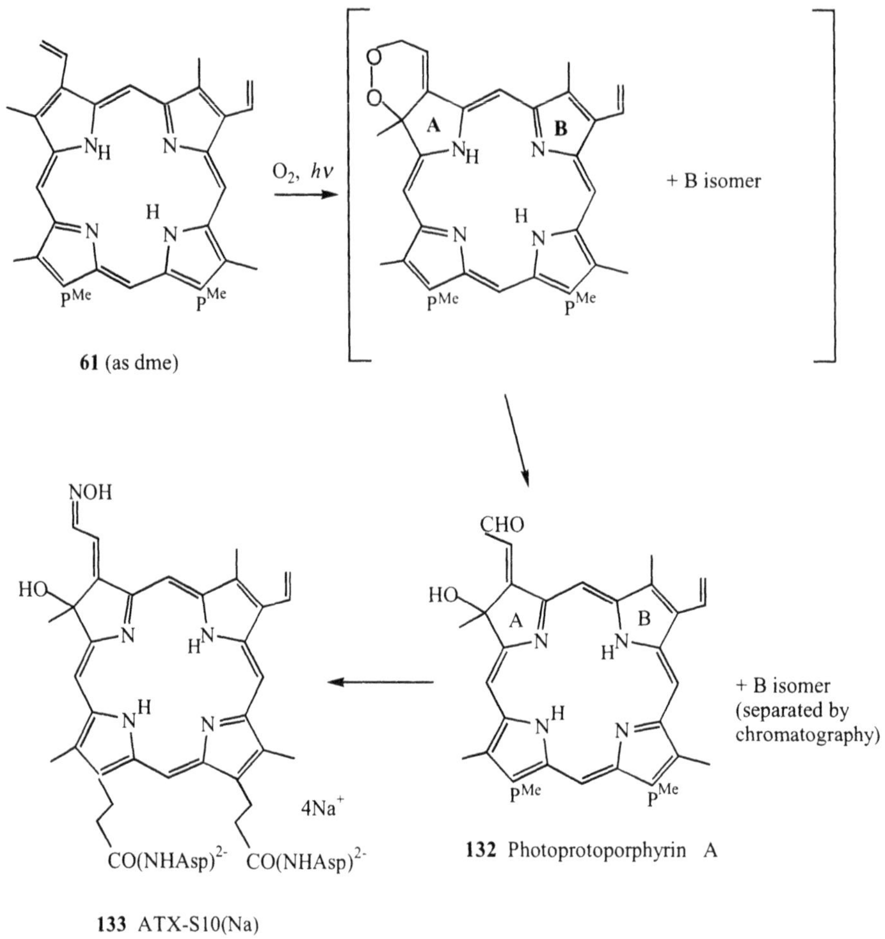

O_2, $h\nu$

+ B isomer

61 (as dme)

133 ATX-S10(Na)

132 Photoprotoporphyrin A

+ B isomer (separated by chromatography)

FIGURE 9.6 CHLORIN PHOTOSENSITISER (**133**) FROM PHOTOPROTOPORPHYRIN (**132**). (S. NAKAJIMA, I. SAKATA AND T. TAKEMURA, *CHEM. ABS.*, 1997, **126**, 311893).

9.5 CHLORINS AND BACTERIOCHLORINS BY TOTAL SYNTHESIS

9.5.1 By Reduction

There are two main preparative procedures for reducing a porphyrin to the corresponding chlorin (see structures **107**, **108**, **109**): (i) metal-acid reduction, usually of the iron complex. Typical reaction conditions involve molten sodium in pentyl alcohol at reflux; and (ii) treatment with diimide, HN=NH, generated from *p*-tosyl hydrazide and potassium carbonate in pyridine. In cases where it can be recognised the former reaction generates predominantly the thermodynamically more stable transoid arrangement at the reduced β,β-bond, the latter the cisoid arrangement. In both reactions, further reduction may occur to give bacteriochlorins and isobacteriochlorins: these substances may be dehydrogenated (eg with *o*-chloranil) to the chlorin if that is the synthetic objective; or, if the tetrahydroporphyrins are required, they may be isolated by acid-fractionation and

77 5,10,15,20-Tetrakis(*m*-hydroxy phenyl)chlorin (m-THPC)

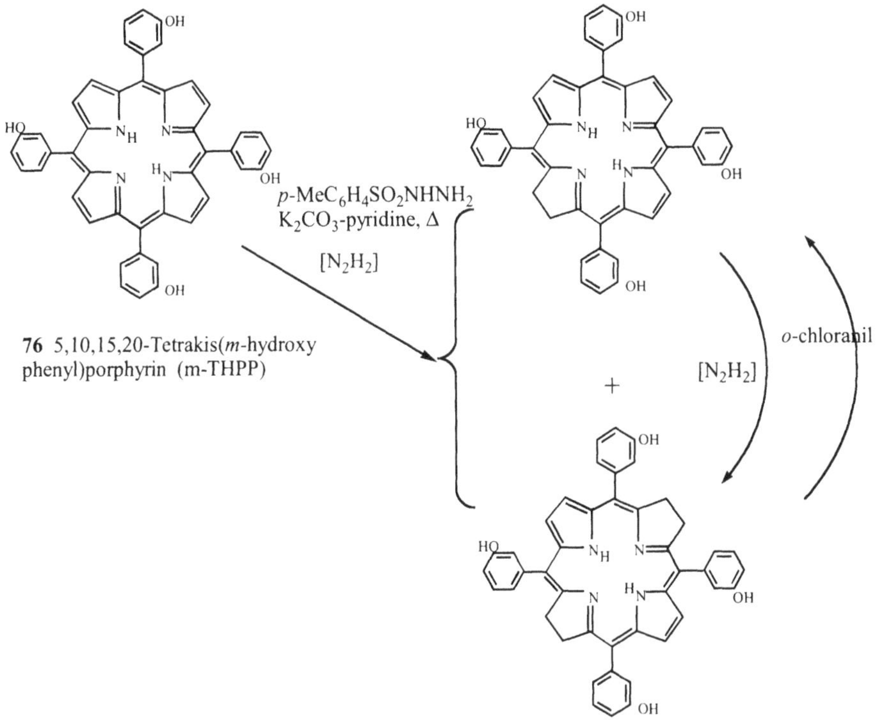

p-$MeC_6H_4SO_2NHNH_2$
K_2CO_3-pyridine, Δ
$[N_2H_2]$

76 5,10,15,20-Tetrakis(*m*-hydroxy phenyl)porphyrin (m-THPP)

+

$[N_2H_2]$

o-chloranil

134 5,10,15,20-Tetrakis(*m*-hydroxy phenyl)bacteriochlorin (m-THPBC)

FIGURE 9.7 SYNTHESIS OF M-THPC (**77**) AND M-THPBC (**134**) BY DIIMIDE REDUCTION OF THE CORRESPONDING PORPHYRIN (**76**). (R. BONNETT, R.D. WHITE, U-J. WINFIELD AND M.C. BERENBAUM, *BIOCHEM. J.*, 1989, **261**, 277–280).

chromatography. Diimide reduction of the metal-free porphyrin gives mainly chlorin and bacteriochlorin: with the zinc complex, the isobacteriochlorin is the predominant tetrahydro product.

Thus, diimide reduction of m-THPP (**76**), a photosensitiser of demonstrated potency in PDT (section 8.4.2), gives the corresponding chlorin, m-THPC (**77**), and bacteriochlorin, m-THPBC (**134**) (Figure 9.7). Similar reactions occur with the *o*- and *p*-isomers (**93**, **94**).

This series (**76**, **77**, **134**) represents the only example so far where the porphyrin nucleus with the same substitution pattern has been subjected to physical and biological examination at all three reduction levels, as shown in Table 9.1.

Although in methanol the molar absorption coefficients and the λ_{max} values of Band I increase in the sequence m-THPP → m-THPC → m-THPBC, the triplet quantum yields and singlet oxygen quantum yields do not alter much on reduction. However the photodynamic effect increases with each reduction step, the magnitude

TABLE 9.1 COMPARISON OF PHOTOPHYSICAL AND PHOTOBIOLOGICAL DATA FOR M-THPP (**76**), M-THPC (**77**) AND M-THPBC (**134**).

	m-THPP (**76**)	m-THPC (**77**)	m-THPBC (**134**)
PHOTOPHYSICAL DATA (MeOH)			
λ_{max} Band I (nm)	644	650	735
ε_{max} (l mol^{-1} cm^{-1})	3400	29600	91000
λ_{max}, fluorescence (nm)	649, 715	653, 720	612, 653, 746
for excitation at (nm)	415	415	500
E_s (kJ mol^{-1})	185	183.5	161.5
ϕ_f	0.12	0.089	0.11
ϕ_T	0.69	0.89	0.83
ϕ_Δ (air)	0.46	0.43	0.43
BIOASSAY DATA (Mouse tumour implant model, section 13.1.4.1)			
λ_{max} in calf foetal serum (nm)*	648	652	741
Dose of photosensitiser (μmol kg^{-1})	6.25	0.75	0.39
Depth of tumour photonecrosis (mm)	4.6 ± 0.45 (13)	5.41 ± 0.39 (19)	5.12 ± 1.21 (8)

*Wavelength of tumour irradiation, fluence 10 J cm^{-2}, 24 hours after photosensitiser administration in DMSO, i.p.

(Photophysical data from R. Bonnett, P. Charlesworth, B.D. Djelal, S. Foley, D.J. McGarvey and T.G. Truscott, *J. Chem. Soc., Perkin Trans.* **2**, 1999, 325–328 . Photobiological data from R. Bonnett, R.D. White, U-J. Winfield and M.C. Berenbaum, *Biochem. J.*, 1989, **261**, 277–280).

of the effect being roughly parallel to ε_{max} of Band I, at which irradiation occurs. Thus, with the bioassay results in Table 9.1 being set for a tumour necrosis depth of *ca.* 5mm, the drug doses required under the same experimental conditions are m-THPP, 6.25; m-THPC, 0.75; and m-THPBC, 0.39 μmol kg^{-1}. Both m-THPC and m-THPBC are being developed commercially as photosensitisers. Because of the greater ease of dehydrogenation and greater difficulty in manufacture of the bacteriochlorin, the main emphasis has been on m-THPC, which has the proprietary name of 'Foscan' (section14.3.2).

Within two years of the announcement of the discovery of this substance and its effectiveness in animal models, the first clinical results (HB Ris, Berne, 1991) had appeared describing four cases of mesothelioma, a cancer of the pleural cavity, often associated with asbestos exposure, which is essentially untreatable. The results were said to be promising, although it might be thought that this difficult condition was not the best place at which to start. m-THPC is now in Phase III clinical trials for cancers of the upper aerodigestive tract, and other conditions (Panel 14.2). It is a very potent sensitiser, and clinicians used to working with HpD or Photofrin have had some difficulty getting used to the low levels of sensitiser and fluence required. Long-lived skin photosensitisation lasting for several days after treatment is still a problem, although less so than with Photofrin.

9.5.2 By Other β,β′-Addition Reactions

cis-Hydroxylation of the β,β′ band of a porphyrin to give a dihydroxychlorin can be effected by osmylation. The reaction (like hydrogenation above) can occur twice: the free base then gives the tetrahydroxybacteriochlorin, whereas osmylation of the zinc(II) complex gives mainly the analogous isobacteriochlorin.

Thus osmylation of octaethylporphyrin (**135**) in dichloromethane followed by reduction of the intermediate cyclic osmate with hydrogen sulphide gives the dihydroxychlorin (**136**, λ_{max} 643 nm, ε 41 800) and the tetrahydroxybacteriochlorin (**137**, λ_{max} 715 nm, ε 53 000). Both of these compounds show PDT activity *in vivo.* Treatment of the diol (**136**) under highly acidic conditions leads, via a pinacol-pinacolone transformation, to the β-oxochlorin (**138**), which is inactive in the *in vivo* bioassay. However, reduction of (**138**) with sodium borohydride in ethanol gives the hydroxy compound which is converted to the corresponding bromide (**139**) with 50% HBr-HOAC. (Figure 9.8).

Compound (**139**) is a secondary benzylic bromide, and reacts readily with nucleophiles, provided that they are not sterically encumbered, (since C-2 has an adjacent *gem*-diethyl group). Thus reaction with ethylene glycol, glycerol, D-glucose and D-mannitol occurs at the most accessible primary alcohol sites to give compounds **140, 141, 142** and **143**, respectively, which have graded amphiphilic character, **140** having one hydroxy function, **143** having five. (**142** is the 6-*O*-**D**-glucose derivative, while **143** is the 1-*O*-**D**-mannityl ether). These are all chlorins (eg **141**, λ_{max} 642 nm, λ 49 000): **141–143** are diastereoisomeric mixtures. Activity in tumour photonecrosis increases with the number of hydroxy groups, as the strong hydrophobicity of (**135**) is gradually ameliorated: (the selectivity of these compounds is referred to in section 13.1.4.2).

In an analogous way, osmylation of β,β′ positions is also observed with *meso*-tetraphenylporphyrin (**27**) to give the corresponding chlorindiol (λ_{max} 644 nm) and the bacteriochlorin tetraol (two isomers separated, both λ_{max} 708 nm).

Osmylation has also been applied to naturally derived systems, most usefully to chlorins, which specifically give dihydroxybacteriochlorins. Thus osmylation of 3-formyl-3-devinylpurpurin 18 methyl ester (**144**) gives the bacteriochlorin (**145**), which on pinacol-pinacolone transformation gives the β-oxobacteriochlorins (**146, 147**) resulting from the two possible alkyl migrations (Figure 9.9).

9.5.3. By Meso-β Cyclisation

Formylation of nickel(II) etioporphyrin (**148**) under Vilsmeier-Haack conditions gives the *meso*-formyl compound, which on Wittig reaction and demetallation gives the *meso*-acrylic ester (**149**) (Figure 9.10). Under mild acid conditions such *meso*-acrylic systems undergo a reversible thermal cyclisation to the neighbouring β-position (here, **149→150**).

This type of cyclisation was used by Woodward in his synthesis of chlorophyll *a*: a mechanistic rationalisation in arrow pushing terms is included in Figure 9.10.

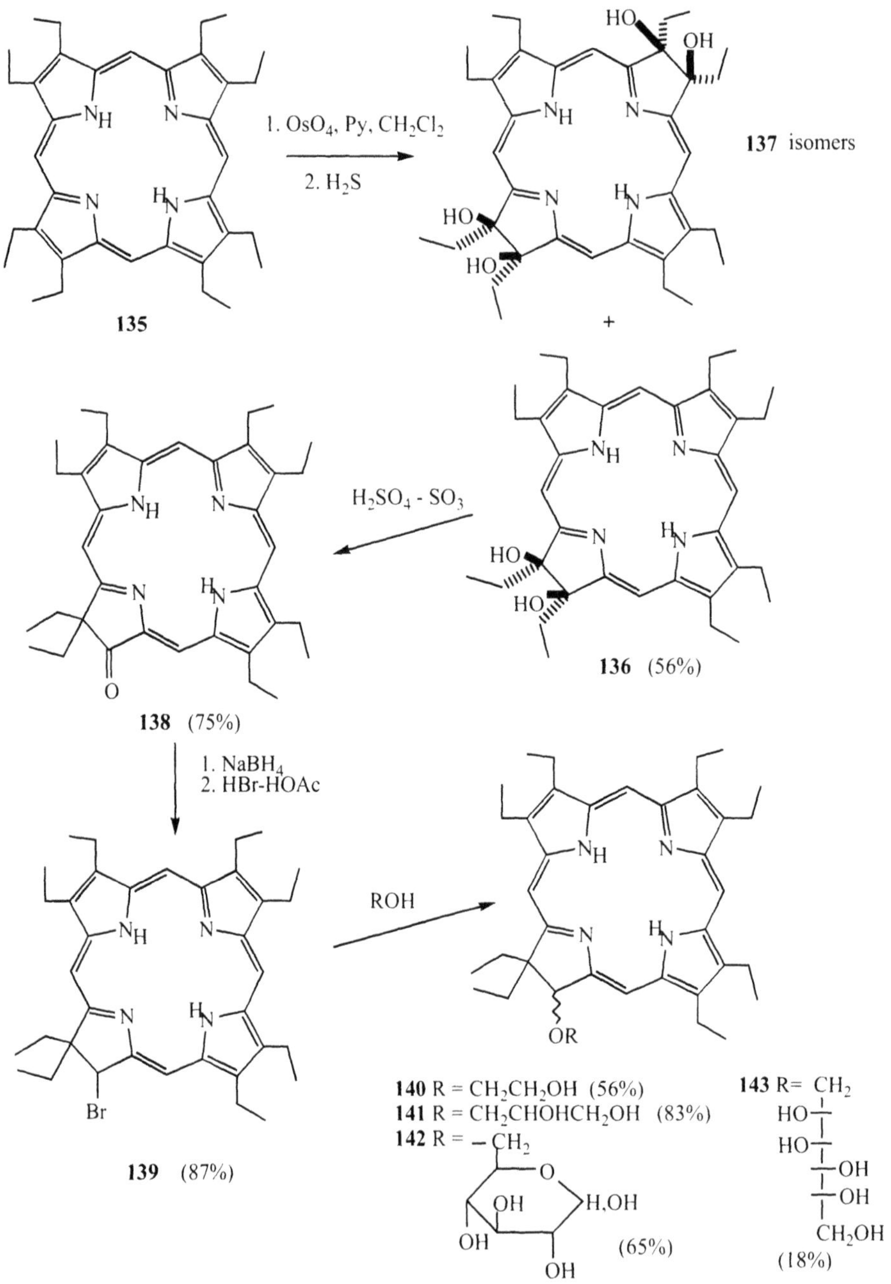

1. OsO_4, Py, CH_2Cl_2

2. H_2S

135

137 isomers

+

H_2SO_4 - SO_3

136 (56%)

138 (75%)

1. $NaBH_4$
2. HBr-HOAc

139 (87%)

ROH

140 R = CH_2CH_2OH (56%)

141 R = $CH_2CHOHCH_2OH$ (83%)

142 R = $-CH_2$ (65%)

143 R = CH_2 (18%)

FIGURE 9.8 CHLORINS OF GRADED AMPHIPHILIC CHARACTER PREPARED FROM OCTAETHYLPORPHYRIN (**135**) FOLLOWING OSMYLATION. (K.R. ADAMS, M.C. BERENBAUM, R. BONNETT, A.N. NIZHNIK, A. SALGADO AND M.A. VALLES, *J. CHEM. SOC., PERKIN TRANS.* **1**, 1992, 1465–1470).

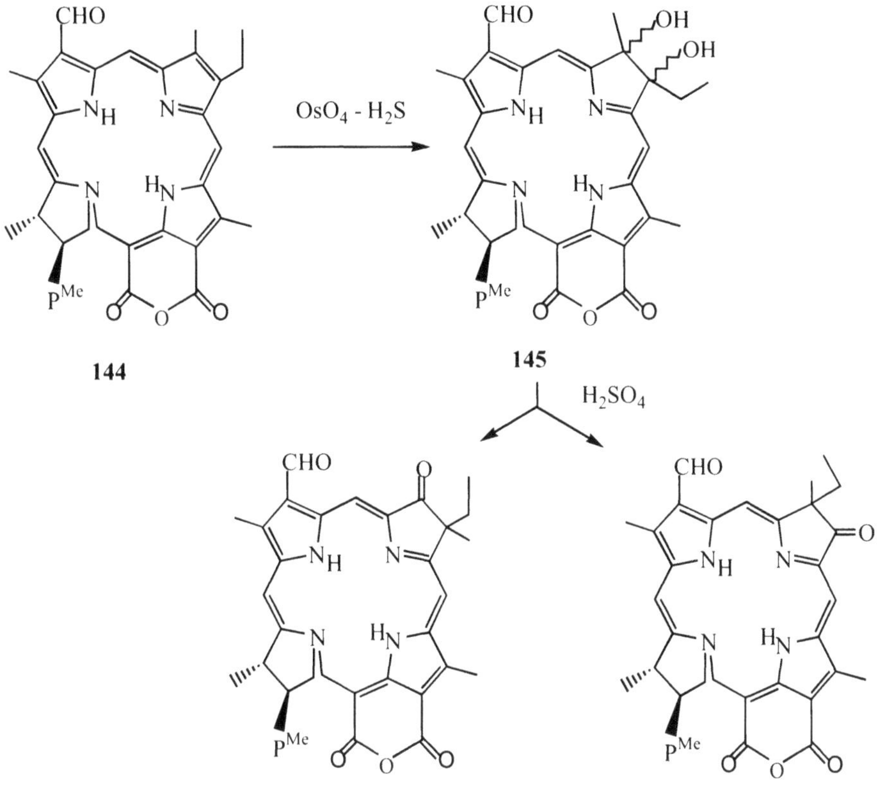

144

OsO_4 - H_2S

145

H_2SO_4

146 λ_{max} 786 nm

147 λ_{max} 777 nm

FIGURE 9.9 OSMYLATION OF 3-FORMYL-3-DEVINYLPURPURIN 18 METHYL ESTER (**144**) AND PINACOL-PINACOLONE REARRANGEMENT OF THE RESULTING DIOL (**145**). (A.N. KOZYREV, R.K. PANDEY, C.J. MEDFORTH, G. ZHENG, T.J. DOUGHERTY AND K.M. SMITH, *TETRAHEDRON LETTERS*, 1996, **37**, 747–751).

In the present case cyclisation occurred selectively to the β-position bearing the ethyl group, presumably to maximise the relief of steric strain, to give the regioisomer (**150**) in 92% yield. Because (**150**) is a chlorin possessing an electron-withdrawing function at a *meso*-position adjacent to the reduced ring, the spectrum / structure falls into the purpurin class (Panel 9.1). Metallation of (**150**) with tin(II) chloride gives the $Cl_2Sn(IV)$ complex (**78**), which has λ_{max} (CH_2Cl_2) 659 nm (ε 30 300). The fluorescence yield is low (heavy atom quenching): on the other hand, the singlet oxygen quantum yield is substantial (ϕ_Δ = 0.6). This material, referred to as SnEt2, but with the proprietary name 'Purlytin', is reported to be in clinical trials for cancer and for age-related macular degeneration (section 14.3.3).

The benzochlorin iminium salts (**79**) have already been mentioned (section 7.2.6). Benzochlorin synthesis is shown in Figure 9.11. The starting point here is, for example, copper(II) octaethylporphyrin which is subjected to a vinylogous Vilsmeier-Haack reaction (3-dimethylaminoacrolein and phosphorus oxychloride) to give the *meso*-acrylic aldehyde derivative (**151**).

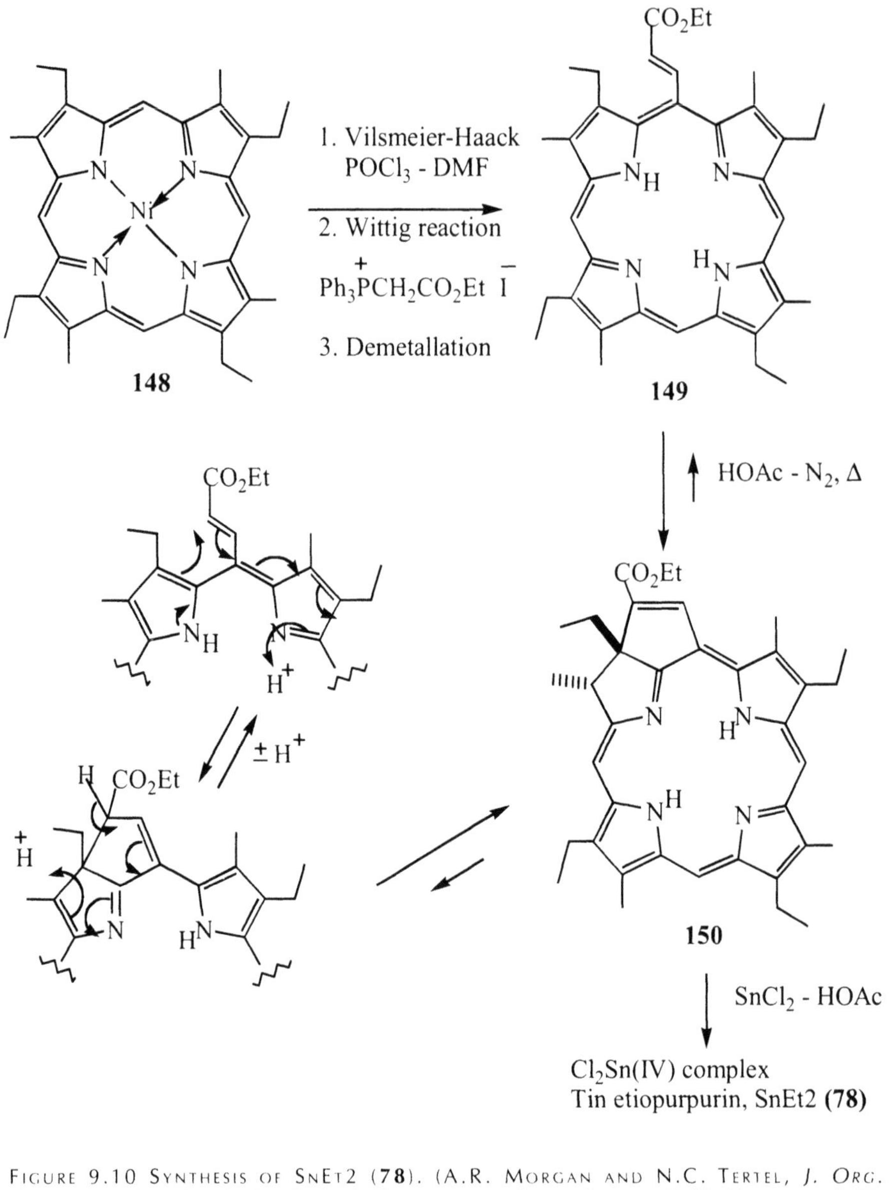

FIGURE 9.10 SYNTHESIS OF SNET2 (**78**). (A.R. MORGAN AND N.C. TERTEL, *J. ORG. CHEM.*, 1986, **51**, 1347–1350. A.R. MORGAN, G.M. GARBO, R.W. KECK AND S.H. SELMAN, *CANCER RES.*, 1988, **48**, 194–198).

Treatment of (**151**) with H_2SO_4/TFA causes cyclisation on to the neighbouring β-position with a Wagner-Meerwein shift of the alkyl group (in mechanistic terms this cyclisation to give (**152**) requires a reduction step, the nature of which is not really clear). Demetallation of the product (**152**) with concentrated sulphuric acid gives the [*c,d*]benzochlorin (**153**), which possesses a "blocked" chlorin system. Direct sulphonation of this generates the monosulphonic acid (**154**).

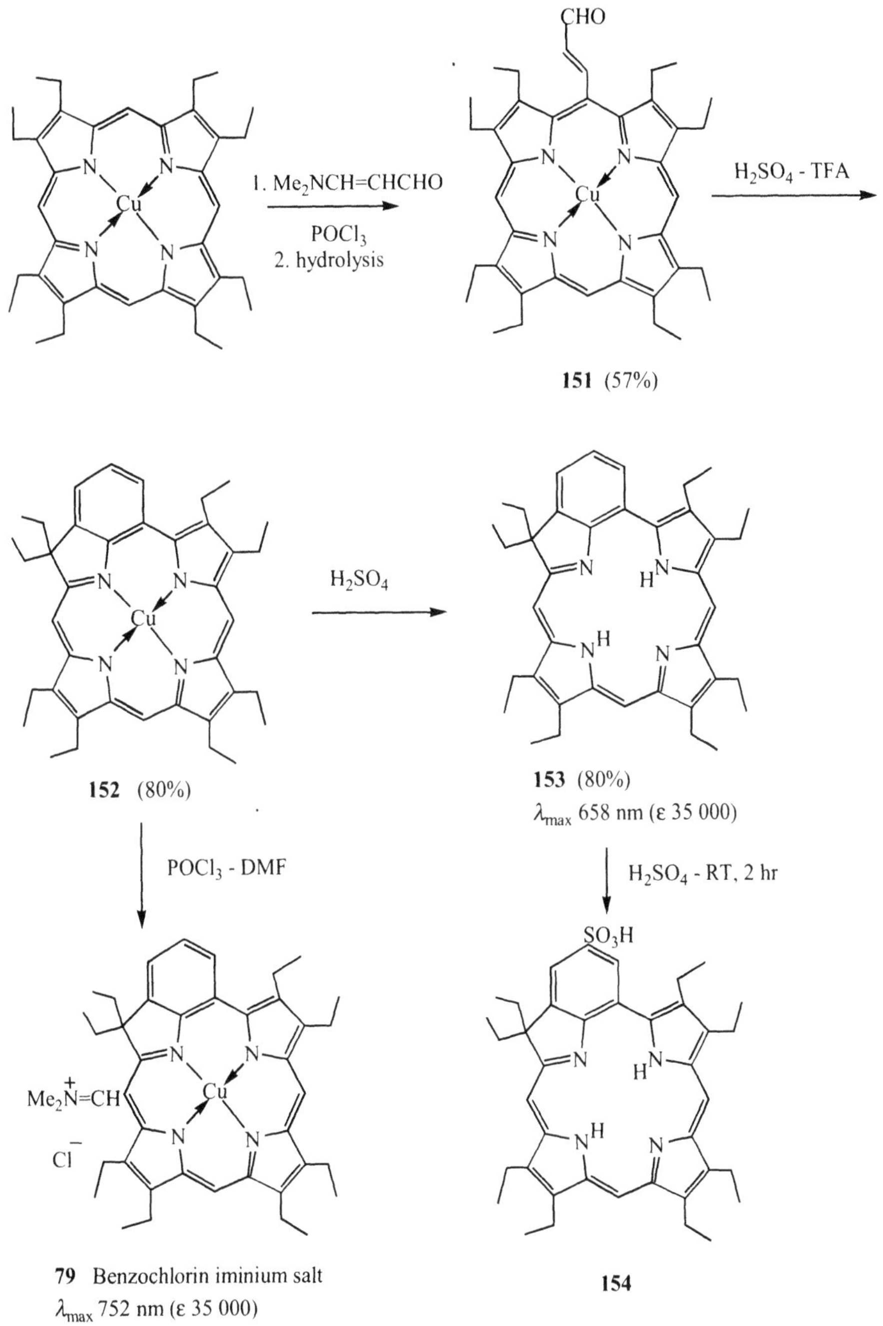

1. $Me_2NCH=CHCHO$
$POCl_3$
2. hydrolysis

CHO

H_2SO_4 - TFA

151 (57%)

H_2SO_4

152 (80%)

153 (80%)
λ_{max} 658 nm (ε 35 000)

$POCl_3$ - DMF

H_2SO_4 - RT, 2 hr

SO_3H

$Me_2\overset{+}{N}=CH$
Cl^-

79 Benzochlorin iminium salt
λ_{max} 752 nm (ε 35 000)

154

FIGURE 9.11 BENZOCHLORIN SYNTHESIS (M.G.H. VICENTE, I.N. REZZANO AND K.M. SMITH, *TETRAHEDRON LETTERS*, 1990, **31**, 1365–1368; A.R. MORGAN *ET AL.*, *PHOTOCHEM. PHOTOBIOL.*, 1992, **55**, 133–136; D. SKALKOS AND J.A. HAMPTON, *MED. CHEM. RES.*, 1992, **2**, 276–281).

Vilsmeier-Haack formylation of (**152**) allows the intermediate iminium salt (**79**) to be isolated. This is remarkable in two respects: it is apparently quite stable, and it shows a considerable bathochromic shift.

Both (**153**) and (**154**) are reported to cause significant tumour regression in PDT experiments.

BIBLIOGRAPHY

H. Scheer (ed.) *Chlorophylls* CRC Press, Boca Raton 1991 and especially K.M. Smith (Chapter 1.6) on synthesis and P.H. Hynninen (Chapter 1.7) on chemical reactions.

F-P. Montforts, B. Gerlach and F. Höper, 'Discovery and synthesis of less common natural hydroporphyrins', *Chem. Rev.*, 1994, **94**, 327–347. Discussion and references for chlorins (eg bonellin, factor I, haem *d*, tunichlorin) and isobacteriochlorins (eg sirohydrochlorin = factor II, haem d_1) and synthetic model compounds.

A.E. Wick, D. Felix, K. Steen and A. Eschenmoser, 'Claisen rearrangement with alkyl and benzyl alcohols with acetal of N,N-dimethylacetamide', *Helv. Chim. Acta*, 1964, **47**, 2425–2429 and references therein. Examples and mechanism of amideacetal Claisen rearrangement (Figure 9.4).

M. Gosmann and B. Franck, 'Synthesis of fourfold enlarged porphyrin with an extremely large, diamagnetic ring current effect', *Angew. Chem. Intl. Edn.*, 1986, **25**, 1100–1101. Use of vinylogous Vilsmeier-Haack reaction using $Me_2NCH = CHCHO$ in the synthesis of the [26] porphyrin (80). See also Panel 7.3.

CHAPTER 10

PHTHALOCYANINES AND NAPHTHALOCYANINES

"On melting the greater part of the *o*-cyanobenzamide, from which the phthalimimide had in part sublimed as white crystals, it was seen that the transformation was not completely smooth. Some water and ammonia were liberated, and in the residue, in addition to the main product, some phthalimide was detectable. When the green melt was dissolved in alcohol and filtered, there remained on the filter a small amount of a blue substance."

Translated from A. Braun and J. Tcherniac, 1907

The first occasion on which phthalocyanine was encountered. The copper complex was encountered (but not correctly formulated) by de Diesbach in 1927. The structural identification by R.P. Linstead followed from the accidental discovery of iron phthalocyanine in a phthalimide manufacturing plant at Scottish Dyes Ltd., Grangemouth, in 1928. The iron came from a crack in the glaze on the reaction vessel.

"The results agree with those required by Linstead's formula, but it is impossible to determine the number of hydrogen atoms in the molecule by this method. It is, however, clearly established that the unit cell contains two centro-symmetric molecules. An important deduction immediately follows regarding the symmetry of the central group. The metal and the four surrounding isoindole nitrogen atoms must all be strictly in one plane. For the metal atom, being unique, must coincide with the centre of symmetry"

J.M. Robertson (1935)

Determination of the structure of phthalocyanine, and of its nickel, copper and platinum complexes, by X-ray crystallography. This was the first 'unknown' complex organic structure to be solved in this way.

10.1 STRUCTURAL CONSIDERATIONS

Substitution of nitrogen for a *meso* carbon bridge in a porphyrin gives an *azaporphyrin* (**155**): replacement of all four *meso* carbon bridges by nitrogens gives a *porphyrazine* (**156**). If we fuse benzo rings to the β,β′-bonds of a porphyrin we arrive at benzoporphyrins such as *tetrabenzoporphyrin* (**157**). Replacement of all four *meso* carbon bridges of this by nitrogens gives a substance used extensively in everyday life, although we may not realise it (blue / green pigments, toner for the photocopier, laser imprinted surface layer of the compact disc), and it has a special name of its own, *phthalocyanine* (**30**). With increasing *meso*-nitrogen substitution and fusion of benzene rings, the macrocycles give more stable metal complexes (copper(II) phthalocyanine dissolves in cold concentrated sulphuric acid without change, and can be precipitated on pouring into ice-water), and become less basic. The electronic spectrum changes dramatically so that the Q band now becomes (for phthalocyanines) the strongest band in the spectrum, and appears in the 670–700nm region. The metal-free phthalocyanines have two strong bands in this region (Q_x and Q_y, Panel 8.1) whereas the metallated phthalocyanines usually have only one. These Q bands are *very* intense (*ca.* 100 000–400 000) and sharp, a property which is exploited in laser address systems.

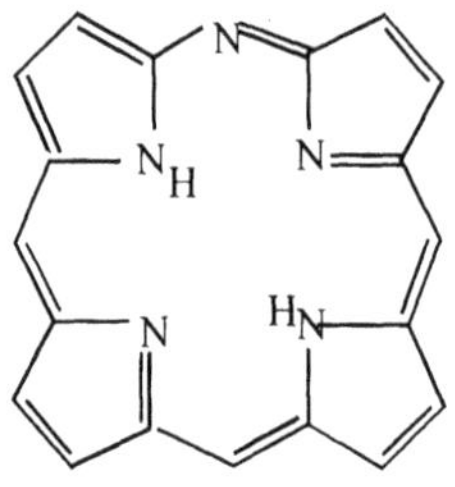

155 5-Azaporphyrin

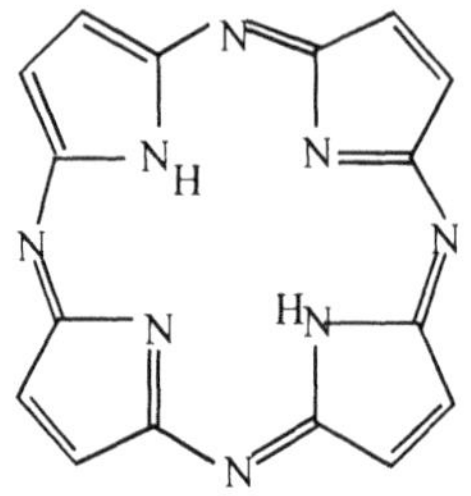

156 Porphyrazine

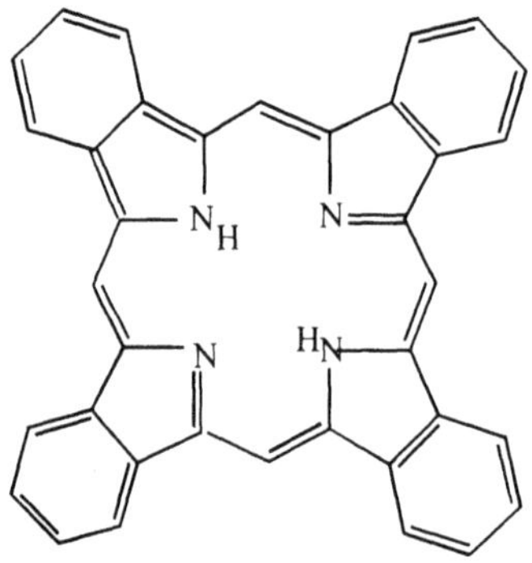

157 Tetrabenzporphyrin

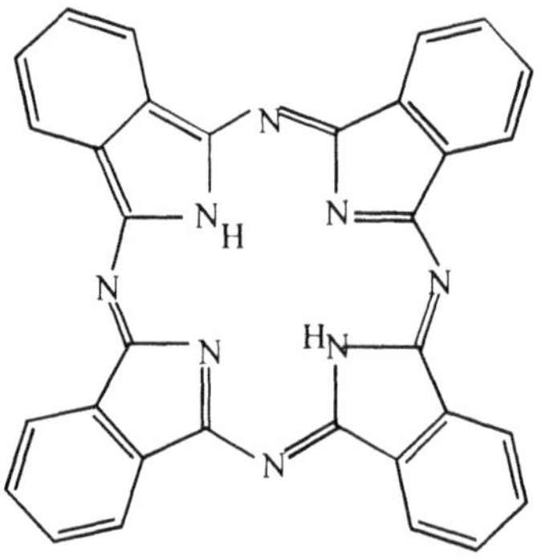

30 Phthalocyanine

Various elaborations of the phthalocyanine system are known, for example with heteroaromatic rings instead of benzenoid rings, and with more extensive aromatic systems. Figure 10.1 shows the effect of step-by-step fusion of additional benzene rings on the visible spectrum of a series of vanadyl complexes. Two general characteristics emerge: (i) as the aromatic system becomes larger the Q bands move further into the red, ending up, indeed, in the near infrared and (ii) at the same time the redox potential changes so that the systems become more easily oxidised. The latter feature means that compounds in this series beyond naphthalocyanine (**31**) are not seriously in contention as PDT sensitisers.

An important structural consideration which applies to all the compounds in Figure 10.1 is that they are large *flat* molecules, a geometry which is very unlike that of most of the chemical constituents of living things. Thus the phthalocyanine nucleus is a disc about 14Å in diameter (*exo*-C – *exo*-C), while naphthalocyanine has a corresponding diameter of about 18Å. Living things are dependant on such discoid molecules with a diameter of 8.5Å (ie β-C-β-C in porphyrins), but how they deal with much larger structures of this sort is still a matter for investigation.

As for nomenclature, the benzo(aza)porphyrins are named by an extension of the rules already given (Panel 4.2) as for example in dibenzo[*b*,*l*]-5-azaporphyrin. Phthalocyanine (tetrabenzo[*b*, *g*, *l*, *q*]-5,10,15,20-tetraazaporphyrin) is an exception, and is numbered as shown.

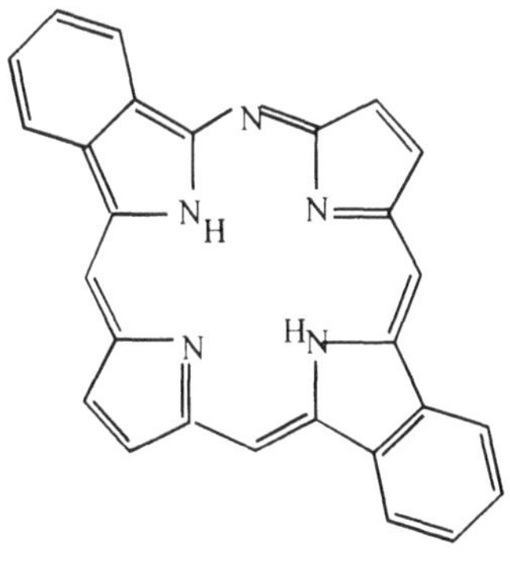

Dibenzo[*b*,*l*]-5-azaporphyrin

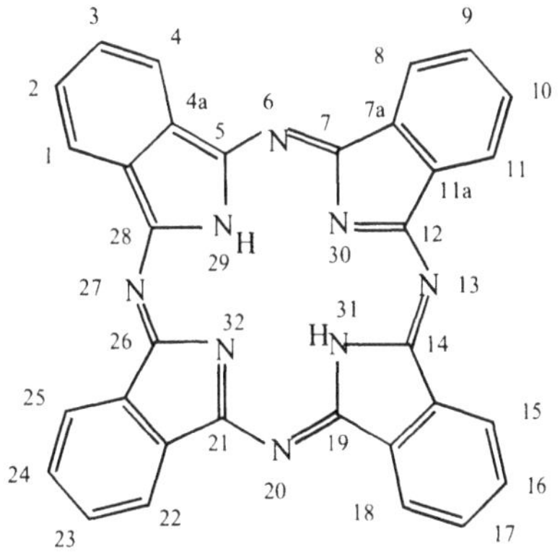

30 Phthalocyanine: usual numbering

Occasionally it is convenient to base the numbering of the phthalocyanine system on that of the phthalic precursor, thus:

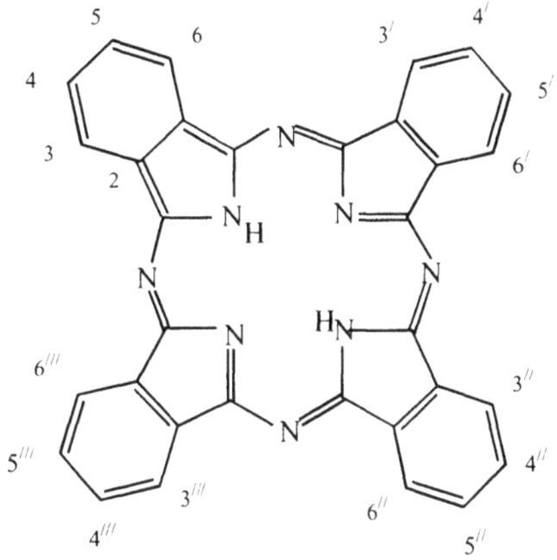

10.2 SPECTRA

Figure 10.1 illustrates the effect of increasing benzenoid annellation: here, for purposes of comparison, vanadyl complexes have been shown, and the starting point is vanadyl porphyrazine. As mentioned above, as the aromatic system becomes larger the Q bands move progressively into the infrared, but with little change in molar absorbance. However, the Q band is more intense than the B band for all the benzo-fused compounds shown in this figure; and although phthalocyanine is a brilliant blue,

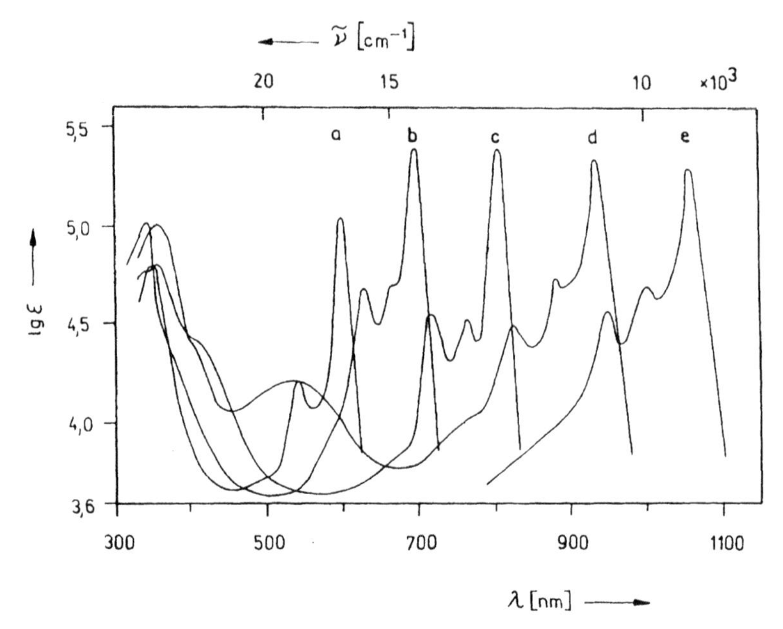

FIGURE 10.1 ELECTRONIC SPECTRA IN TOLUENE OF VANADYL PORPHYRAZINES WITH INCREASING BENZENOID ANNELLATION. REPRINTED FROM W. FREYER AND LE QUOC MINH, *MONATSCH. CHEM.*, 1986, **117**, 475-489 WITH PERMISSION. KEY:

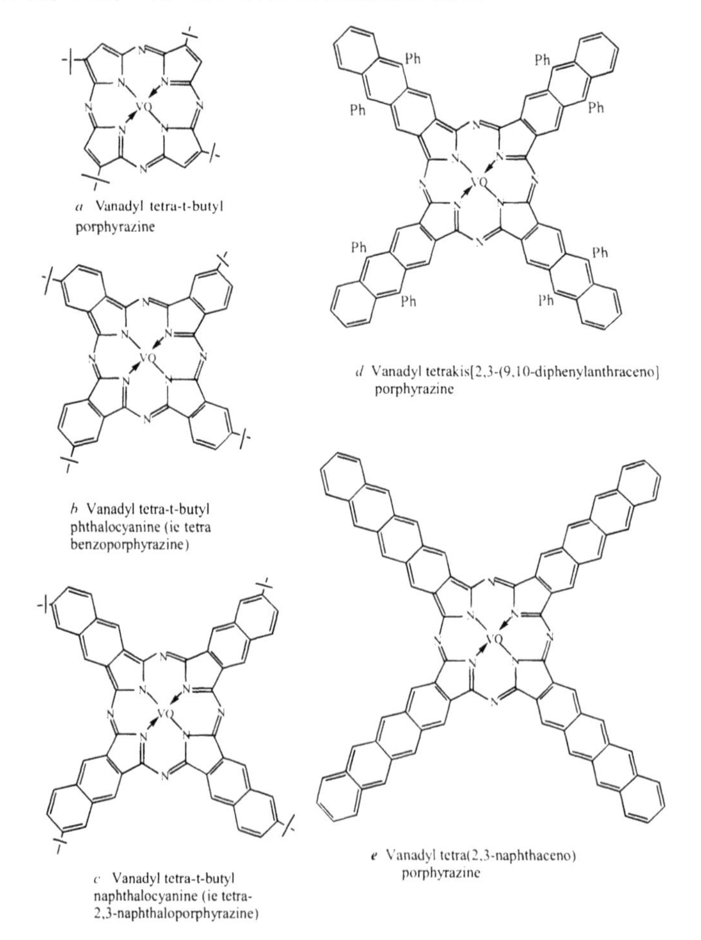

the more extended systems, with strong absorption in the infrared, have a rather non-descript yellow grey colour.

Figure 10.2 displays the spectra of examples in this series which have been considered for PDT applications.

10.3 5-AZAPORPHYRINS

A recent synthesis of this rather uncommon system starts with bilirubin (**51**) (Figure 10.3). Dehydrogenation with ferric chloride, esterification, and cyclisation in the presence of zinc ions as template gives the oxonium salt (**163**) which is converted into 5-azaprotoporphyrin dimethyl ester (**158**) with ammonia. Application of the vinylporphyrin photooxygenation chemistry already discussed (section 9.4.3) generates the azachlorin (**164**, isomers). 5-Azaprotoporphyrin dimethyl ester (**158**) has Band I at 624 nm (ε 22 700) (Figure 10.2a) in chloroform, while (**164**, mixed isomers) has λ_{max} 674 nm (ε 66 000): both compounds have $\phi_\Delta \sim 0.6$. These examples show the hyperchromic effect of the *meso*-aza function on Band I (cf protoporphyrin dimethyl ester, Band I 631nm, ε 5 600).

The parent porphyrazine (**156**) has been prepared by reductive cyclotetramerisation (see section 10.5.2) of maleodinitrile and of succinimidine. The best route (Figure 10.4) appears to be from succinonitrile: unstable hexa and tetrahydro intermediates can be detected, but the chlorin can be isolated and oxidised in a separate step to (**156**).

10.4 BENZOPORPHYRINS

In a sense the benzoporphyrins are a halfway house between porphyrins and phthalocyanines. Although the spectra of monobenzoporphyrins do not show much change with respect to porphyrins in the Band I region, the *opp*-dibenzoporphyrins (eg **159** in *o*-$C_6H_4Cl_2$, λ_{max} 644 nm, ε 24 800) and the tetrabenzoporphyrins (eg **160** in 1-$C_{10}H_7Cl$, λ_{max} 674 nm, ε 47 800) do show bathochromic and hyperchromic effects, and in each the Q-band region as a whole is dominated by two strong bands (Figure 10.2 b,c).

The benzoporphyrins are much less common substances than either the porphyrins or the phthalocyanines. Panel 10.1 gives a selection of synthetic routes.

In vitro and *in vivo* experiments have shown that the *opp*-dibenzoporphyrins and the tetrabenzoporphyrins are biologically active where administered in liposomes. Zinc(II) tetrabenzoporphyrin (**157**, zinc complex) was found to be promising, showing high tumour and plasma levels and a high tumour / muscle ratio (17 at 24 hours) *in vivo*. Water-soluble sulphonated *meso*-aryltetrabenzoporphyrins have been reported to be active in tumour photonecrosis, and to have low toxicity.

10.5 PHTHALOCYANINES

10.5.1 General

The phthalocyanines are the only macrocycles in this series which are produced commercially in tonnage amounts (Zeneca Specialities, Ciba Speciality Chemicals). Their principal applications are as pigments (typically insoluble stable colourants of symmetrical planar structure in this case, used as a solid or slurry) and dyes (typically

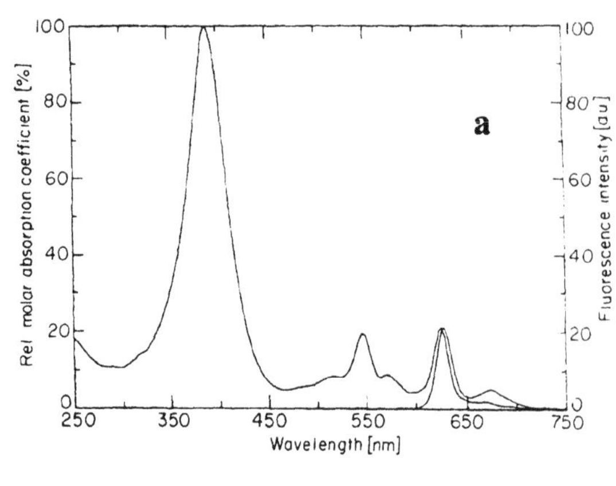

A. 5-Azaprotoporphyrin dimethyl ester (**158**) in $CHCl_3$. ε_{389} = 110 000. The corrected fluorescence spectrum is also shown (λ_{Ex} = 405 nm). (Reprinted from *J. Photochem. Photobiol. B: Biol.*, **23**, by K. Schiwon, M-D. Brauer, B. Gerlach, C.M. Müller and F-P. Montforis, Photophysical and photochemical properties of azaporphyrin and azachlorin derivatives, 239–243, copyright 1994, with permission from Elsevier Science).

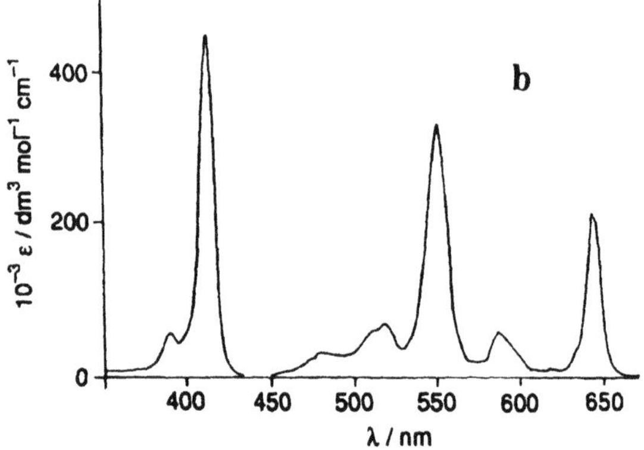

B. 2,3,12,13-Tetramethyldibenzo[*g,q*]porphyrin (**159**) in *o*-dichlorobenzene. ε for 450–650 nm is reduced to one tenth of that shown. (Reproduced from R.Bonnett and K.A. McManus, *Chem. Commun.*, 1994, 1129–1130 with permission).

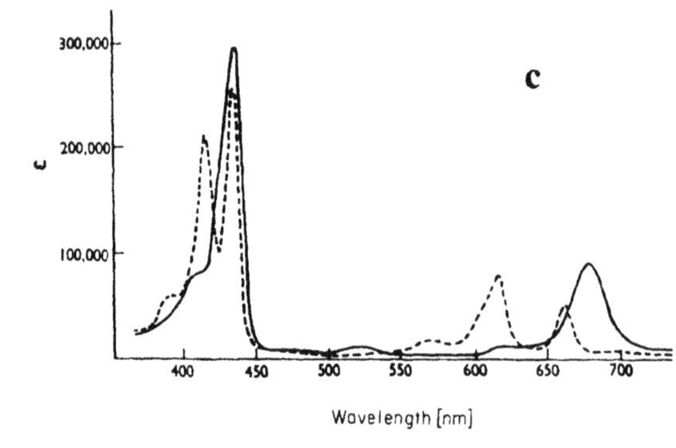

C. Tetrakis(3, 6-dimethylbenzo)porphyrin (**160**): dotted line — free base in pyridine containing 2% trifluoroacetic acid; continuous line — dication in trifluoroacetic acid. (Reproduced from C.O. Bender, R. Bonnett and R.G. Smith, *Chem. Commun.*, 1969, 345–346 with permission).

Figure 10.2 Electronic spectra of aza and annellated systems. (Scales differ: this series of spectra illustrates the variation in presentation encountered in the literature structures **158–162** follow).

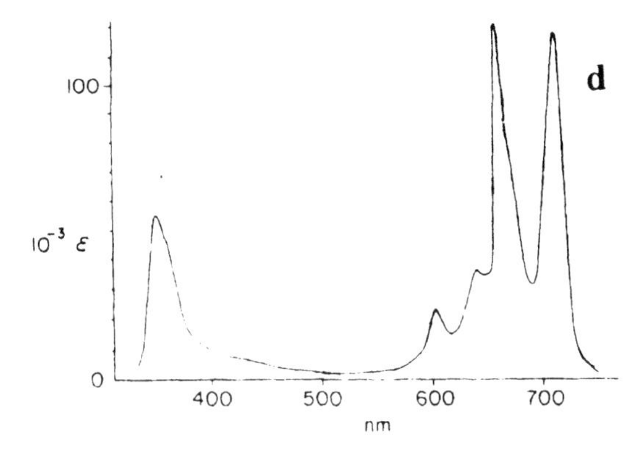

d. Phthalocyanine (**30**) in 1-chloronaphthalene. (Reproduced from G.E. Ficken and R.P. Linstead, *J. Chem. Soc.*, 1952, 4846–4854 with permission).

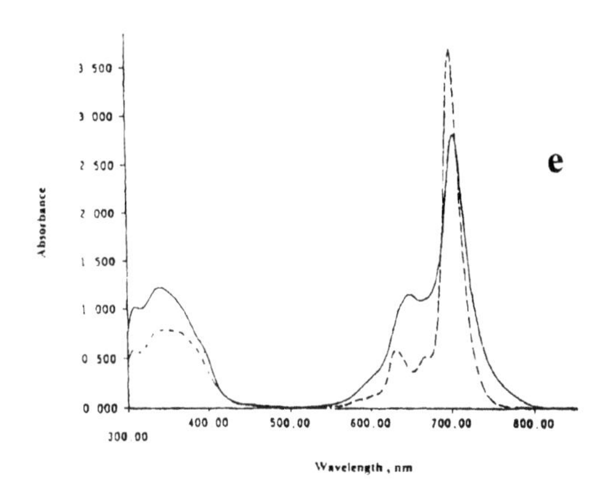

e. Zinc(II) octa(decyl)phthalocyanine (**161**). Full line: 2.4 x 10^{-7} M in cyclohexane, ε_{MAX} = 115 000. Broken line: same concentration in cyclohexane: ethanol = 94:6, ε_{MAX} = 158 000. (Reproduced from M.J. Cook, I. Chambrier, S.J. Cracknell, D.A. Mayes and D.A. Russell, *Photochem. Photobiol.*, 1995, **62**, 542–545 with permission).

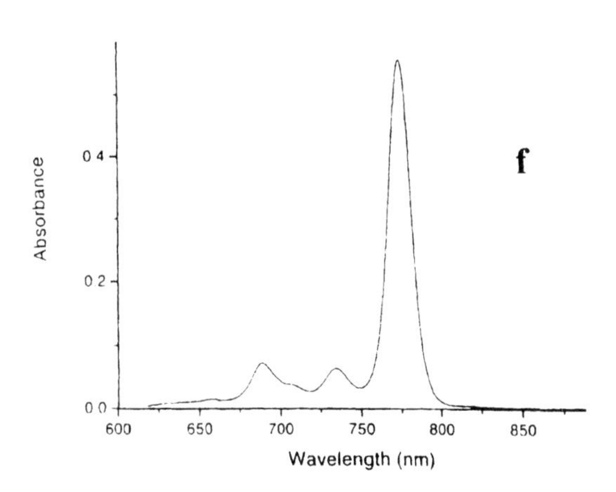

f. Di[diisobutyl(octadecyl)siloxy]silicon(IV) naphthalocyanine (**162**) in tetrahydrofuran at 1.3 x 10^{-6} M. (Reprinted from *J. Photochem. Photobiol. B: Biol.*, **42**, by M. Soncin, A. Busetti, R. Biolo, G. Jori, G. Kwag, Y-S. Li, M.E. Kenney and M.A.J. Rodgers, Photoinactivation of amelanotic and melanotic melanoma cells sensitized by axially substituted Si-naphthalocyanines, 202–210, copyright 1998, with permission from Elsevier Science).

Figure 10.2 (Continued).

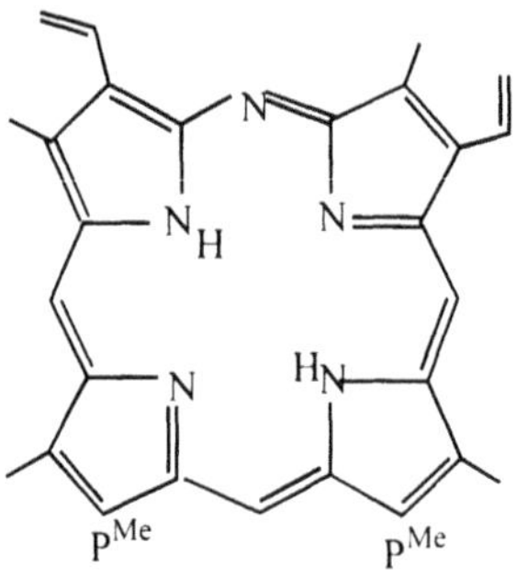

N NH N HN P^{Me} P^{Me}

158

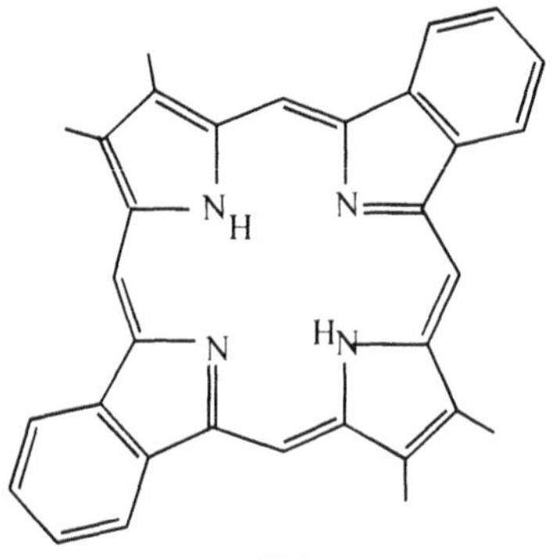

NH N N HN

159

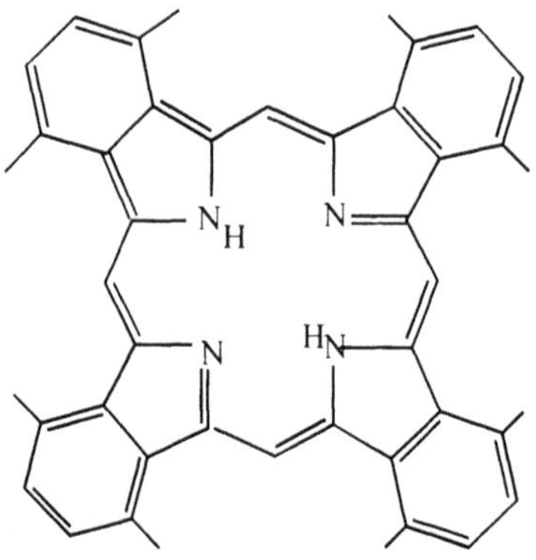

NH N N HN

160

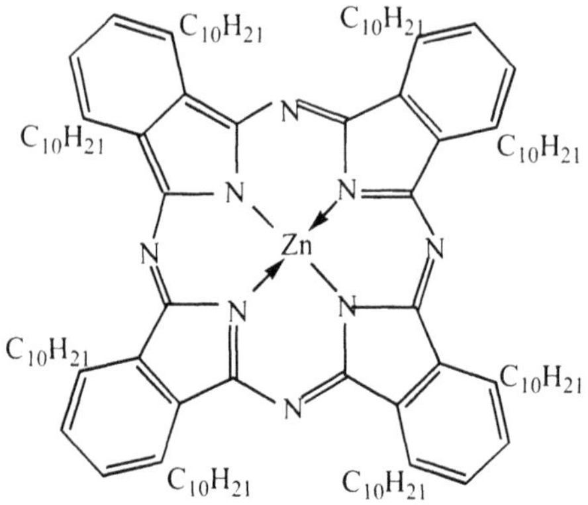

$C_{10}H_{21}$ $C_{10}H_{21}$ $C_{10}H_{21}$ $C_{10}H_{21}$ N N N Zn N N N N N $C_{10}H_{21}$ $C_{10}H_{21}$ $C_{10}H_{21}$ $C_{10}H_{21}$

161

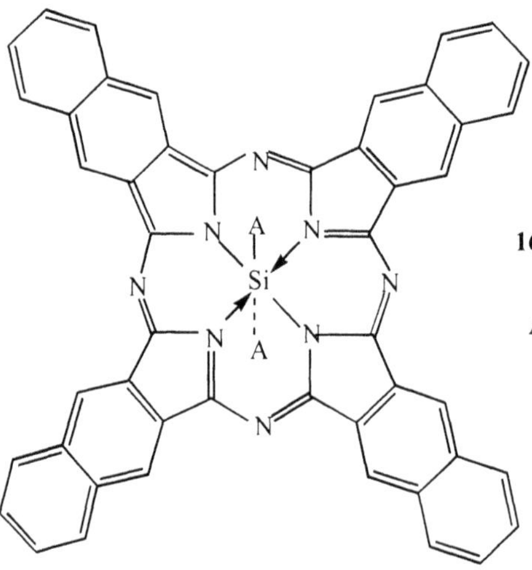

N N N A Si A N N N N N

162

$A = (Me_2CHCH_2)_2Si(C_{18}H_{37})$–O–

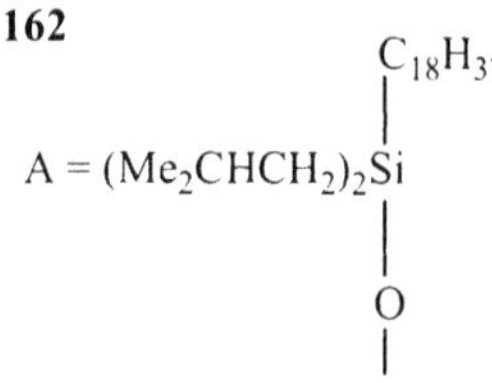

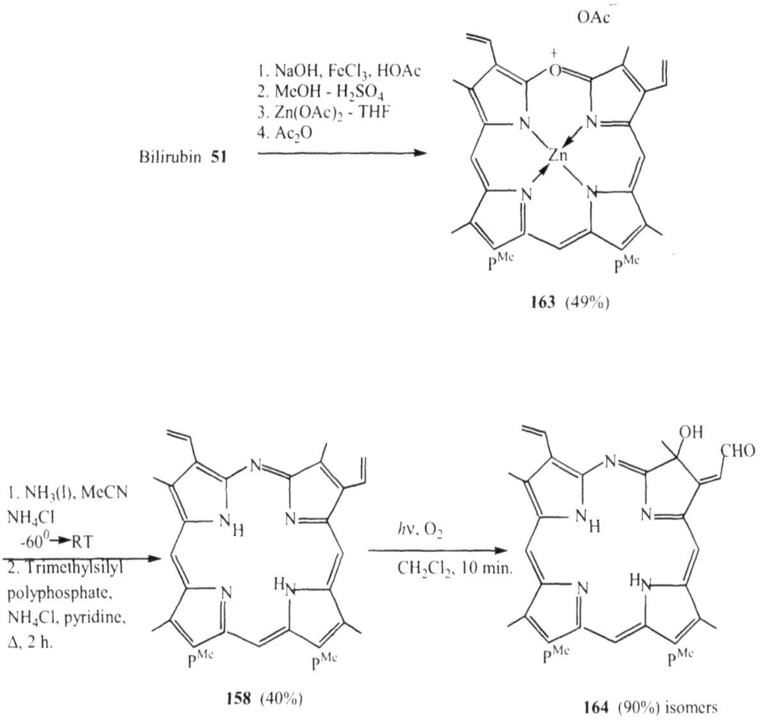

Bilirubin **51**

1. NaOH, $FeCl_3$, HOAc
2. MeOH - H_2SO_4
3. $Zn(OAc)_2$ - THF
4. Ac_2O

OAc^-

163 (49%)

1. NH_3(l), MeCN, NH_4Cl, $-60^0 \rightarrow$ RT
2. Trimethylsilyl polyphosphate, NH_4Cl, pyridine, Δ, 2 h.

158 (40%)

$h\nu$, O_2
CH_2Cl_2, 10 min.

164 (90%) isomers

FIGURE 10.3 CONVERSION OF BILIRUBIN INTO 5-AZAPORPHYRIN DERIVATIVES (F-P. MONTFORTS AND B. GERLACH, *TETRAHEDRON LETTERS*, 1992, **33**, 1985–1988).

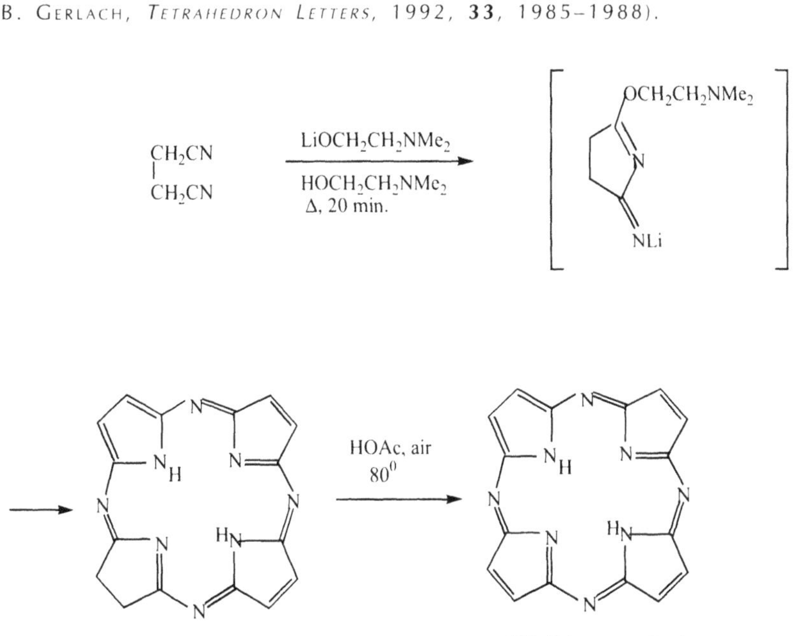

CH_2CN | CH_2CN

$LiOCH_2CH_2NMe_2$
$HOCH_2CH_2NMe_2$
Δ, 20 min.

$OCH_2CH_2NMe_2$ ... NLi

HOAc, air
80^0

Tetraazachlorin (18%)

λ_{max} nm (ε) (PhCl) 345 (35 500), 520 (26 300), 619 (5 500), 645 (6 000) and 678 (53 700).

156 Tetraazaporphyrin (= porphyrazine)

λ_{max} nm (ε) (PhCl) 335 (50 100), 545 (39 800) and 617 (56 200).

FIGURE 10.4 SYNTHESIS OF PORPHYRAZINE (**156**). (R.P. LINSTEAD AND M. WHALLEY, *J. CHEM. SOC.*, 1952, 4839–4846; E.A. MAKAROVA, G.V. KOROLYOVA AND E.A. LUKYANETS, RSC CONFERENCE ON PHTHALOCYANINES, EDINBURGH, 1998).

Panel 10.1 Syntheses of Benzoporphyrins

1. *By 4 x 1 syntheses giving tetrabenzoporphyrins*

18% λ_{max} 632 nm (ε 151 000)

(R.P. Linstead and F.T. Weiss, *J. Chem. Soc.*,1950, 2975-2981).

81% λ_{max} (pyridine) 642 nm (ε 127 000)

(C.O. Bender, R. Bonnett and R.G. Smith, *J. Chem. Soc. C*, 1970, 1251–1257)

2. *By self-condensation of benzopyrromethenes*

31% λ_{max} ($CHCl_3$) 641 nm (ε 29 500)

(R. Bonnett and K.A. McManus, *J. Chem. Soc., Perkin Trans. 1*, 1996, 2461-2466)

continued next page

3. *By Dieckmann cyclisation of porphyrin diesters*

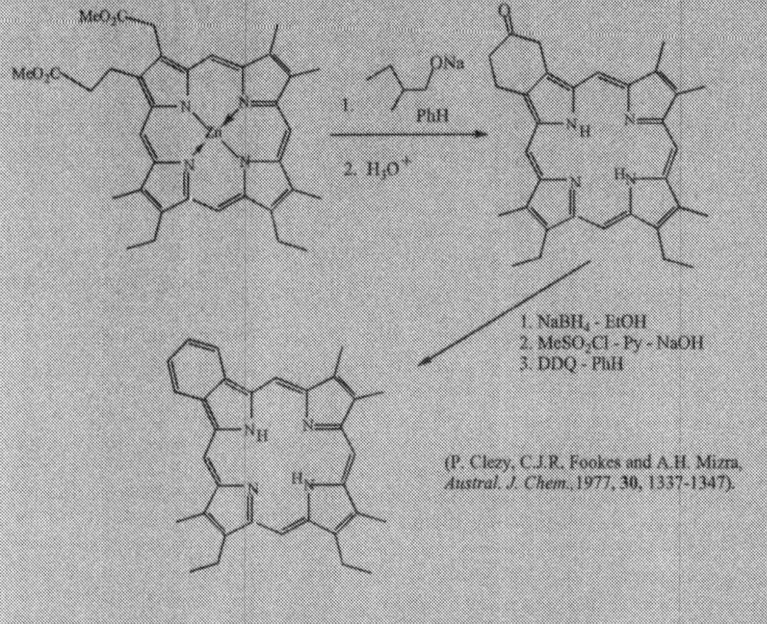

1. ONa, PhH
2. H_3O^+

1. $NaBH_4$ - EtOH
2. $MeSO_2Cl$ - Py - NaOH
3. DDQ - PhH

(P. Clezy, C.J.R. Fookes and A.H. Mizra, *Austral. J. Chem.*,1977, **30**, 1337-1347).

4. *By Diels-Alder annellation of a porphyrin*

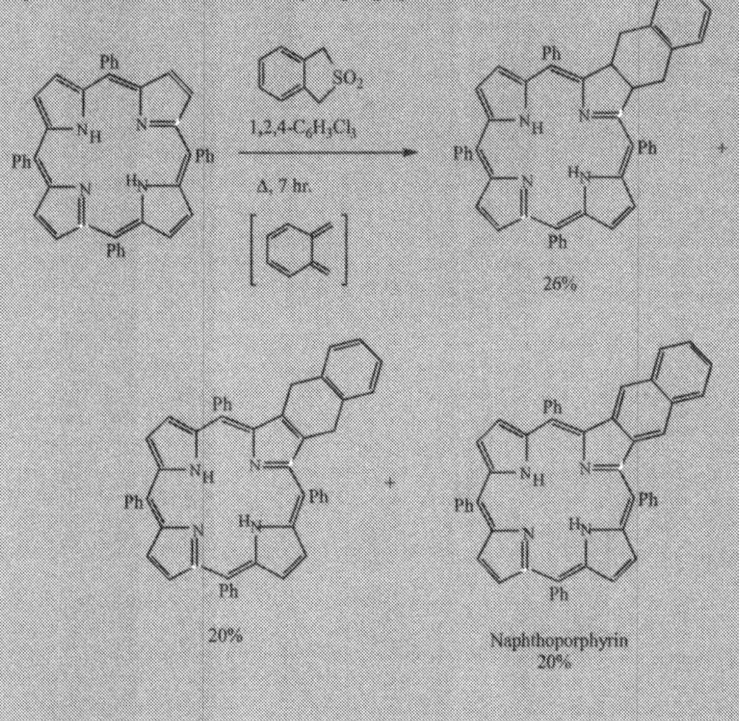

SO_2

1,2,4-$C_6H_3Cl_3$

Δ, 7 hr.

26%

20%

Naphthoporphyrin
20%

DDQ treatment would be expected to convert the chlorin and the porphyrin to the naphthoporphyrin

(A.C. Tomé, P.S.S. Lacerda, M.G.P.M.S. Neves and J.A.S. Cavaleiro, *Chem. Commun.*,1997, 1199-1200).

continued ***next page***

5. *From tetrahydroisoindole precursors*

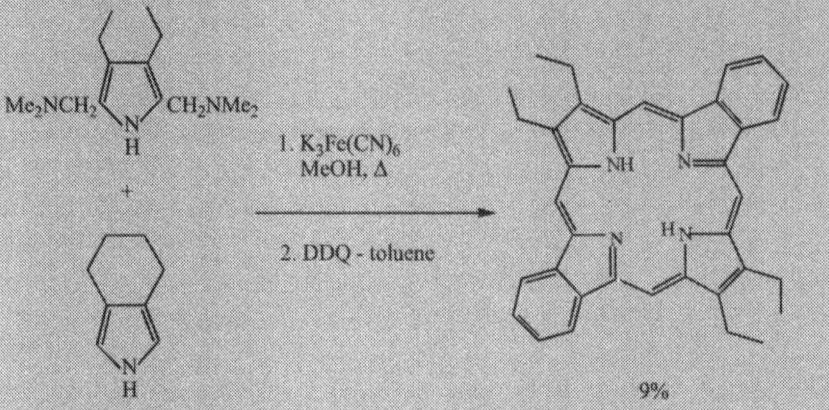

Me_2NCH_2 N H CH_2NMe_2 + N H

1. $K_3Fe(CN)_6$ MeOH, Δ

2. DDQ - toluene

NH N N HN

9%

(L.T. Nguyen, M.O. Senge and K.M. Smith, *J. Org. Chem.*, 1996, **61**, 998-1003).

6. *Using a masked isoindole synthon*

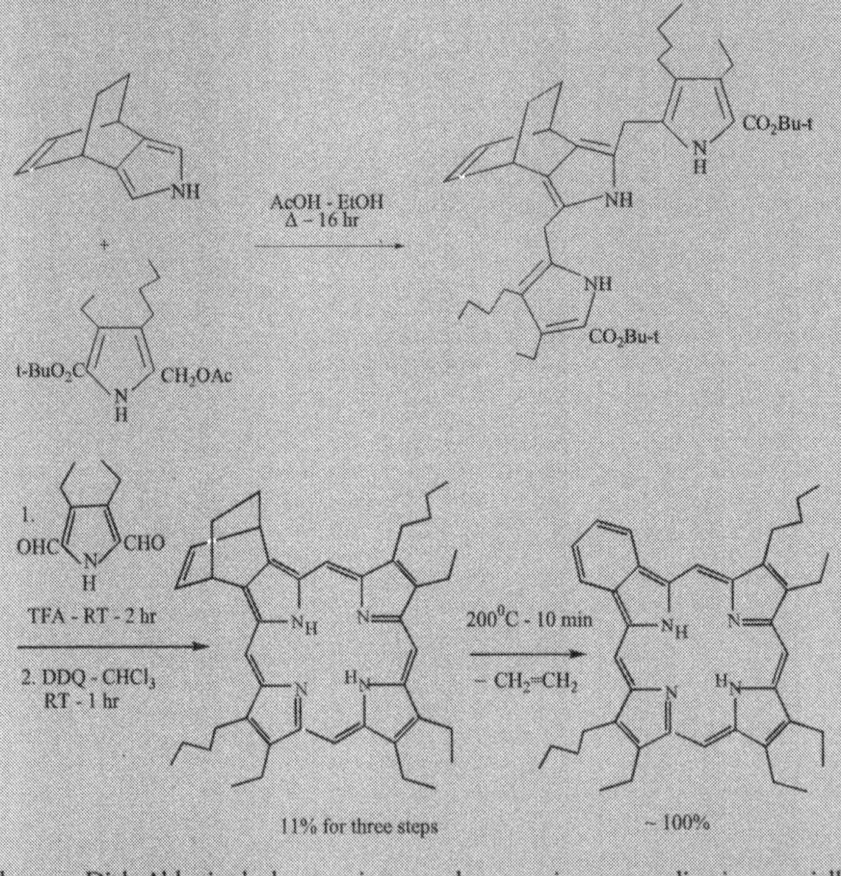

NH + t-BuO$_2$C N H CH_2OAc

AcOH - EtOH Δ - 16 hr

CO_2Bu-t N H NH NH CO_2Bu-t

1. OHC N H CHO TFA - RT - 2 hr

2. DDQ - $CHCl_3$ RT - 1 hr

NH N N HN

11% for three steps

200°C - 10 min

− CH_2=CH_2

NH N N HN

~ 100%

The retro Diels-Alder in the last step is a very clean reaction, proceeding in essentially quantitative yield. (S. Ito, T. Murashima, H. Uno and N. Ono, *Chem. Commun*, 1998, 1661–1662)

less symmetrical because of solubilising substituents, used in solution ie in the dye bath) to confer a range of blue and green hues. Copper phthalocyanine (Monastral Blue) is the typical blue pigment: it is the largest volume colourant in this series being produced in amounts of ~ 50,000 tonnes / year. Various crystal forms exist, and of these two are used commercially. The α-form is a purple-blue (paints) and the β-form is a greener blue (inks applications). Green colours are produced by aromatic chlorination (to give up to $CuPcCl_{16}$). Nucleophilic substitution of the halogen with thiolate (eg ArS^-, 140°C, N-methylpyrrolidone) gives $CuPc(SAr)_{16}$ which has λ_{max} 800 nm (ε ~ 200 000), is nearly colourless, and is the basis of security applications (eg invisible barcoding).

Water solubilisation is achieved by sulphonation, eg

$$CuPc \xrightarrow[\text{2. } NH_3 \text{ 3. } H_2O]{\text{1. } ClSO_3H} CuPc(SO_3H)_2(SONH_2)$$

CI Direct Blue

Reactive dyes are produced, for example, using chlorotriazine derivatives:

$$CuPc(SO_3H)_x(SO_2Cl)_y \xrightarrow[\text{2. hydrolysis}]{\text{1. } NH_2CH_2CH_2NHCOMe} CuPc(SO_3H)_x(SO_2NHCH_2CH_2NH_2)_y$$

$$\xrightarrow{\text{2,4-dichloro-6-methoxy-1,3,5-triazine}} CuPc(SO_3H)_x\left(SO_2NHCH_2CH_2NH\text{-}C_3N_3(Cl)(OMe)\right)_y$$

Reactive dyes of this general sort ('Procion' from Zeneca, 'Cibacron' from Ciba) suffer nucleophilic substitution of the remaining halogen by hydroxy groups in cellulose (cotton), leading to covalent attachment of the dye.

In recent years new applications have become important which make use of the electroactive properties of the phthalocyanines. Thus in electrophotography (eg photocopying, laser printing) the photoconductive properties of titanyl phthalocyanine and phthalocyanine are employed: the toner contains copper phthalocyanine. The move to 'high tec' has also involved electrochromic display, ink jet and optical data (compact disc) applications. Compact discs on which the information is stored in laser-ablated features in a stable phthalocyanine layer are expected to be long-lived. Patent activity in this general area is vigorous, especially from Japanese electronic companies.

10.5.2 Synthesis

Nearly all the syntheses of phthalocyanines are of the 4 × 1 type (Class C in Panel

8.2). This puts a severe symmetry restriction on the products which can be obtained. Because of potential electroactive (preceding para.) and photodynamic therapy applications, a reliable step-wise synthetic route is highly desirable. This has not yet emerged, because the necessary chemistry to handle the *meso*-nitrogen bridge in two ring and three ring intermediates (well established for the *meso*-carbon bridge in Type A and Type B syntheses of porphyrins, Panel 8.2) is not yet known. Various ingenious strategies for circumventing the inflexibility of the Type C synthesis have been proposed, and are summarised in Panel 10.2.

Iron phthalocyanine was discovered in 1928 by accident while phthalimide was being prepared (at Scottish Dyes, Grangemouth) in an iron reactor which had cracked glazing (see also the chapter heading). Chemical study, initially at Scottish Dyes but especially by Linstead who laid the foundation to the chemistry of these compounds, resulted in the formulation (**30**) in 1934, and the structure was confirmed by X-ray analysis by Robertson the next year (chapter heading). This was the first occasion on which a complex (albeit planar) organic structure had been solved by this technique, in itself an important milestone.

The syntheses which emerged from these and derived studies are shown in Figure 10.5. Industrially it is preferable on cost grounds to start with phthalic acid or anhydride, but in the laboratory, or where purer materials are required, the phthalonitrile or diiminoisoindoline routes are preferable.

The diiminoisoindolines are prepared from the *o*-dinitriles by nucleophilic cyclisation (eg potassamide in liquid ammonia), thus:

NH_2^- C≡N C≡N — 1. NH_2^- / NH_3 2. H^+ source → NH NH NH

while the *o*-dinitriles may be obtained by dehydration of the corresponding diamides (eg Figure 10.6), by nucleophilic substitution:

MeO MeO Br Br — CuCN, Δ solvent → MeO MeO CN CN

or by Diels-Alder reaction:

O — BuLi, RBr 2x → R O R — NCCH=CHCN → R O R CN CN — 1. $LiN(SiMe)_2$ THF, -78^{0}C 2. H_2O → R R CN CN

Panel 10.2 Synthetic Approaches to Phthalocyanines lacking D_{4h} Symmetry of Peripheral Substitution

1 *Mixed synthesis (crossed-condensation).*

This can be made to work quite well, but gives a complex mixture which must often be separated by chromatography which can be tedious and inefficient. However the reaction can be directed to some extent by appropriate choice of reactant ratio.

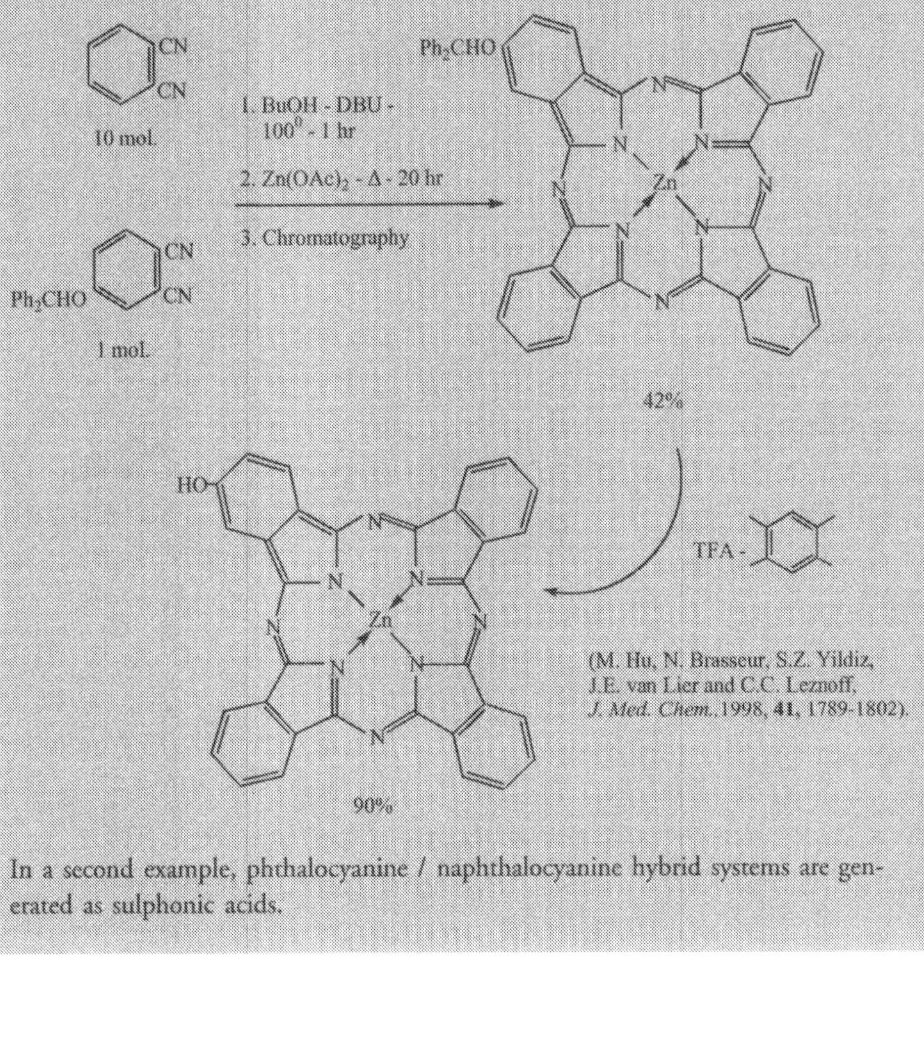

CN
CN
10 mol.

Ph_2CHO CN
CN
1 mol.

1. BuOH - DBU - 100^0 - 1 hr
2. $Zn(OAc)_2$ - Δ - 20 hr
3. Chromatography

Ph_2CHO

42%

TFA -

HO

90%

(M. Hu, N. Brasseur, S.Z. Yildiz, J.E. van Lier and C.C. Leznoff, *J. Med. Chem.*,1998, **41**, 1789-1802).

In a second example, phthalocyanine / naphthalocyanine hybrid systems are generated as sulphonic acids.

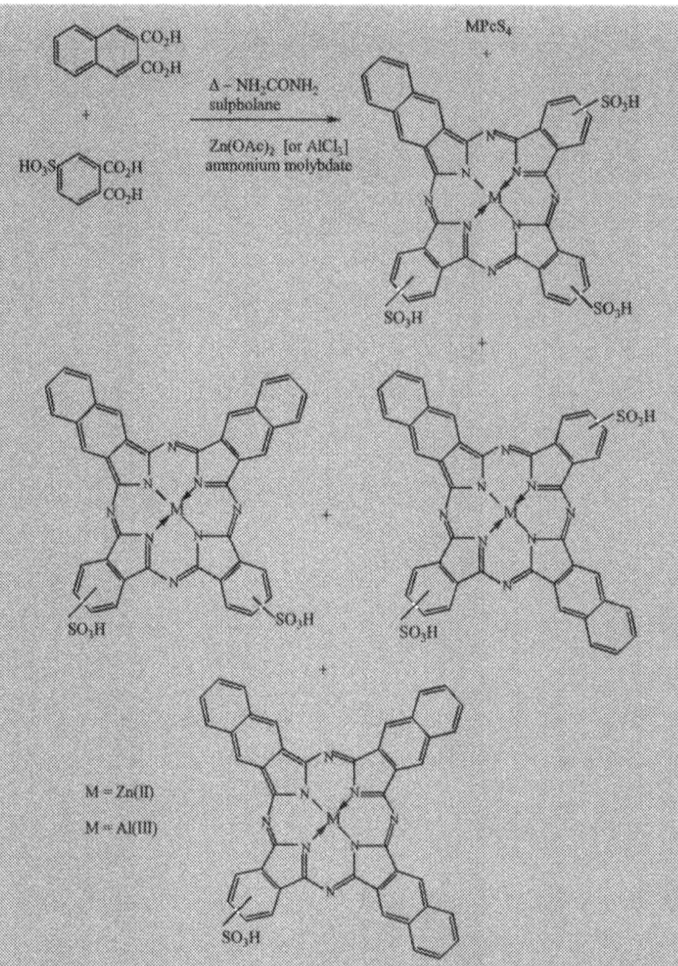

Δ - NH_2CONH_2
sulpholane

$Zn(OAc)_2$ [or $AlCl_3$]
ammonium molybdate

$MPcS_4$

M = Zn(II)

M = Al(III)

In bioassay *in vitro* and *in vivo*, the di and tri sulphonic acids are the most active. (P. Margaron, R. Langlois, J.E. van Lier and S. Gaspard, *J. Photochem. Photobiol. B Biol.*, 1992, 14, 187-199).

2. *Mixed synthesis - minor monomer substituted to facilitate separation.*

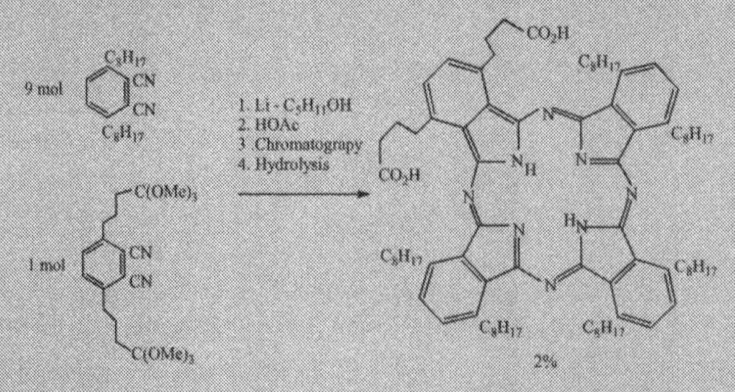

(N.B. McKeown, I. Chambrier and M.J. Cook, *J.Chem. Soc.,Perkin Trans.1*, 1990, 1169-1177).

3. *Solid phase synthesis.*

Using immobilised phthalonitrile (Tr = trityl).

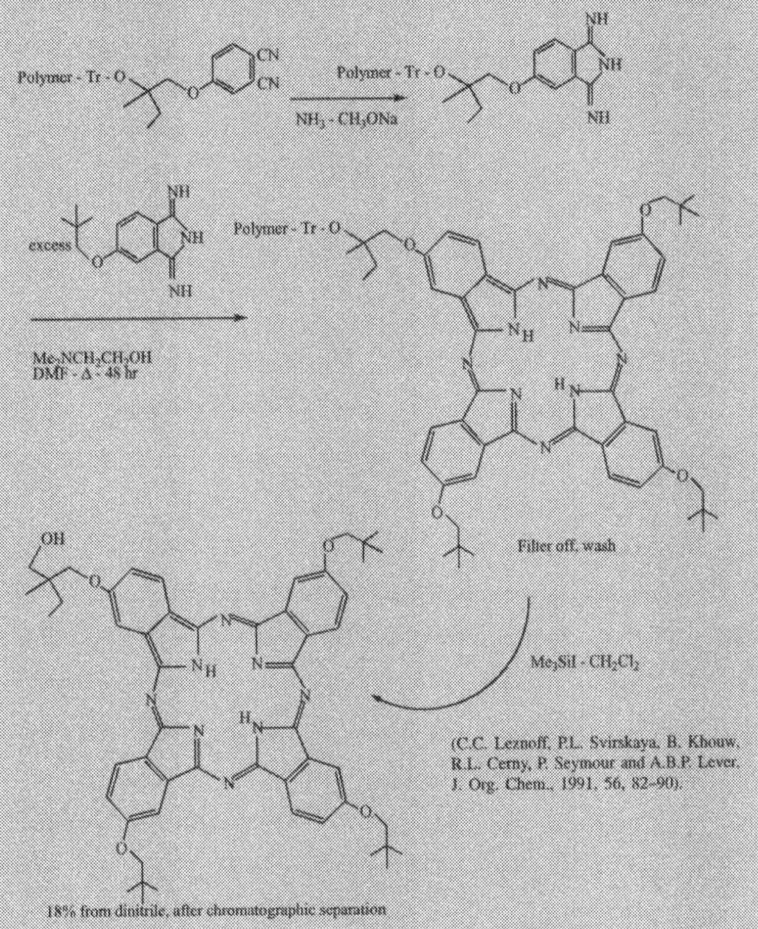

Polymer - Tr - O

NH_3 - CH_3ONa

excess

$Me_2NCH_2CH_2OH$
DMF - Δ - 48 hr

Filter off, wash

Me_3SiI - CH_2Cl_2

(C.C. Leznoff, P.L. Svirskaya, B. Khouw, R.L. Cerny, P. Seymour and A.B.P. Lever, J. Org. Chem., 1991, 56, 82–90).

18% from dinitrile, after chromatographic separation

4. *From boron(III) subphthalocyanine by macroring expansion.*

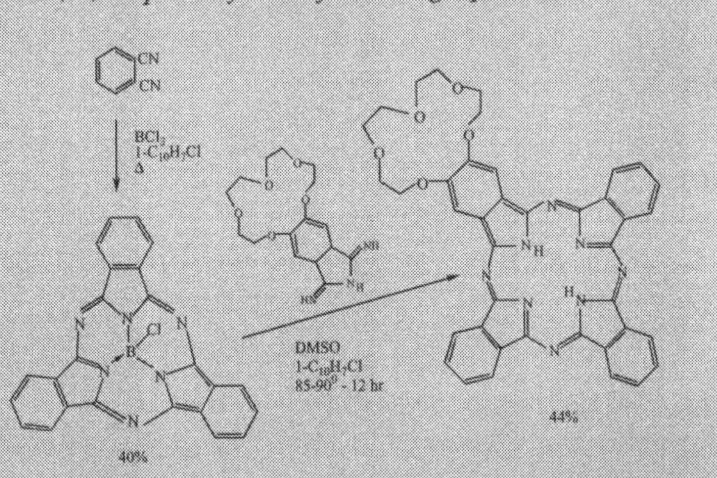

BCl_3
1-$C_{10}H_7Cl$
Δ

DMSO
1-$C_{10}H_7Cl$
85-90° - 12 hr

40%

Chloroboron(III) subphthalocyanine

44%

A. Meller and O. Ossko, *Monatsch. Chem.*, 1972, **103**, 150–155. N. Kobayashi, R. Kondo, S. Nakajima and T. Osa, *J. Am. Chem. Soc.*, 1990, 112, 9640–9641. E. Musluoglu, A. Gürek, V. Ahsen, A. Gül and O. Bekaroglu, *Chem. Ber.*, 1992, **125**, 2337–2339.

5. *Via half-phthalocyanines.*
-rational synthesis of *adj*-disubstituted phthalocyanines.
Treatment of phthalonitrile with lithium methoxide in methanol gives an azabenzopyrromethene derivative.

CN CN —LiOMe - MeOH, Δ→ MeO N N N NLi

(first observed with 4-nitrophthalonitrile: S.W. Oliver and T.D. Smith, *J. Chem. Soc., Perkin Trans 2*, 1987, 1579-1582).

Although this material is fragile, and cannot be purified, it can be used to direct subsequent condensation steps with a substituted phthalonitrile to give predominantly the *adj*-disubstituted phthalocyanine.

MeO N N N NLi —1. Ph_2CHO CN CN, octanol, 100^0C, several hr; 2. $Zn(OAc)_2$ - Δ, 8 hr→

Ph_2CHO $OCHPh_2$ N N N N N N N N Zn —TFA→ HO OH N N N N N N N N Zn

30% *adj* - substituted 88%

(Hu et al., *J. Med. Chem.*, 1988, **41**, 1789–1802).

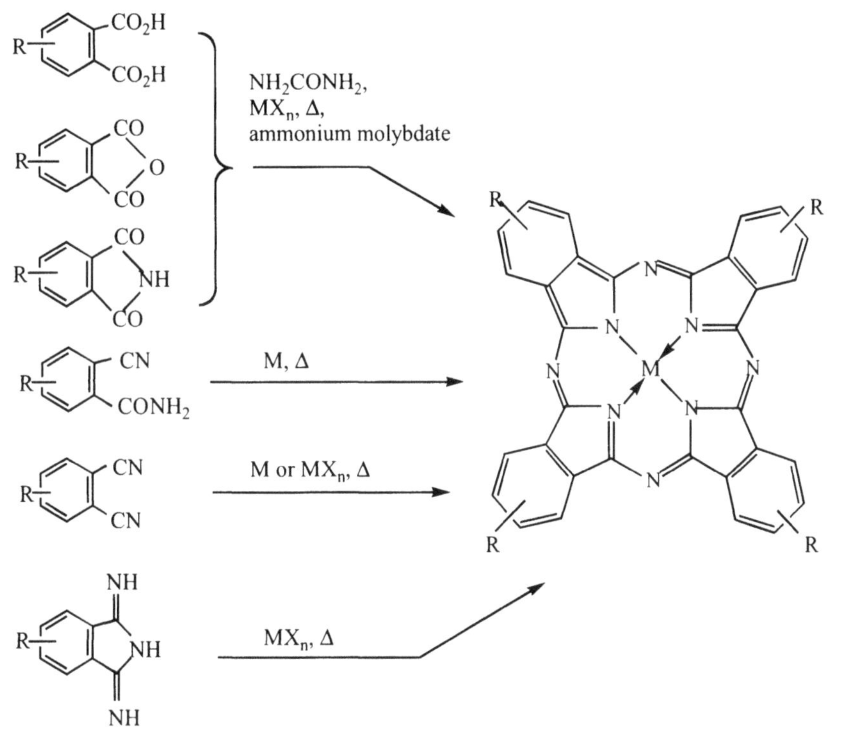

FIGURE 10.5 ROUTES TO METALLOPHTHALOCYANINES.

An actual example is shown in Figure 10.6, starting from *o*-xylene: this also illustrates a convenient route to metal-free compounds using sodium (or lithium) isopentoxide in the macrocyclisation step (Na or Li being readily removed, Panel 8.1). Alternative routes to the metal-free compound are (i) to treat the phthalonitrile with ammonia gas in 2(N,N-dimethylamino)ethanol and (ii) to reflux the diiminoisoindoline in this solvent: yields are usually 80–90%.

The following general points apply to this type of synthesis.

(i) The reactions are carried out in a melt or a solvent, typically a high boiling solvent. Recently, however, Leznoff has described what appears to be a general room temperature procedure. For many years lithium or sodium isopentoxide in isopentanol under reflux has been employed in this reaction (as in Figure 10.6). With higher alcohols the reaction is observed to occur slowly at room temperature as shown in Figure 10.7. The reaction does not proceed with $C_5H_{11}OLi/C_5H_{11}OH$ under these conditions: the reason for the difference in behaviour is not clear. Interestingly, (**166**) is obtained as a single Type isomer, presumably for steric reasons (see (v) below).

(ii) Alkyl substitution (as in Figure 10.6) renders the products soluble in organic solvents, and amenable to chromatographic purification in the usual way (section 10.5.3.1). In the absence of substituents, the phthalocyanines and their metal complexes are usually very sparingly soluble, and have to be purified by

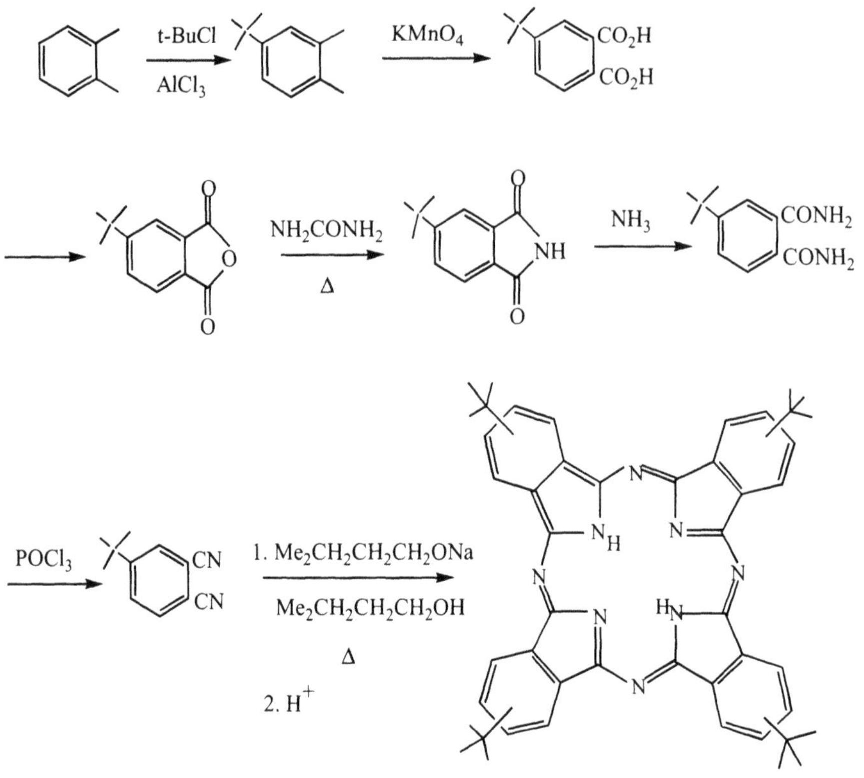

t-BuCl
$AlCl_3$
$KMnO_4$
CO_2H
CO_2H
O
O
O
NH_2CONH_2
Δ
O
NH
O
NH_3
$CONH_2$
$CONH_2$
$POCl_3$
CN
CN
1. $Me_2CH_2CH_2CH_2ONa$
$Me_2CH_2CH_2CH_2OH$
Δ
2. H^+
N
N_H
N
N
N
N
HN
N
165

Figure 10.6 Synthesis of tetra-t-butylphthalocyanine (**165**). A mixture of Type isomers (Panel 4.2) is obtained.

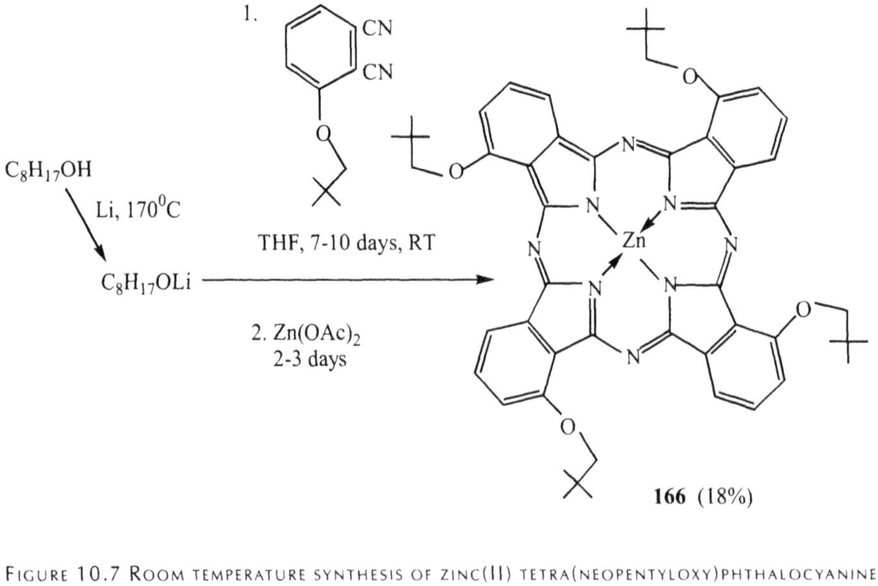

1.
CN
CN
O
$C_8H_{17}OH$
Li, 170^0C
$C_8H_{17}OLi$
THF, 7-10 days, RT
2. $Zn(OAc)_2$
2-3 days
O
O
O
O
N
N
N
N
Zn
N
N
N
N
166 (18%)

Figure 10.7 Room temperature synthesis of zinc(II) tetra(neopentyloxy)phthalocyanine (C.C. Leznoff, M. Hu and K.J.M. Nolan, *Chem. Commun.*, 1996, 1245–1246).

washing out the impurities. Since the system is quite rugged, small amounts may be purified by sublimation, the impurities charring.

(iii) The reaction is a *reductive* cyclotetramerisation probably proceeding in most cases by an intermediate which we can formulate as A or its equivalent eg B. The dianion of the product requires four of these monomeric units plus two electrons, thus:

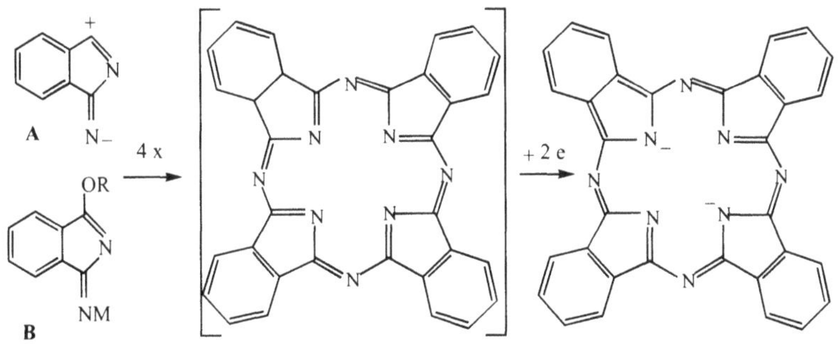

4n π e in macro ring

Phthalocyanine dianion

The electron source is the metal, the metal ion, the solvent (eg $RCH_2OH \rightarrow RCHO$) or the catalyst (eg ammonium molybdate).

(iv) Yields are very variable, but copper(II) complexes are generally formed in good or very good yields. The reaction can be extended to aromatic and heteroaromatic 1,2-dinitriles in general.

(v) As a class C reaction (Panel 8.1), if the monomer is unsymmetrically substituted then four type isomers will be formed in statistical distribution (I 12.5%, II 12.5%, III 50%, IV 25%) resulting in the Type III isomer (Panel 4.2) being the major product. This mixture of type isomers results, provided that steric interactions do not intervene: for example, with 3,4-di(phenylethynyl)-phthalonitrile, the cyclotetramerisation gives only the Type I isomer for reasons of steric repulsion. Figure 10.7 shows another example.

10.5.3 PDT Photosensitisers

10.5.3.1 Zinc(II) phthalocyanine

There have been several studies on this substance, which is available off the shelf, though it is poorly soluble, even in organic solvents. For example, the effect of zinc(II) phthalocyanine, administered in a liposomal preparation, on induced or transplanted rhabdomyosarcoma in hamsters has been studied by Shopova and Jori. With a 24 hour drug-light interval, the photosensitiser concentration was greater in the tumour than in the liver; and it was greater in the induced tumour than in the transplanted one. However, the photodestruction of tumour was similar for both models, in spite of the higher uptake by the induced tumour.

In general, the solubility of porphyrin and phthalocyanine macrocycles in *organic* solvents can be increased by the substitution of alkyl groups larger than methyl at the periphery (Panel 8.1.1). For this reason, octaethylporphyrin (**135**) has been a common substrate for chemical and PDT studies in the porphyrin series (cf structures **136–143, 152–154**). In the phthalocyanine series octalkyl derivatives can be readily prepared from 3,6-dialkylphthalonitriles (above). Zinc(II) octapentyl phthalocyanine (**167**), administered as an emulsion, is more effective and more selective than zinc(II) phthalocyanine in *in vivo* bioassay.

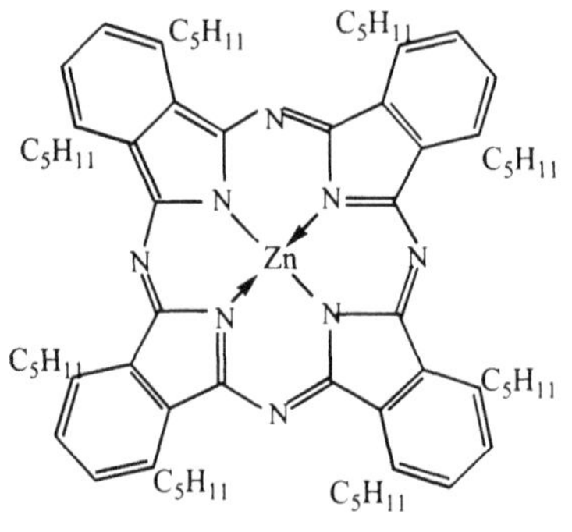

167

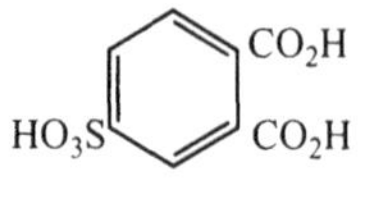

168

10.5.3.2 Sulphonic acids

The sulphonic acids derived from the direct sulphonation (H_2SO_6-SO_3; $ClSO_3H$) of aluminium and gallium phthalocyanines have been considered in section 7.2.4.2. Such electrophilic substitution occurs at both *endo* and *exo* positions, the distribution of positional isomers depending on the substrate.

If sulphonation at just the *exo* position is required, this can be achieved by ring synthesis starting with 4-sulphophthalic acid (**168**). Phthalocyanine synthesis following Figure 10.4 gives phthalocyanine tetra *exo* sulphonic acid (mixture of Type isomers). Mixed syntheses (4-sulphophthalic and 4-phthalic acid) gives a mixture of lower sulphonic acids which may with some effort be separated by reverse phase chromatography (cf Figure 7.2).

In biological assay, the uptake / activity of these compounds appears to be $S_{2a}>S_{2o}$ and $S_2>S_1>S_3>S_4$. Aggregation effects increase in the sequence $S_4<S_1$.

The aluminium phthalocyanine sulphonic acids have been used in extensive clinical trials against various human cancers (section 14.3.6 and Panel 14.6).

Various other polar derivatives of phthalocyanine have been prepared for PDT applications. These include anionic (carboxylic acid, phosphonic acid), cationic (eg pyridinium quaternary salts) and neutral [eg polyethylene glycol ("pegylated") derivatives].

10.5.3.3 Hydroxyphthalocyanines

In the tetraphenylporphyrin series, substitution by hydroxy functions has led to increased potency in PDT (sections 8.4.2(ii) and 9.5.1), and it was clearly of interest to determine if a similar effect occurred in the phthalocyanine series. Tetrahydroxylated phthalocyanines are made in the usual way (Figure 10.5) using ether protection (eg Ph_2CH-O): mono (**169**) di(**170, 171**) and tri (**172**) hydroxy compounds are available by the routes indicated in Panel 10.2 (items 1 and 5).

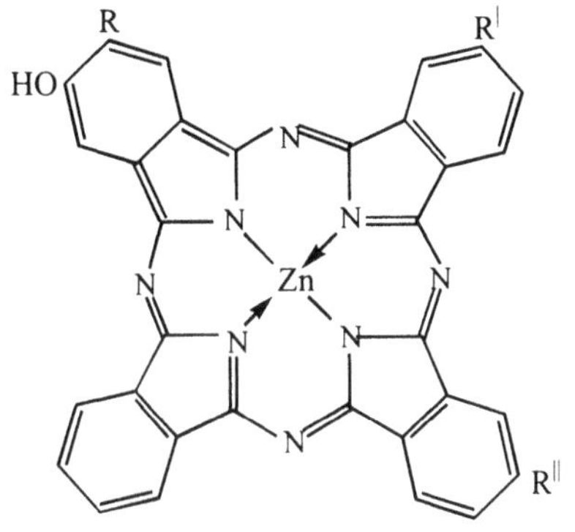

169 2-Hydroxy ZnPc. R = R^{I} = R^{II} = H

170 2,3-Dihydroxy ZnPc. R = OH, R^{I} = R^{II} = H

171 2,9-Dihydroxy ZnPc. R^{I} = OH, R = R^{II} = H

172 2,9,16-Trihydroxy ZnPc. R^{I}=R^{II}= OH, R=H

For biological assay the compounds (**169-172**) were emulsified with Cremophor. Evaluation *in vitro* (MTT assay) against the EMT-6 mammary tumour cell line showed that the monohydroxy compound (**169**) was the most active, activity falling off in the sequence **169** > **170** > **171** > **172**. However, *in vivo* the zinc(II) 2,9-dihydroxyphthalocyanine (**171**) proved to be the most effective in causing complete tumour regression, being more potent than Photofrin, but *less* potent than zinc(II) phthalocyanine. The dihydroxy compound (**171**) was given at a dose of 2 μmol kg^{-1} (1.2 mg kg^{-1}), with a drug-light interval of 24 hours, and a fluence of 400 J cm^{-2}. This energy dose is considerably greater than that required with m-THPP (*ca* 10 J cm^{-2}). However, it is interesting that in (**171**) the hydroxy functions are about 9Å apart, whereas in the less active (**172**) the *opp* hydroxy functions are 12Å apart. This result argues against the tubulin hypothesis described in section 7.2.4.4.

10.5.3.4 Axial ligand variation

There is some evidence that the hydrophobicity of the phthalocyanines can be adjusted for the purpose of improving biological interactions by varying the axial ligands. Thus the silicon phthalocyanine with a quaternary salt side chain (**173**) is poorly taken up by V79 cells in culture, whereas the corresponding tertiary amine (**174**) shows high uptake and photoinactivation.

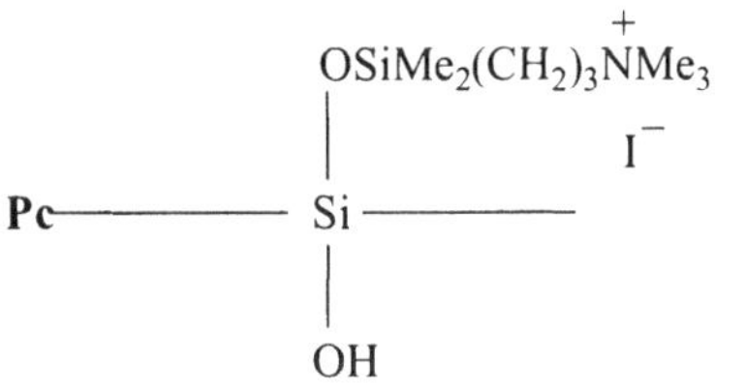

Pc—Si(OH)—$OSiMe_2(CH_2)_3\overset{+}{N}Me_3$ I^-

173

Pc—Si(OH)—$OSiMe_2(CH_2)_3NMe_2$

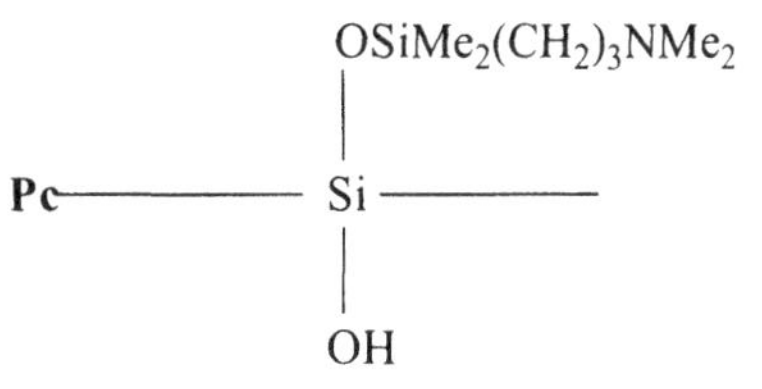

174 λ_{max} 668 nm (ε 230 000)

Axial ligand changes can improve solubility: for example, **174** is soluble in organic solvents. Although the possibility of ligand exchange in the inoculum or in the biological system needs to be borne in mind, compound (**174**), also referred to as Pc4 in a screening programme with some related compounds (Kenney, Ben-Hur) is showing promise in virucidal (blood purification) PDT applications (section 14.5.3.5).

10.6 NAPHTHALOCYANINES

The naphthalocyanines come in two varieties, derived from naphthalene-1,2-dinitrile or naphthalene-2,3-dinitrile, and it is the latter that have been extensively studied in the PDT area. As with phthalocyanine, naphthalocyanine is a very sparingly soluble compound, but solubility in organic solvents can be enhanced by bulky alkyl substituents. The synthetic strategy is also the same as that in the phthalocyanine series, ie a 4 × 1 approach, starting with the 2,3-dinitrile. The initial problem is to synthesise this, and one approach is shown in Figure 10.8.

Initial results of PDT studies with naphthalocyanines were not encouraging. For example, silicon(IV) naphthalocyanine with hydrophilic axial ligands (polyethylene glycol) were water soluble, and accumulated in fibrosarcoma in mice, but showed little selectivity or photodynamic activity. Studies with sulphonated naphthalocyanines (mixture made by direct sulphonation) showed them to be less potent PDT sensitisers than the corresponding phthalocyanines, and to be susceptible to photodegradation.

More recently these compounds have begun to look more promising, and with a strong absorption in the 770–800 nm region they have a potential application in the treatment of melanoma, because the screening absorption of melanin, strong throughout the visible, has begun to decrease significantly in the near infrared.

Two recent results hold promise in this area. Shopova and Wöhrle have demonstrated that zinc(II) naphthalocyanine (**31**, zinc complex) localises in and photoinactivates pigmented melanoma cells. Kreimer-Birnbaum, Jori, Kenney & Rodgers have studied the effect of silicon(IV) naphthalocyanines substituted with trialkylsiloxy ligands on melanotic and amelanotic (ie cells which have lost their pigmentation) melanoma cells. The two compounds studied were NPc-Si$[OSi(isoBu)_2C_{18}H_{37}]_2$ (**162**) and NPc-i$[OSi(C_{10}H_{21})_3][OSiMe_2CH_2CH_2CH_2NMe_2]$ (**175**).

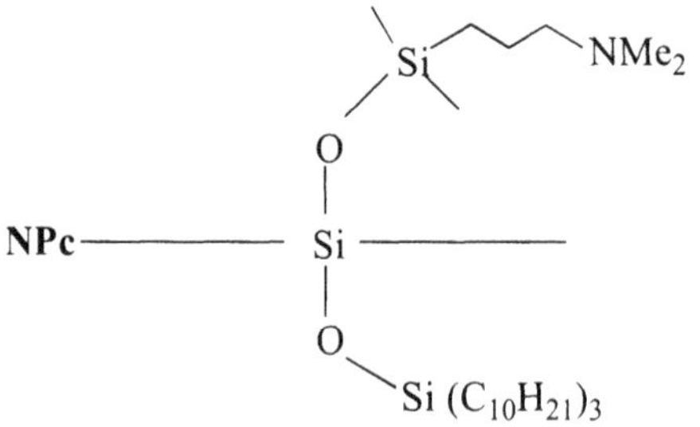

Si NMe$_2$

O

NPc Si

O

Si $(C_{10}H_{21})_3$

175

The second compound (**175**), which has λ_{max} 775 nm ε 400 000, was found to be the more effective. An analysis of the spectra of melanotic cells showed that at this wavelength the melanin absorbance is about 10% that of (**175**). Melanotic cells were less susceptible to photodynamic killing than amelanotic cells for two reasons: (i) the optical filtering effect of the melanin and (ii) the triplet lifetime, which for (**175**) was observed to be lower in melanotic cells (17 μs) than in amelanotic cells (40 μs). Morphological studies on irradiated cells suggested that cell death correlated with damage to lysosomes and the cytoplasmic membrane.

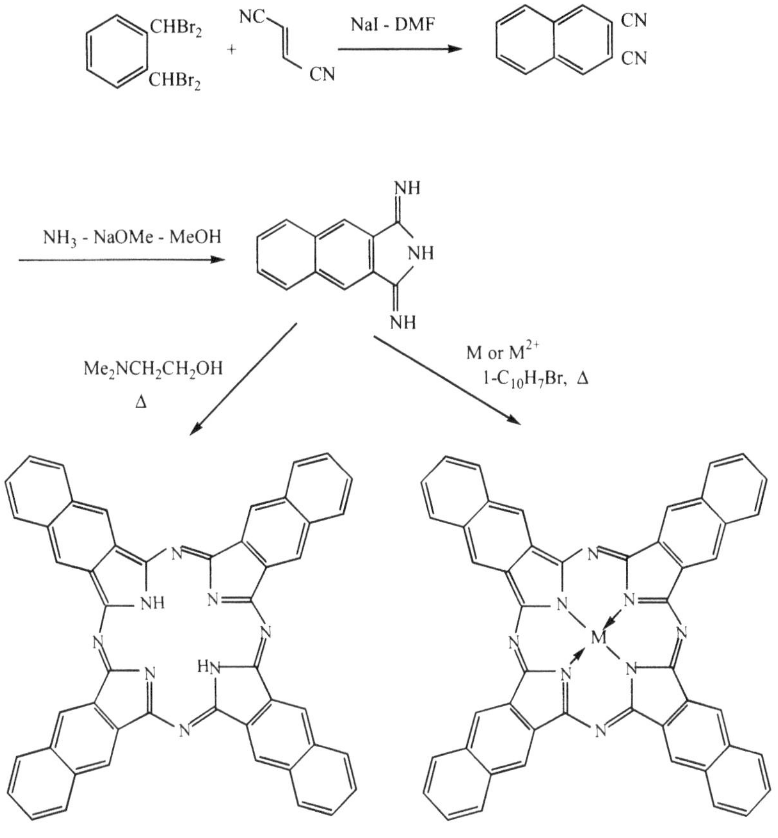

31 Naphthalocyanine

Metallo naphthalocyanine

FIGURE 10.8 SYNTHETIC APPROACH TO NAPHTHALOCYANINE. THE FIRST STEP IS VIEWED AS A DIELS-ALDER REACTION OF DIBROMO-*O*-QUINONEDIMETHANE, FOLLOWED BY ELIMINATION OF 2HBR. (E.I. KOVSHEV, V.A. PUCHNOVA AND E.A. LUKYANETS, *J. ORG. CHEM. USSR* (ENGL. TRANS.), 1971, 7, 364).

The effect of structural variation in the macrocyclic ligand has also been studied. Of a series of substituted amino derivatives examined against Lewis lung carcinoma implants *in vivo* the tetrabenzamido compound proved to be the most effective.

BIBLIOGRAPHY

V.D Poole, 'Azaporphins; Benzoporphins; Benzoazaporphins; Phthalocyanines and related structures', in *Rodd's Chemistry of Carbon Compounds*, Volume IVB (Ed. S. Coffey), Elsevier, Amsterdam 1977, pp.329–340.

A.H. Jackson, 'Azaporphyrins' in *The Porphyrins* (ed. D. Dolphin), Academic Press, New York, 1978. Vol.1, pp.365–388. Includes phthalocyanines.

C.C. Leznoff and A.B.P. Lever (eds) *Phthalocyanines: properties and applications*, VCH Publishers, New York. Vols. 1–4 (1989, 1993, 1993 and 1996). Volume 1 (1989) has chapters on synthesis (C.C. Leznoff) and photobiology (I. Rosenthal, E. Ben-Hur). Volume 3 (1993) has a chapter on redox properties (A.B.P. Lever, E.R. Milaeva, G. Speier).

N.B. McKeown, *Phthalocyanine Materials: Synthesis, Structure and Function*, Cambridge University Press, 1998. Not about PDT, but a useful introduction to the electroactive applications of phthalocyanines.

CHAPTER 11

OTHER PHOTOSENSITISERS

In section 4.3.2(vi) reasons were advanced to explain the major emphasis placed on PDT sensitisers related to the porphyrins. However, these reasons are not overwhelming, and certainly do not prevent consideration of photosensitisers based on other chromophores, as this chapter will show. Indeed, histochemistry is based on the differential staining of tissues, including tumour tissues, with a variety of organic dyes. It has to be said, however, that in many cases the proposed PDT sensitisers, for example the xanthenes, have simply been selected from what was available on the shelf, and have not been subjected to thoughtful structural variation and bioassay to sort out the best therapeutic candidate.

11.1 CYANINE DYES

The cyanine dyes were originally developed in another part of applied photosensitisation — that of photographic sensitisers: any clinical application may be regarded as a bonus.

11.1.1 Merocyanine 540

Merocyanine 540 (**26**) has already been referred to in section 4.3.2 (v). It has been advanced as a phototherapeutic agent against malignant haemopoietic cells, for the extracorporeal purging of bone marrow in leukemia cases. Although it localises in leukemia cells, the phototoxicity does not appear to be sufficiently selective. Comparative studies in two myeloid leukemia cell lines in the mouse have revealed :(i) that m-THPC (77) is significantly more potent and more selective in photokilling

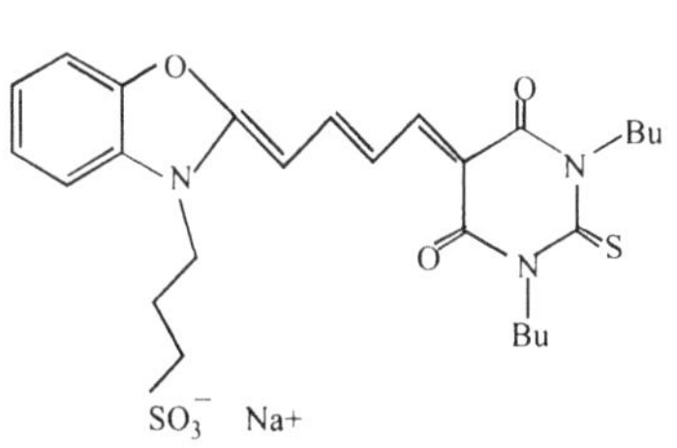

SO_3^- Na+

26 Merocyanine 540

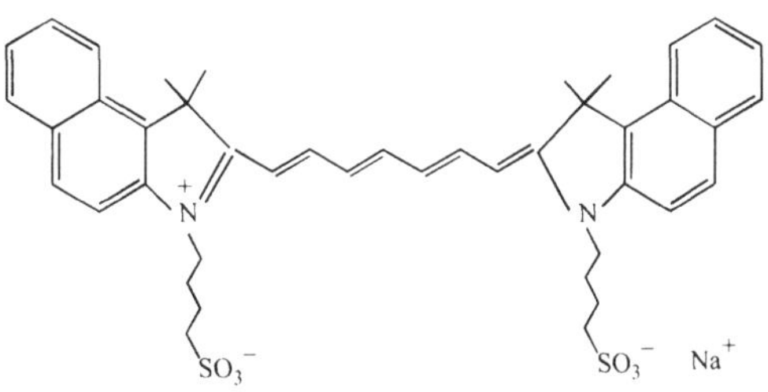

SO_3^- SO_3^- Na^+

176 Indocyanine green

leukemia cells than is merocyanine 540 and (ii) that both m-THPC and merocyanine 540 will induce apoptosis (programme cell death, section 13.4) in leukemia cells, with the m-THPC being observed to kill virtually all the cells in this way.

11.1.2 Indocyanine Green

The significant point about indocyanine green (**176**) (Merck Index no. 4992) is that it has already been approved (since 1956) for clinical administration, for example, in the determination of plasma volume. This means that, if it were a useful PDT sensitiser, the route to regulatory approval would be much easier because of existing clinical experience. The PDT effect on keratinoytes in cell culture has been investigated: the dye is taken up by the cells, and a photodynamic effect is observed. This effect is quenched by sodium azide, but a singlet oxygen mechanism seems rather unlikely since the ϕ_T value is only ~0.01. The dye has an extended delocalised system, with λ_{max} at 805 nm, but the absorption in the visible region is weak, which is an additional advantage.

11.2 HYPERICIN

Hypericin (**20**, section 4.3.2.(i)) is a naturally-occurring extended quinone: it has λ_{max} 590 nm (ε 41 600) in ethanol, and ϕ_Δ = 0.36 in that solvent. In aqueous media ϕ_Δ falls to 0.02, possibly due to aggregation.

This substance has PDT activity against mammary carcinoma implants in athymic mice. In a single clinical case (mesothelioma) where it was applied superficially, it was not effective by itself, but appeared to act synergistically with HpD given subcutaneously.

Photovirucidal activity has also been recorded.

11.3 PHENOTHIAZINES

Many of the phenothiazine dyes, typified by methylene blue (**22**) and toluidine blue (**23**), are commercially available, and there is a considerable literature on the use of these dyes, and especially toluidine blue, to stain and visualise cancerous lesions *in vivo* as an aid in diagnosis.

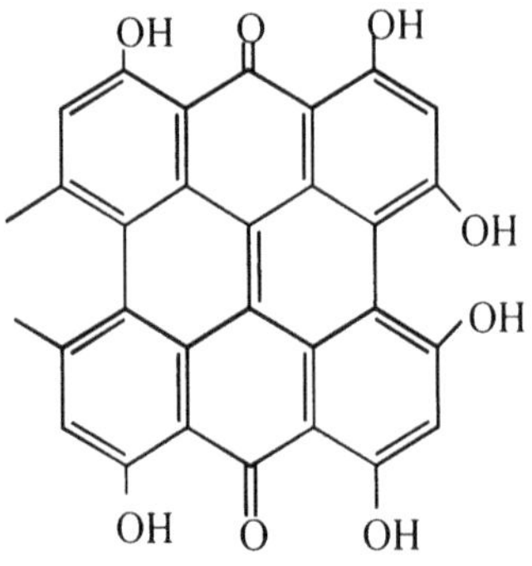

20 Hypericin

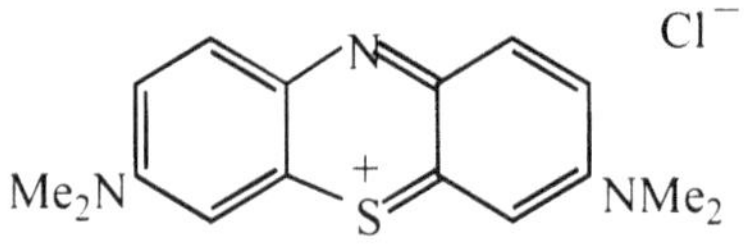

22 Methylene blue

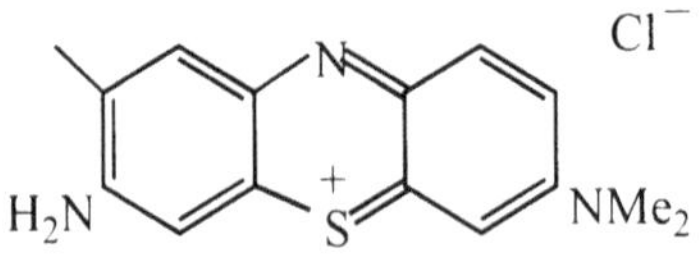

23 Toluidine blue

Phenothiazinium dyes have photomicrobicidal properties, and have activity as PDT agents in *in vitro* experiments with cancer cell lines.

Nevertheless, in spite of being good singlet oxygen sources (such that methylene blue, ϕ_Δ 0.51, Table 2.1, is frequently employed in aqueous systems, for example in searching out tryptophan at or near the active site of an enzyme), the phenothiazinium dyes have not emerged as good PDT agents against cancer. This is because when injected they tend to be rapidly excreted and lack selectivity.

Recent studies on a more lipophilic phenothiazinium salt (**177**) have shown more promise. Administered i.v. at 10 mg kg^{-1} to Wistar rats with fibrosarcoma implants,

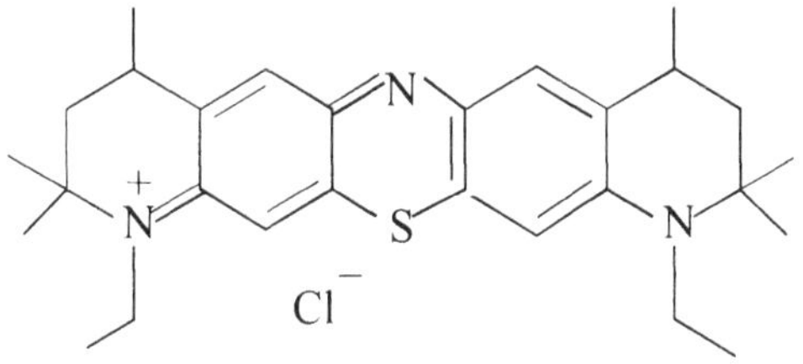

177

it was found that tumour / skin and tumour / muscle ratios (determined by fluorescence measurements) were 9 and 4 respectively, after four hours. Fluorescence microscopy indicated that (**175**) was predominanty localised in vessel walls.

11.4 PORPHYCENES

The porphycenes are isomers of porphyrins that, like the porphyrins, possess an 18 πe aromatic system. The compounds were first described in 1986 by Vogel and his colleagues (Cologne), and are synthesised by McMurry coupling of bipyrrole dialdehydes, as shown in Figure 11.1 for the parent compound (**178**).

Although the syntheses are often low-yielding from uncommon starting materials, the novelty of the system provides interest from the intellectual property point of view and many novel compounds have been prepared with PDT in mind. Cytopharm Inc (Menlo Park, California) in association with Glaxo-Wellcome are thought to be developing compounds of this series for PDT applications (section 14.3.9).

The porphycene system has a porphyrin-like spectrum, but Band I is much more intense and is shifted to the red. Thus (**178**) shows λ_{max} 630 nm (ε 51 900) in benzene. Photophysical measurements show that ϕ_{Δ} varies between about 0.15 and 0.40 depending on substitution pattern. Reduction gives 2,3-dihydroporphycene,

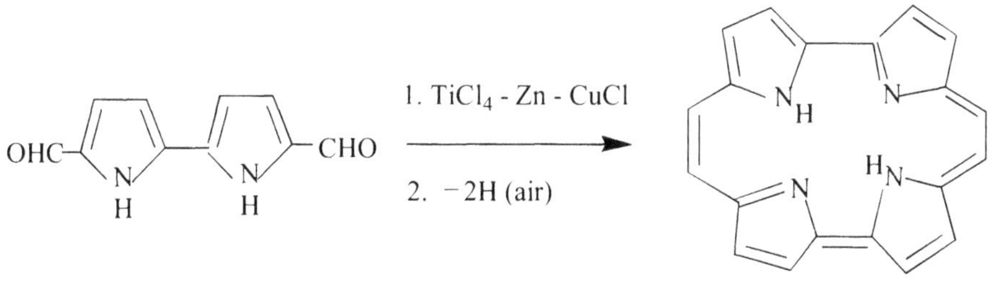

178 Porphycene

FIGURE 11.1 SYNTHESIS OF PORPHYCENE (178). (E.VOGEL, M. KÖCHER, H. SCHMICKLER AND J. LEX, *ANGEW CHEM. INT EDN. ENGL.*, 1986, **25**, 257–259).

an analogue of chlorin; but Band I does not intensify as it does in the porphyrin series.

2,7,12,17-Tetraphenylporphycene (**179**), which is the porphycene analogue of TPP (but with β-substitution), has λ_{max} 659 nm (ε 84 600) and a singlet oxygen quantum yield of *ca.* 0.25. In the light of the favourable experience with m-THPP (**76**) (section 8.4.2) it would be interesting to investigate the corresponding 2,7,12,17-tetrakis(*m*-hydroxyphenyl) derivative.

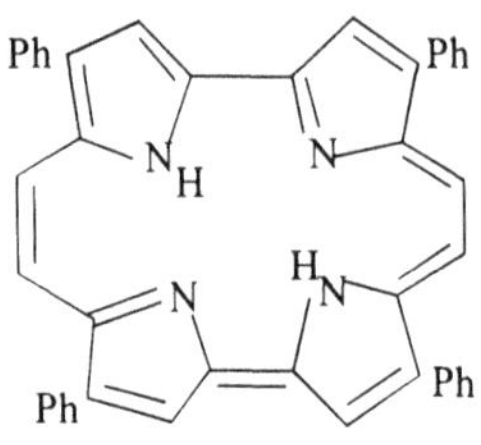

179

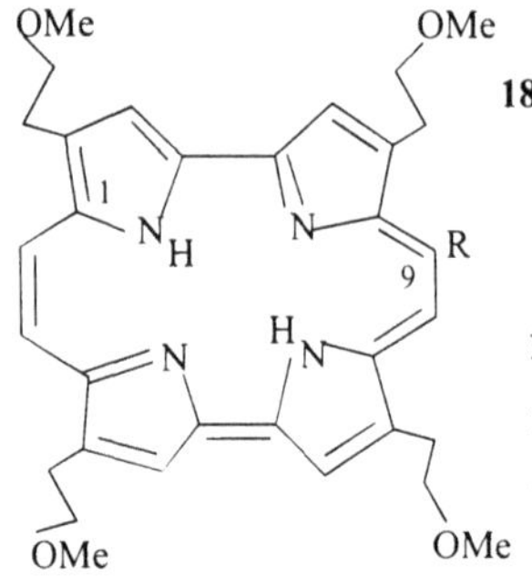

180 R=$NHCO(CH_2)_3CONHCH(CO_2H)(CH_2)_2CO_2H$
hydrophilic, injected in PBS

181 R=$NHCO(CH_2)_3CO_2H$

182 R=OC_6H_{13}

183 R=$NHCO(CH_2)_{16}CH_3$

181,182,183 are injected in liposomal preparations

Porphycene (**178**) is, of course, hydrophobic: for PDT applications compounds with substituents which will confer overall amphiphilic character are needed (section 7.2.4). Several such compounds, including (**180–183**), are being evaluated: they are all 2,7,12,17-tetrakis(2-methoxyethyl) compounds with substituent variation at C-9. Some results of pharmacokinetic experiments in animal models are shown in Table 11.1. The more hydrophilic compounds show lower tumour uptake with lower selectivity, but distribution occurs rapidly. On the other hand, the more hydrophobic compounds (eg **183**) equilibrate more slowly, but the concentrations eventually reached are higher and there is more selectivity.

TABLE 11.1 HYDROPHOBICITIES, PLASMA UPTAKE AND TUMOUR / MUSCLE RATIOS FOR SOME PORPHYCENES. (A. SEGALLA, F. FEDELI, E. REDDI, G. JORI AND A. CROSS, *INTL. J. CANCER*, 1997, **72**, 329–336).

Structure	Retention time[a] (min)	Plasma $t_{1/2}$[b] (hr)	[Tumour]/[Muscle]
180	1.0	0.02	2.0 at 20 min.
181	1.3	0.02	2.8 at 20 min.
182	6.6	0.37	19.0 at 3 hr.
183	9.7	6.26	26.3 at 24 hr.

a Reverse phase chromatography: higher retention = more hydrophobic
b Time after injection at which the plasma concentration of the photosensitiser falls to half of its initial value

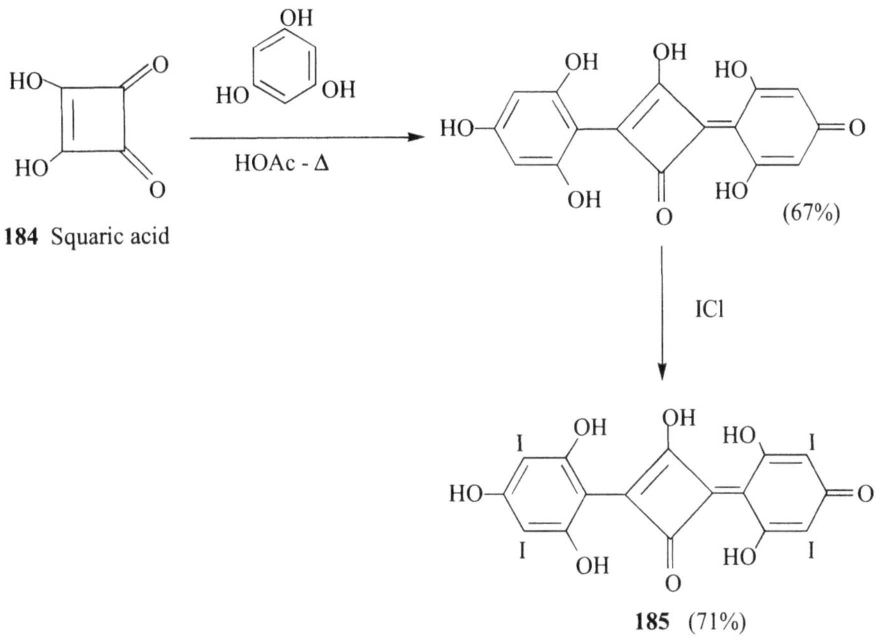

184 Squaric acid

HOAc - Δ

(67%)

ICl

185 (71%)

FIGURE 11.2 SYNTHESIS OF A SQUARAINE DERIVATIVE. (A. TREIBS AND K. JACOB, *ANGEW. CHEM.INTL.EDN. ENGL.*, 1965, **4**, 694. D. RAMAIAH, A. JOY, N. CHANDRASEKHAR, N.V. ELDKO, S. DAS AND M.V. GEORGE, *PHOTOCHEM.PHOTOBIOL.*, 1997, **65**, 783-790).

11.5 SQUARAINES

The squaraine dyes are derived from squaric acid (**184**) by condensation with π-excessive aromatic compounds. Figure 11.2 shows an example where condensation occurs with two molecules of phloroglucinol (1,3,5-trihydroxybenzene). Iodination of the product gives the tetraiodosquaraine derivative (**185**).

These compounds have been suggested as PDT photosensitisers, although no biological studies have yet appeared. The compounds have a complex acid-base behaviour, and the spectrum depends very much on the pH, as shown in Table 11.2. The species of (**185**) present at near neutrality is the squaraine monoanion, which has λ_{max} 617 nm (ε 63 000). The non-halogenated squaraines have low ϕ_{isc}, but because of the heavy atom (iodine) substitution the triplet of (**185**) is readily formed, and the triplet lifetime is 36 μs. ϕ_{Δ} is 0.47. Clearly this chromophore has potential in PDT studies, especially if the band position can be shifted further into the red by appropriate substitution.

11.6 TEXAPHYRINS

The texaphyrins have an expanded coordination with five nitrogen atoms, and it is metal complexes rather than the free base which appear to be making the running. The basic system is formed by condensing a diformyltripyrrane with a diamino compound. When the diamine is an *o*-phenylenediamine, or a similar molecule, the initially formed bis-azomethine is easily oxidised to the aromatic texaphyrin (**186**) (Figure 11.3).

TABLE 11.2 SPECTROSCOPIC BEHAVIOUR OF SQUARAINE IONS IN MeOH – H_2O.

Species	pH	Structure	λ_{max} nm (ε)
SqH+	0.5	(i) 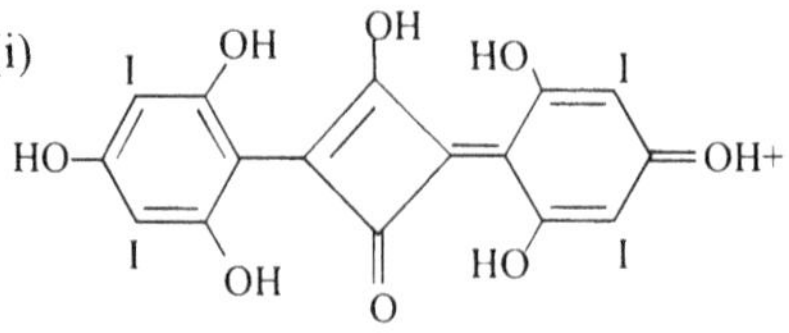	563 (98 000)
		(ii) aggregate	720
Sq	3.5	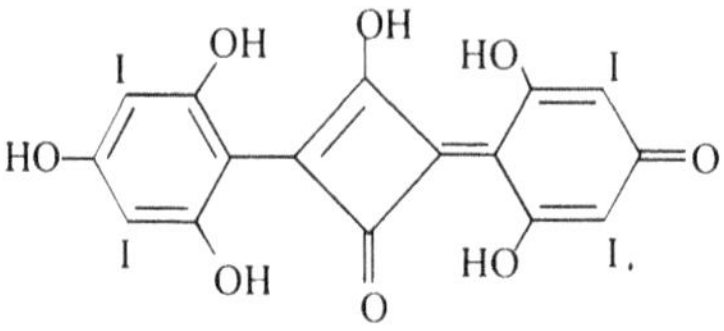	508 (45 000)
Sq^-	6.3	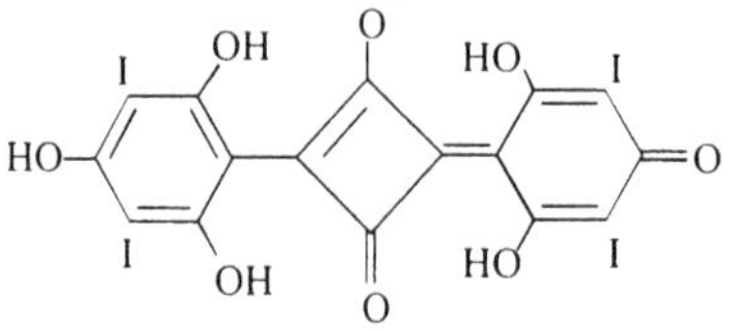	617 (63 000)
Sq^{--}	8.0	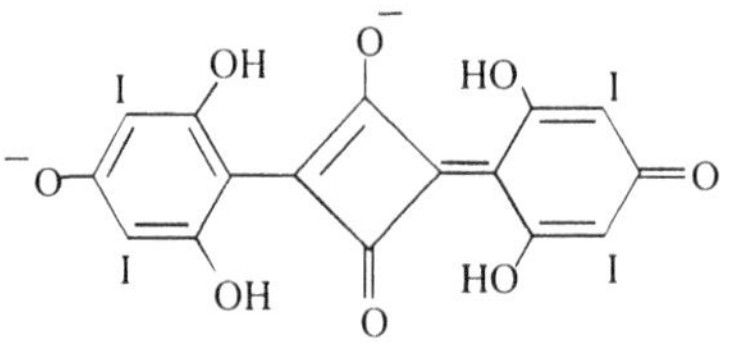	562 (27 000)

The texaphyrins readily form complexes with metal ions, including those with larger ionic radii than can be readily accommodated by porphyrins. X-ray studies on a lutetium(III) complex (Lu radius 0.98Å) show that the metal is eight coordinate, and is 0.27Å above the mean plane of the macrocyclic nitrogens. On the other hand, a lanthanum(III) complex (La^{3+} radius 1.27A) is ten-coordinate, and here the metal is about 0.9Å above the mean plane of the macrocyclic nitrogens, and the macrocycle adopts a saucer-shaped conformation.

As far as bioactivity is concerned, gadolinium(III) complexes are promising as contrast agents in magnetic resonance imaging, and lanthanum(III) and lutetium(III) complexes show phototumoricidal properties *in vivo*.

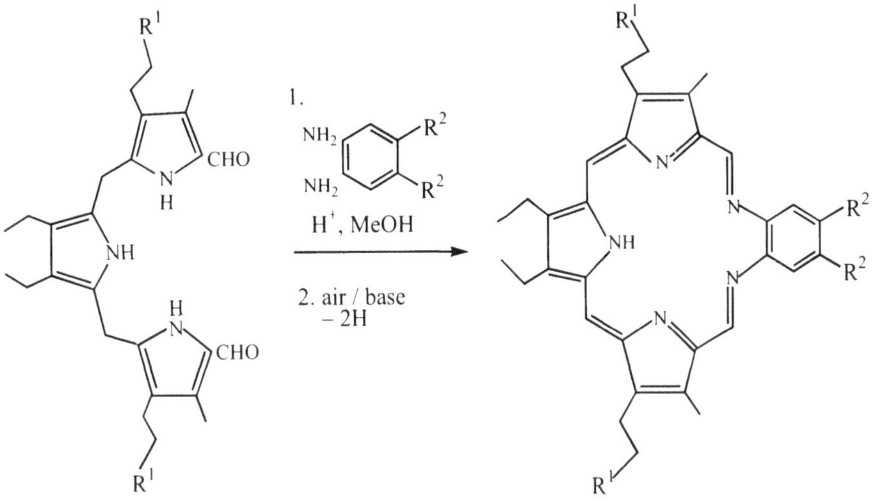

1. NH_2 / NH_2 (o-phenylenediamine, R^2, R^2), H^+, MeOH
2. air / base − 2H

186

FIGURE 11.3 SYNTHESIS OF THE TEXAPHYRIN SYSTEM (J.L. SESSLER, G, HEMMI, T.D. MODY, T. MURAI, A. BURRELL AND S.W. YOUNG, *ACC.CHEM.RES.*, 1994, **27**, 43–50).

These complexes have a strong absorption band in the 600–900 nm region, the position of which can be varied by suitable choice of substituent (including metal).

The texaphyrin nucleus (**186**) is hydrophobic like the other macrocyclic nuclei we have considered earlier. In order to make the photosensitising drug more water-soluble, short -$(CH_2CH_2O)_n$- chains have been attached. This process, which is used elsewhere in pharmaceutical chemistry to increase water solubility, relies on the hydrogen bonding capacity of the polyether chain. Often rather long chains derived from polyethylene glycols are used (say, of average molecular weight 2000, but preparations of polyethylene glycol (PEG) are commercially available of various average molecular weight), and the process is referred to as 'pegylation'. In the present case, two triethylene glycol chains are used: the free base is metallated with lutetium acetate in $MeOH/Et_3N$ to give the lutetium texaphyrin derivative (**187**). This is being commercially developed under the proprietary name of "Lutrin" (section 14.3.8 and Panel 14.8) by Pharmacyclics in Sunnyvale, California. It has attractive properties, with λ_{max} 732 nm (ε 42 000), selectivity for tumour tissue, and a short drug-light interval (3 hours).

The electronic absorption spectrum in phosphate buffered saline, containing 10% foetal bovine serum (FBS) to mimic the biological situation, is shown in Figure 11.4.

11.7 XANTHENES

The xanthene dyes, for example, fluorescein (**188**), eosin (**24**) and rose bengal (**25**), are excellent photosensitisers for singlet oxygen formation (Table 2.1) and are perhaps the most commonly used compounds where singlet oxygen is required in

187

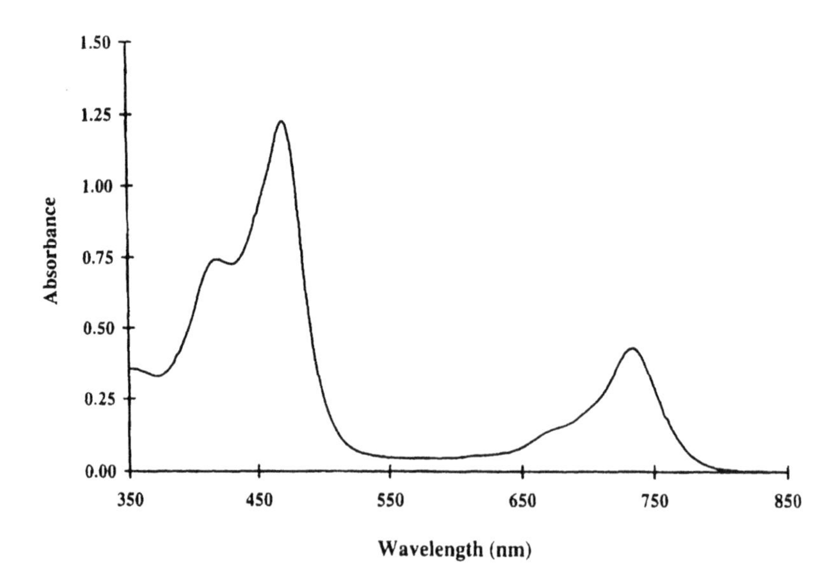

FIGURE 11.4 ELECTRONIC ABSORPTION SPECTRUM OF THE TEXAPHYRIN COMPLEX (**187**); 20 μG ML^{-1} IN PBS CONTAINING 10% FBS. (REPRODUCED WITH PERMISSION FROM S.W. YOUNG, K.W. WOODBURN, M. WRIGHT, T.D. MODY, Q. FAN, J.S. SESSLER, W.C. DOW AND R.A. MILLER, *PHOTOCHEM. PHOTOBIOL.*, 1996, **63**, 892–897).

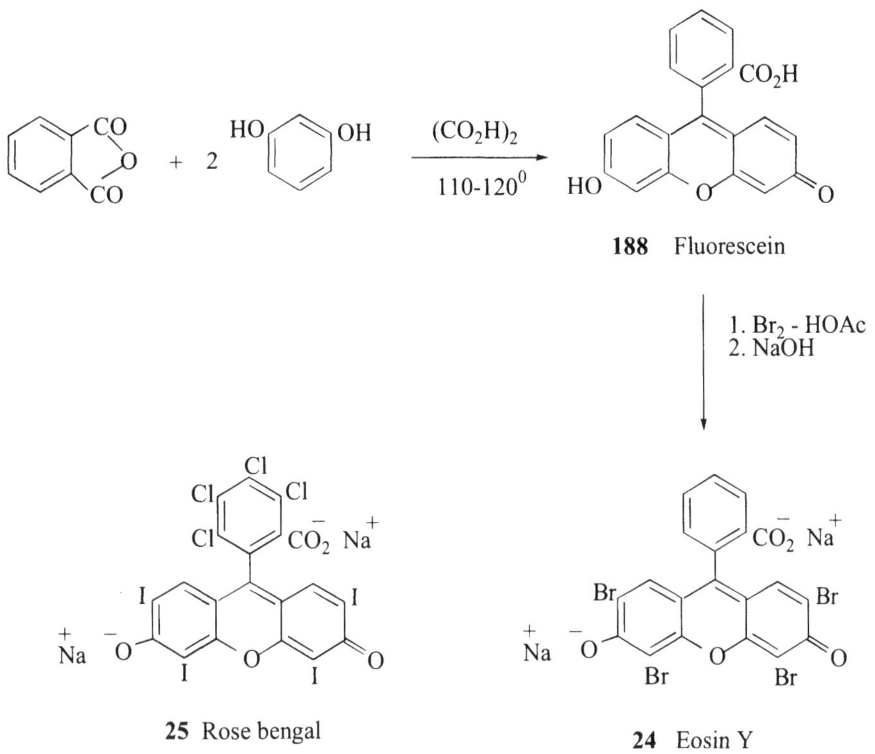

$(CO_2H)_2$ 110-120°

188 Fluorescein

1. Br_2 - HOAc
2. NaOH

25 Rose bengal

24 Eosin Y

FIGURE 11.5 SOME COMMON XANTHENE PHOTOSENSITISERS.

organic synthesis (eg Panel 3.3.1). Fluorescein is prepared by the condensation of phthalic anhydride and resorcinol, and eosin Y is made by direct bromination of the two activated aromatic rings (Figure 11.5).

The earliest clinical experiments in this field were carried out by Jesionek and von Tappeiner with eosin (in 1903 and 1905, section 1.2.4) and in 1974 Dougherty started his PDT studies with fluorescein. However, he quickly moved on to HpD because it was much more effective.

The available xanthenes are too water soluble to give a good clinical result, except perhaps by topical application. Given systemically they do not seem to show the localisation and selectivity required. It is conceivable that they could do so if they were suitably substituted with lipophilic groups.

11.8 CONJUGATED PHOTOSENSITISERS

It is the biochemical usage of *conjugated* photosensitisers that is required here. Conjugation in this sense may be defined as the covalent attachment of the principal molecule to a second molecule, which may be small or large, and which confers some particular (usually beneficial) property on the principal molecule. Thus, in Panel 5.2 we saw how conjugation of bilirubin, which is virtually insoluble in water, with the uronic acid *D*-glucuronic acid, gave the diglucuronide, which is water soluble, and which can be readily secreted into the bile.

Similar solubilising potential has been applied in developing PDT photosensitisers. Thus, m-THPC (77) although it is freely soluble in methanol, is almost insoluble in water at pH7 (section 9.5.1). However, when polyethylene glycol residues (mean molecular weight about 2000) are attached to the four phenolic functions, the substance becomes water soluble. Although the photosensitiser dose (in mg kg^{-1}) must necessarily increase, such conjugates are of interest because of the greater tumour specificity which they show.

Another application of conjugation is to improve targetting by attachment to a system which fits a receptor site on the tumour cell, giving conjugates which are regarded as the *third generation photosensitisers* (section 7.1). Covalent attachment to monoclonal antibodies continues to receive attention, although a practical PDT system which can be used clinically has not appeared as yet. However, the search continues. In cancer chemotherapy, such an approach has reached the clinic in the form of Mabthera, marketed by Roche for the treatment of the most common type of non-Hodgkin's lymphoma.

On the premise that breast carcinoma cells possess an excess of estrogen receptors, Montforts and his group have prepared the estradiol derivative (**189**, see also Figure 9.4), in the hope that selectivity will be improved. In a conceptually related, but

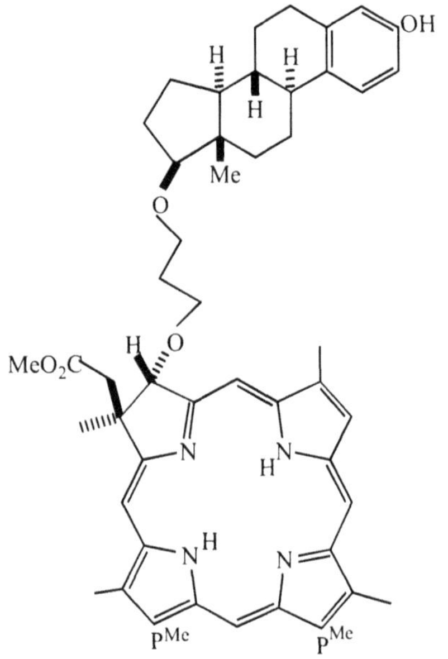

189

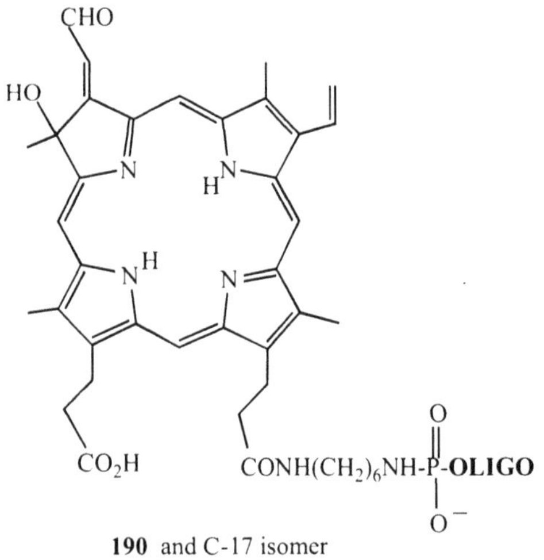

190 and C-17 isomer

OLIGO = TTCTTCTCCTTTCT
(see Panel 5.1 for structures of nucleotide bases)

biochemically very different area, Brault and his colleagues have prepared oligonucleotide conjugates of chlorins (eg **190**, prepared from photoprotoporphyrin A, **132**, section 9.4.3). Such oligonucleotide conjugates form specific complexes with suitable single-stranded and double-stranded polynucleotides, and site-directed photodamage to nucleic acid structures has been observed using such oligonucleotide conjugates and red light (668 nm). This sort of work appears to be relevant to the current search for systems with photovirucidal activity (section 14.5.3.5)

BIBLIOGRAPHY

M. Wainwright, 'Non-porphyrin photosensitisers in biomedicine', *Chem. Soc. Rev.*, 1996, **25**, 351–359.

J.Y. Chen, N.K. Mak, J.M. Wen, W.N.Leung, S.C. Chen, M.C. Fung and W.H.Cheung, 'A comparison of the photodynamic effects of Temoporfin (mTHPC) and MC540 on leukemia cells: efficacy and apoptosis', *Photochem. Photobiol.,* 1998, **68**, 545–554. References on preclinical studies on merocyanine 540 in the treatment of leukemia.

S. Frickweiler, R-M. Szeimies, W. Bäumler, P. Steinbach, S. Karrer, A.E. Goetz, C. Abels, F. Hofstädter and M. Landthaler, 'Indocyanine green: intracellular uptake and phototherapeutic effects *in vitro*', *J. Photochem. Photobiol., B: Biol.,* 1997, **38**, 178–183.

H. Koren, G.M. Schenk, R.H. Jindra, G. Alth, R. Ebermann, A. Kubin, G. Koderhold and M. Kreitner, 'Hypericin in phototherapy', *J. Photochem. Photobiol., B: Biol.,* 1996, **36**, 113–119.

Q. Peng, S.B. Brown, J. Moan, J.M. Nesland, M. Wainwright, J.Griffiths, B. Dixon, J. Cruse-Sawyer and D. Vernon, 'Biodistribution of a methylene blue derivative in tumour and normal tissues of rats', *J. Photochem. Photobiol., B: Biol.,* 1993, **20**, 63–71.

D. Mew, V. Lum, C.K. Wat, G.H.N. Towers, C.H.C. Sun, R.J. Walter, W. Wright, M.W. Berns and J.G. Levy, 'Ability of specific monoclonal antibodies and conventional antisera conjugated to hematoporphyrin to label and kill selected cell lines subsequent to light activation', *Cancer Research,* 1995, **45**, 4380–4386.

T. Hasan, 'Photosensitizer delivery mediated by macromolecular delivery systems', in *Photodynamic Therapy:BasicPrinciples and Clinical Practice,* (eds B.W. Henderson and T.J. Dougherty), Marcel Dekker, New York, 1992, pp.187–200.

M.B. Vrouenraets, G.W.M. Visser, F.A. Stewart, M. Stigter, H. Oppelaar, P.E. Postmus, G.B. Snow and G.A.M.S. van Dongen, 'Development of *meta*-tetrahydroxyphenylchlorin monoclonal antibody conjugates for photoimmunotherapy', *Cancer Res.*, 1999, **59**, 1505–1513.

A.S. Boutorine, D. Brault, M. Takasugi, O. Delgado and C. Hélène, 'Chlorin-oligonucleotide conjugates: synthesis, properties, and red-light-induced photochemical sequence-specific DNA cleavage in duplexes and triplexes', *J. Amer. Chem. Soc.,* 1996, **118**, 9469–9476.

CHAPTER 12

PHOTOBLEACHING

" I have now noticed the principal facts respecting the powerful agencies of solar light, in producing, changing, and destroying mineral, vegetable, and animal colours; which agencies as far as we know, or can judge, seem to be principally, if not exclusively, exerted, in promoting, under particular circumstances, and with particular coloured, or colouring, matters, *an abstraction or diminution of their oxygene;* and with other matters and other circumstances, in causing a *new* or *additional combination of it.*"

Edward Bancroft, Fellow of the Royal Society of London, and of the American Academy of Arts and Sciences, of the State of Massachussetts Bay, *Experimental Researches concerning the Philosophy of Permanent Colours,* Cadell and Davies, London,1813, making the first reference to reductive and oxidative pathways in photobleaching.

12.1 DEFINITIONS

Photobleaching is a long established study in the chemistry of dyestuffs and pigments, where resistance to decolorisation by light is (usually*) a highly valued property. In that area, dyes which have such stability to light are said to be *light fast,* or to have *fastness* to light. Such stability refers to the absorption properties of the colorant, and this leads to the common sense definition of photobleaching as the loss of colour (*ie* absorbance) caused by light.

In photochemistry and photobiology, however, the custom has emerged of using the term *photobleaching* to refer to loss of absorbance and/or fluorescence on exposure of the system to light. This definition, which will be used here, can lead to confusion unless it is made clear whether absorption loss or fluorescence loss is being referred to, since the two processes are often not in parallel. The term photobleaching is also sometimes used to refer to reversible processes (particularly those observed at high photon flux with laser sources), but here we shall use the term to refer to irreversible processes, leading to the chemical change, including the total destruction, of the chromophore. Thus, two types of photobleaching can be considered:

*Indigo, used to dye jeans is an exception. This dye is rather susceptible to photobleaching, which was why there was a need a century ago to replace it with light fast compounds. However, the ways of fashion are unpredictable, and the etiolated hue of partially photobleached indigo became very trendy starting in the 60s and continuing.

(i) *photomodification*, where loss of absorbance or fluorescence occurs at some wavelengths, but the chromophore is retained in a *modified* form; and
(ii) *true photobleaching*, where chemical change is deep seated, and results in small fragments which no longer have appreciable absorption in the visible region.

Often these two processes occur concomitantly.

Photobleaching of fluorescence may not run parallel to photobleacing in absorption because fluorescence, although very sensitive, is also very susceptible to quenching, for example arising from aggregation (Panel 12.1). The importance of fluorescence measurements in the present context is that, because they are so sensitive, they can be used at the concentrations encountered *in vivo*, where absorption measurements are usually not really feasible. Moreover, using fluorescence microscopy it is possible to determine fluorescence for individual cells, and for organelles within cells, during the irradiation process.

12.2 POTENTIAL VALUE OF PHOTOBLEACHING IN PDT

Ideally an injected photosensitiser should be specific for tumour tissue: in the real world, we have to settle for selectivity rather than specificity (section 7.2.4) and hence some photosensitiser will be found in normal tissue, where it may cause undesirable photosensitisation (eg for a period of several weeks with Photofrin). If the photosensitiser were moderately susceptible to photobleaching, then after the main task of destroying the tumour under strong directed illumination had been completed, the residual part of the photosensitiser located in normal tissue would be susceptible to photobleaching under ambient light conditions, thus reducing the period during which exposure to strong sunlight would need to be avoided. With a photobleaching sensitiser it might be possible to arrange matters so that the therapeutic dose of light, while damaging the tumour, does not damage the surrounding tissue (because it contains a drug level in the region of the threshold for photodynamic action), but starts off the photobleaching process.

As this implies, photobleaching is expected to affect dosimetry, both of the photosensitiser and of the light. For photobleaching drugs there may be some benefit in ramping the light dose (so that as the concentration of active photosensitiser decreases, the concentration of photons increases), and clearly allowance will need to be made for photobleaching in determining drug dose. Photosensitisation by photoproducts is another consideration. In animal model experiments, a photosensitation threshold is often found, below which no significant effect is observed. As Figure 12.1 shows, this threshold is higher for a readily photobleached sensitiser (such as m-THPBC) than for a less readily photobleached analogue (m-THPC).

12.3 KINETIC STUDIES

Because of the ease with which photobleaching can be followed spectroscopically in solution in this series, there has been an abundance of kinetic studies. For example, Krasnovskii and his colleagues reported rate constants for singlet oxygen quenching and for competing chemical reaction (eg. photobleaching) for some porphyrin

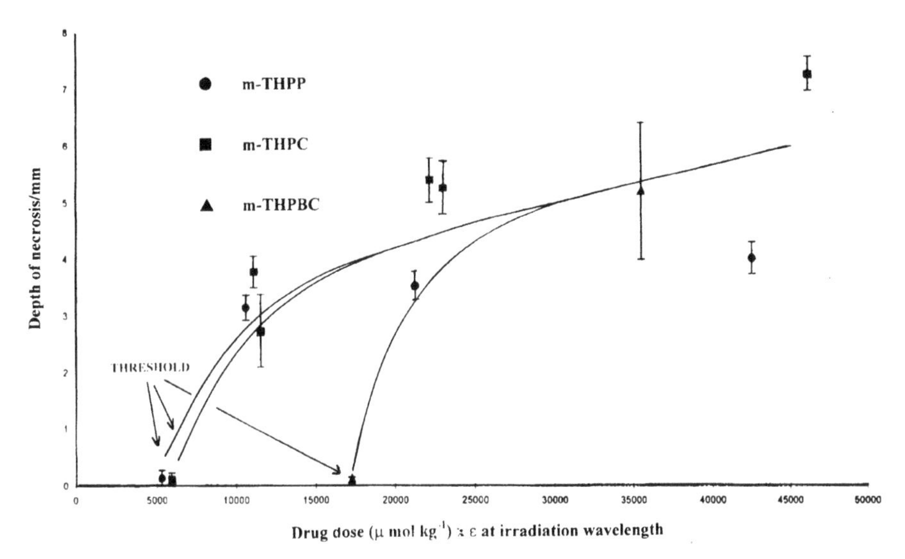

FIGURE 12.1 PLOT OF THE DEPTH OF TUMOUR PHOTONECROSIS AGAINST THE PRODUCT OF ε AT THE WAVELENGTH OF IRRADIATION AND THE DRUG DOSE FOR M-THPP (**76**), M-THPC (**77**) AND M-THPBC (**134**) IN THE SAME ANIMAL MODEL UNDER IDENTICAL CONDITIONS. THE FIGURE SHOWS THAT (I) THE MORE READILY PHOTOBLEACHED DRUG (M-THPBC, SECTION 12.4.2.1) HAS THE HIGHER PHOTOSENSITISATION THRESHOLD AND (II) BIOLOGICAL ACTIVITY FOR THESE COMPOUNDS WHICH HAVE THE SAME PERIPHERAL SUBSTITUTION PATTERN, ROUGHLY PARALLELS ε_{MAX} AT BAND I (THE WAVELENGTH OF IRRADIATION). (ADAPTED FROM R. BONNETT, P. CHARLESWORTH, B.D. DJELAL, S. FOLEY, D.J. MCGARVEY AND T.G. TRUSCOTT, *J. CHEM. SOC., PERKIN TRANS.* **2**, 1999, 325–328).

compounds (Table 12.1). It is clear that these compounds are much better at physically quenching singlet oxygen than at reacting with it chemically.

Because of its importance in relation to HpD (Chapter 6) the photobleaching of haematoporphyrin and its relatives has been examined extensively. Rotomskis found both photobleaching and photomodification occurring at the same time, the latter leading to new absorption at about 640 nm, which was attributed to a chlorin. The rate of photobleaching was found to be sensitive to the degree of aggregation. It was concluded that photobleaching arose from the monomeric material, whereas

TABLE 12.1 RATE CONSTANTS FOR QUENCHING OF SINGLET OXYGEN (k_q) AND IRREVERSIBLE OXIDATION (k_{ox}).

	k_q (M^{-1} s^{-1})	k_{ox} (M^{-1} s^{-1})
Protoporphyrin **61** (as dme)	5×10^5	5×10^3
Mesoporphyrin dme **99**	2×10^6	400
Chlorophyll *a* **17**	7×10^8	2×10^6
TPP **27**	9×10^5	30
Tetraphenylchlorin (TPC)	2×10^6	120
Zinc(II) TPC	4×10^9	2×10^8
Tetraphenylbacteriochlorin **29**	1×10^8	1.5×10^5

(A.A. Krasnovskii, Y.A. Venediktov and O.M. Chernenko, *Biophysics,* 1982, **27**, 1009–1016).

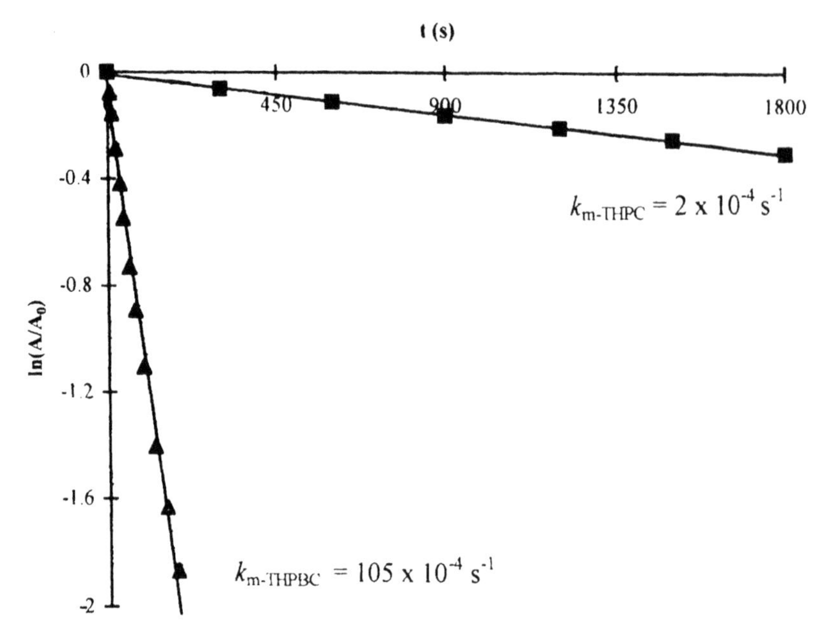

FIGURE 12.2 PHOTOBLEACHING OF M-THPC (**77**) AND M-THPBC (**134**) IN METHANOL IN AIR USING AN ARGON LASER SOURCE (514 NM) UNDER IDENTICAL CONDITIONS, AND FOLLOWING THE DIMINUTION OF BAND I (M-THPC 652 NM, M-THPBC 735 NM). THE SOLUTIONS WERE OPTICALLY MATCHED AT T = 0 ([M-THPC] = 0.04 MM, [M-THPBC] = 0.02 MM). UNDER THESE CONDITIONS THE SENSITISERS ARE MONOMERIC.

photomodification occurred with the aggregate. *Aggregation* is a significant property of porphyrins (and of aromatic systems in general), and is considered in Panel 12.1. In mixed solvent systems, aggregation generally increases when the proportion of the poor solvent increases, and at the same time the rate of photobleaching decreases.

Often photobleaching is due to photooxidation, and under these circumstances the rate of photobleaching follows the redox potential of the system (photobleaching rate: porphyrin < chlorin < bacteriochlorin). Phthalocyanines appear to be rather resistant to photobleaching. Electron donor substituents, and metallation with low valent metals (MgII, ZnII), lower the redox potential, and the rate of photobleaching is expected to increase. Thus the zinc(II) complex of monoaspartylchlorin e_6 (**125**) is photobleached more rapidly than is the free base (buffer pH 7.4, air).

Figure 12.2 shows first order plots for the photobleaching of m-THPC (**77**) and m-THPBC (**134**) as monomers in methanol in air using argon laser irradiation (514 nm): the bacteriochlorin is bleached at about 50 times the rate of the chlorin under these conditions.

The rate of photobleaching is sensitive to solvent and to additives, especially thioethers. As an example of solvent effect, the rate of photobleaching of m-THPC increases by 20 fold in formamide compared with methanol solution, a result possibly related to the high dielectric constant of formamide. Protoporphyrin provides an example of the effect of additives. In organic solvents both photomodification (Figure 12.3) and photobleaching are found. However in aqueous media in the presence

TABLE 12.2 PHOTOBLEACHING QUANTUM YIELDS (ϕ_{PB}) OF SOME PORPHYRINS (5μM) AT 25°C IN 0.1 M PHOSPHATE BUFFER AT pH 7, IN AIR.

	ϕ_{pb}
Haematoporphyrin	4.7×10^{-5}
Uroporphyrin I	2.8×10^{-5}
$TPPS_4$	9.8×10^{-6}
Photofrin	5.4×10^{-5}

(J.D. Spikes, *Photochem. Photobiol.*, 1992, **55**, 797–808.)

of erythrocyte ghosts (representing a natural membrane system) irradiation in air results in rapid photobleaching. Similarly, addition of methionine to aqueous-oil emulsions of protoporphyrin (and also haematoporphyrin and mesoporphyrin) markedly enhances photobleaching. Whitten has suggested the involvement of persulphoxides ($R_2S^+\text{-}OO^-$), but the mechanism deserves further clarification.

Table 12.2 shows the apparent quantum yields for photobleaching for some early PDT sensitisers. The values are all low. Apart from uroporphyrin, these substances are expected to be partially aggregated under the conditions described.

12.4 PRODUCT STUDIES

The excited state of a molecule, with one electron in an antibonding orbital, and an electronic hole in a bonding orbital, is expected to be more easily oxidised and more easily reduced than the ground state. With porphyrins, oxidation is the more commonly encountered: mechanistically this may involve a Type I or a Type II pathway, or both (section 4.4), ie it is a self-sensitised reaction. In the following sections, examples are given to illustrate the types of products formed.

12.4.1 Photomodification

12.4.1.1 Protoporphyrin

The self-photooxygenation of protoporphyrin was first observed by Hans Fischer, but the structures of the products were elucidated much later by Inhoffen and by Whitten and their colleagues.This reaction provides the most closely studied example of porphyrin photomodification, although the products of true photobleaching, a concomitant process which leads to low yields of photomodification products, have usually not been worked out (but see section 12.4.2).

The products which have been identified are shown in Figure 12.3. They consist of an isomeric pair of hydroxychlorin aldehydes (**132** and **191**), called photoprotoporphyrin A and B, respectively; two isomeric porphyrin aldehydes (**192**, **193**); and a porphyrin dialdehyde (**194**). Interestingly, compound (**192**) is spirographis porphyrin, the iron complex of which is the prosthetic group of the oxygen transporting protein of the polychaete *Spirographis spallanzanii.*

In organic solvents such as dichloromethane the photoprotoporphyrins are the major products, the ratio photoprotoporphyrin / porphyrin aldehyde being about 10 : 1. Under these conditions the products appear to arise from singlet oxygen cycloadditions ($4\pi + 2\pi$ to give **132** and **191,** section 3.3.3; and $2\pi + 2\pi$ to give

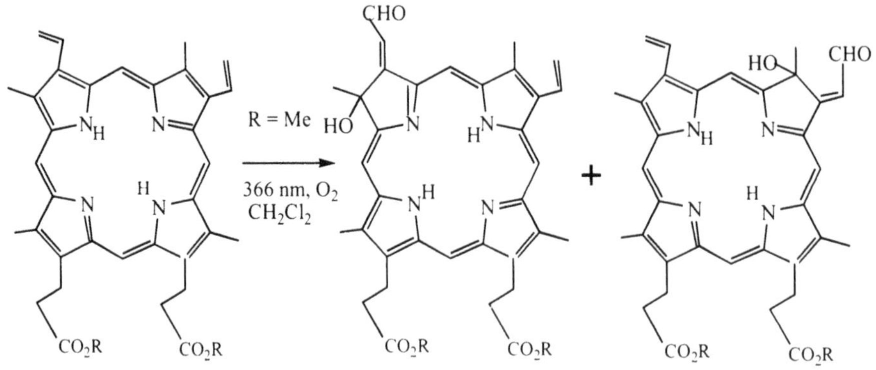

R = Me or H (**61**) → R = Me, 366 nm, O_2, CH_2Cl_2 → **132** Photoprotoporphyrin A + **191** Photoprotoporphyrin B

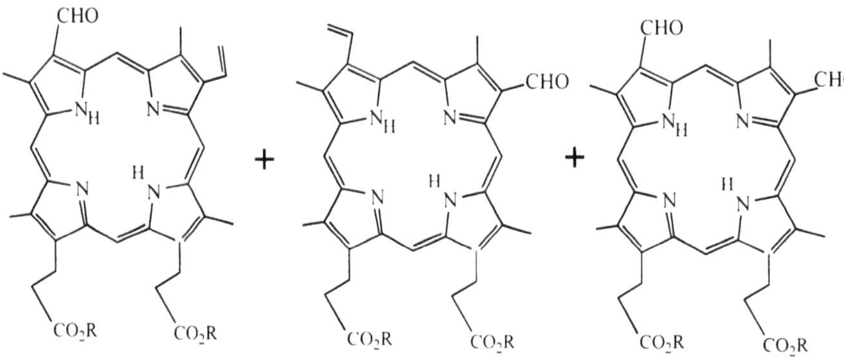

192 + **193** + **194**

FIGURE 12.3 PHOTOOXYGENATION PRODUCTS OF PROTOPORPHYRIN (**61**) AND ITS DIMETHYL ESTER. UNDER THE CONDITIONS SHOWN THERE IS A CONVERSION TO (**132** + **191**) OF 46%, TO (**192** + **193** + **194**) OF 1.4%, AND A RECOVERY OF STARTING MATERIAL OF 21% AFTER 2 H.

the porphyrin aldehydes, section 3.3.2). However in more organised media, such as SDS-H_2O micelles or DPPC-H_2O liposomes, the porphyrin aldehydes become the major products. It is thought that in these more structured systems, which indeed more closely resemble the *in vivo* situation, an intramicellar Type I process

$$\text{Por} \xrightarrow{h\nu} \text{Por}^{*} \xrightarrow{^{3}O_2} \text{Por}^{\bullet +} + O_2^{\bullet -}$$

becomes more important. In this connection, it should be borne in mind that there is a significant difference in oxygen solubility in aqueous and lipophilic environment (Panel 7.2, Table 7.2): in the lipid phase of the micelle, the liposome, or the lipid-rich cellular membrane, the concentrations of the transported hydrophobic sensitiser *and* of oxygen are increased.

Both (**132**) and (**191**) are chlorins, and have been used as starting materials for PDT sensitisers (eg **133**, section 9.4.3 and Figure 9.6; **190**, section 11.8).

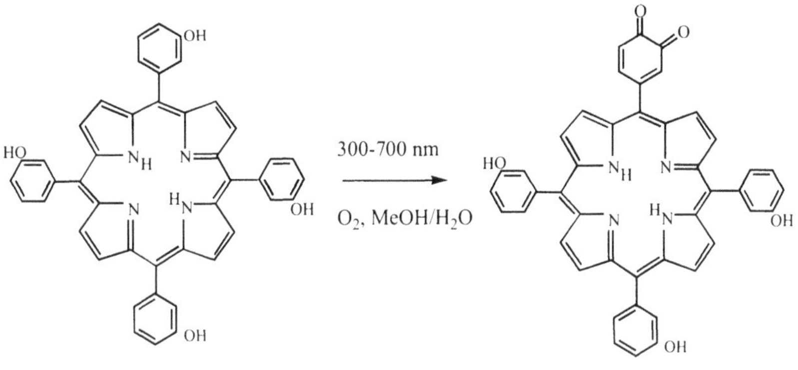

76 m-THPP

195 20%

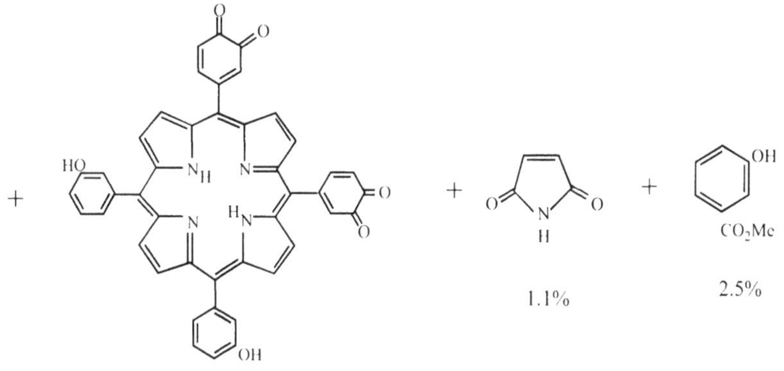

196 5%

FIGURE 12.4 PHOTOOXYGENATION OF M-THPP (**76**). (G. MARTINEZ, 1999).

12.4.1.2 m-THPP

Photooxygenation of m-THPP (**76**) in aqueous methanol (Figure 12.4) gives the mono- and di-*o*-quinones (**195**, **196**), both being less polar and less fluorescent than the starting material. The loss of fluorescence is ascribed to the deactivation of the S_1 state by rapid intramolecular electron transfer to the *o*-quinone.

It is possible that the catechols are formed initially, and are themselves oxidised to the *o*-quinones under the reaction conditions, although direct hydroxylation would suggest the intermediacy of hydroxy radicals. Alternatively, the formation of the *o*-quinone can be rationalised using a singlet oxygen mechanism, thus:

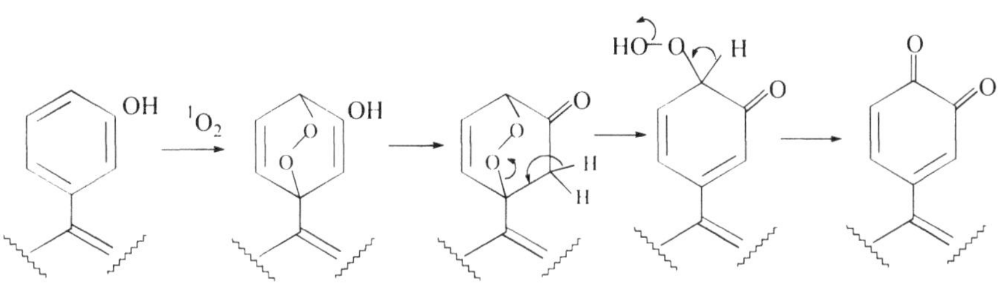

Linear tetrapyrroles resulting from cleavage at one *meso* bridge [analogous to (**197**) obtained on photooxygenation of zinc(II) tetraphenylporphyrin] have not as yet been identified. However, colourless products resulting from repeated *meso*-cleavages have been found (maleimide, methyl *m*-hydroxybenzoate) as shown in Figure 12.4. These fragments are most easily isolated when the photoxygenation is carried out in methanol.

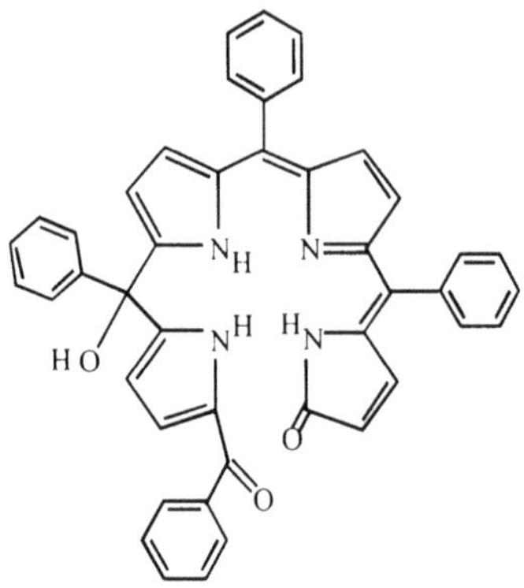

197

12.4.1.3 m-THPC

Although m-THPC (77) is rather readily susceptible to true photobleaching (Figure 12.2 and section 12.4.2.1) it has been possible to detect a variety of photomodification products after low intensity white light irradiation. The products, separated and identified by on-line HPLC-electrospray ionisation tandem mass spectrometry, have been formulated as shown in Figure 12.5. The on-line methodology is very powerful in the examination of small amounts of photoproducts in this way, but is not convenient for the isolation of the products. It seems possible that the compounds formulated as catechols in Figure 12.5 might actually be isolated as the *o*-quinones (as in Figure 12.4).

12.4.1.4 Purpurins

Photooxygenation of the purpurins considered in section 9.5.3 occurs by a $2\pi + 2\pi$ cycloaddition, to give the product of oxidative cleavage of the double bond in the exocyclic ring, thus:

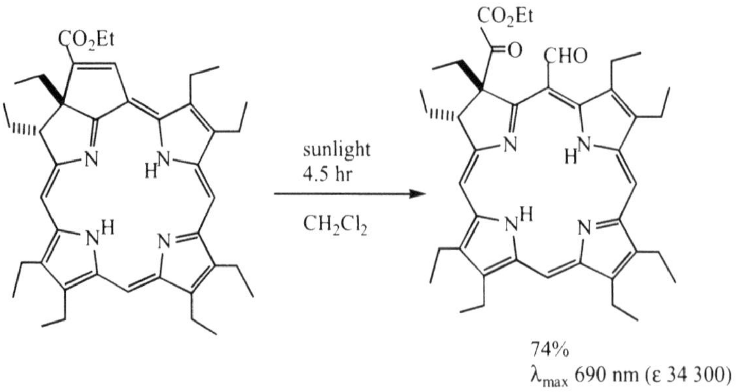

sunlight 4.5 hr, CH_2Cl_2

74%
λ_{max} 690 nm (ε 34 300)

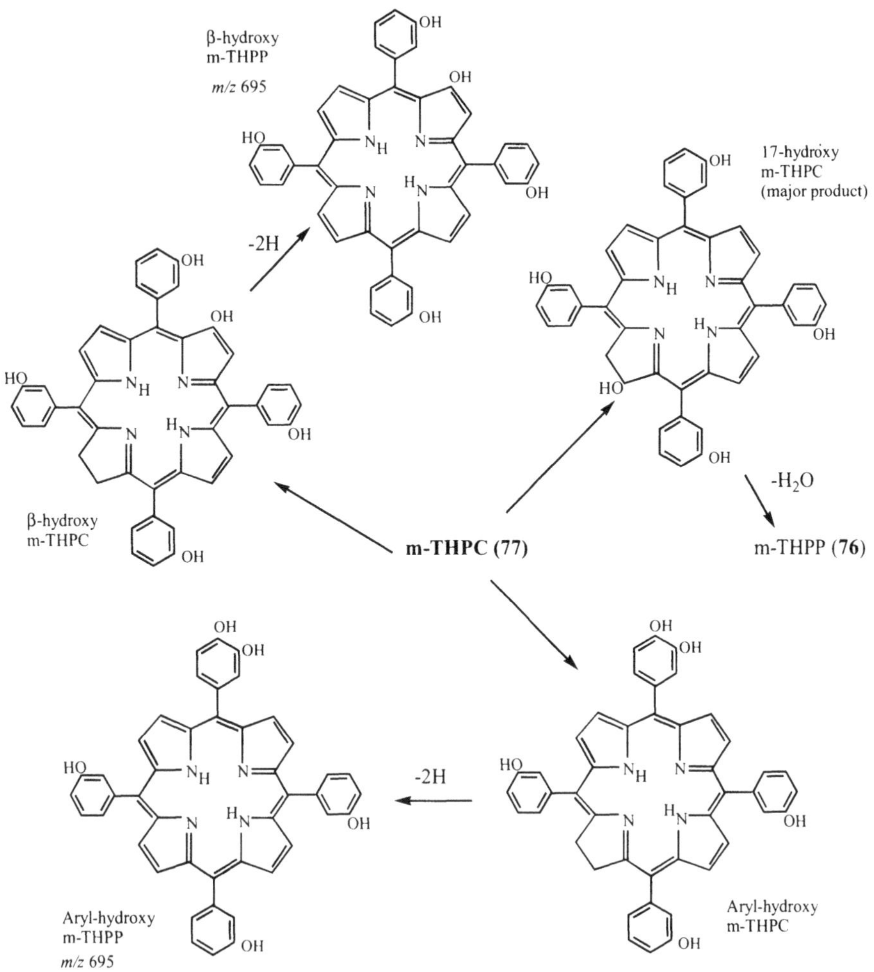

FIGURE 12.5 STRUCTURAL FORMULATION OF THE PRODUCTS, AND PROPOSED MAJOR PATHWAYS, IN THE PHOTOOXYGENATION OF M-THPC IN METHANOL UNDER MILD CONDITIONS (AMBIENT LABORATORY LIGHT, 1 WEEK). THE STRUCTURES REST PRIMARILY ON MS / MS INTERPRETATIONS. (R.M. JONES *ET AL.*, *J. CHROMATOGR. A*, 1996, **722**, 257–265).

Like the reaction leading to the starting purpurin (section 9.5.3), this reaction also finds precedent in Woodward's chlorophyll synthesis.

12.4.1.5 Haematoporphyrin

The above photomodifications are photooxidations, and require oxygen. It was at first thought that haematoporphyrin might eliminate water sufficiently readily under irradiation to form protoporphyrin, which would yield known products (Figure 12.3). The peak emerging at about 640 nm (Figure 12.6) would then be ascribed to photoprotoporphyrins. However, the reaction takes a quite different course.

Irradiation of haematoporphyrin (**2**) in PBS in a water-cooled reactor gives a mixture of diastereoisomeric chlorins. Preparative TLC separates three bands, but each band is still a mixture. The electronic spectrum of the most mobile band is

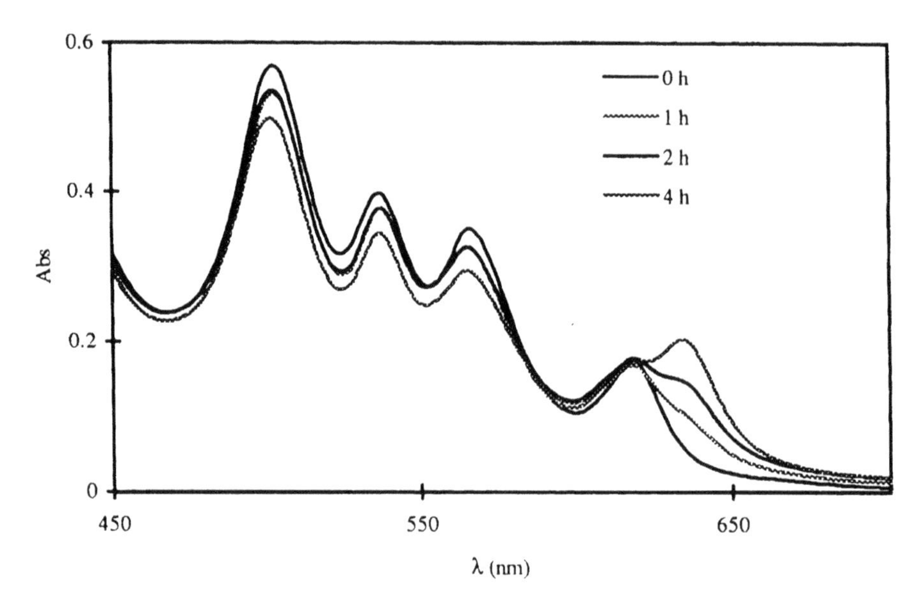

FIGURE 12.6 CHANGES IN THE VISIBLE SPECTRUM DURING THE IRRADIATION OF HAEMATOPORPHYRIN IN PHOSPHATE-BUFFERED SALINE AT pH 7.07. (PHILIPS MLU 300 W SOURCE).

shown in Figure 12.7 which also shows possible structures (**198**, **199**) of the products (NMR and MS evidence). There is also spectroscopic evidence from other studies for a product with λ_{max} 660 nm, and a product with λ_{max} 529 nm (possibly a linear tetrapyrrole).

The photoreduction of m-THPP (**76**) occurs in benzene with pyrrolidine as the hydrogen donor. The major product is m-THPC.

12.4.2 Photobleaching

The photomodification of sensitisers is usually accompanied by some photobleaching, especially under oxidative conditions: in many cases the products have not been identified.

12.4.2.1 m-THPC and m-THPBC

The rapidity of the photobleaching of these two compounds has been illustated in Figure 12.2. Product studies under preparative conditions have led to the isolation of colourless fragments (imides, benzoic acid derivatives) which arise by repeated cleavage at the *meso*-bridges (Table 12.2).

The dehydrogenation of m-THPBC to m-THPC is a minor process, but does occur.

12.4.2.2 Other photosensitisers related to porphyrins

It may be anticipated that other photobleaching processes occurring in the presence of air will also lead to imides or their hydrolysis products. Thus, phthalocyanines would be expected to give phthalimides, and so on.

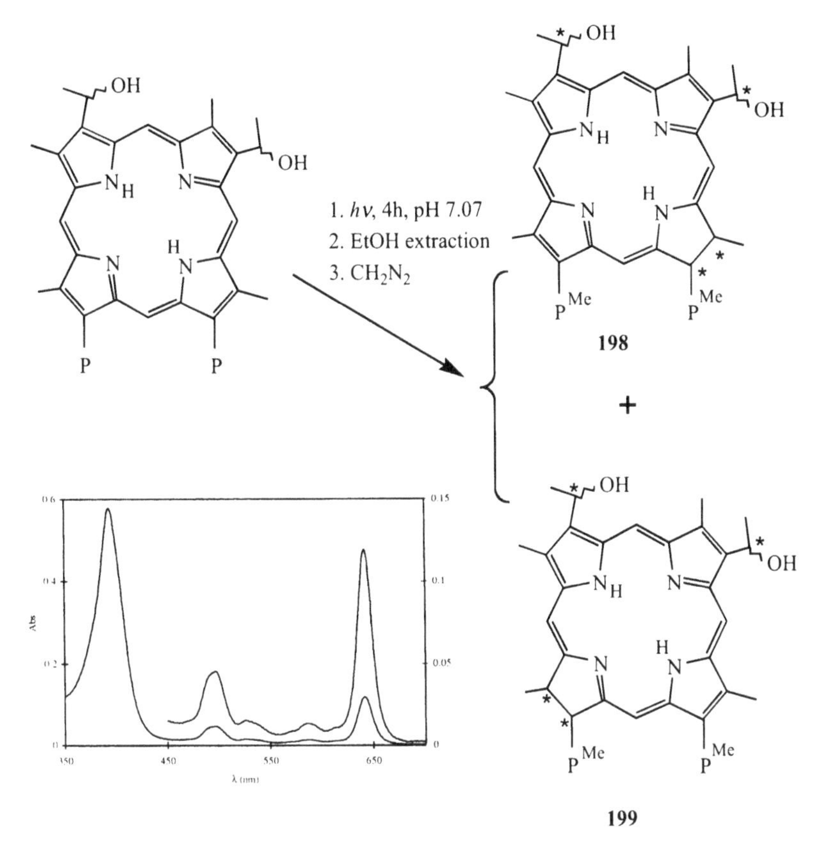

FIGURE 12.7 SPECTRUM OF CHLORIN ESTER (FRACTION 1, LEAST POLAR) IN CHLOROFORM.

Possible formulations of chlorin esters.
3 bands on TLC each a mixture of diastereoisomers. Total yield 8%
plus 28% recovery of haematoporphyrin dme.

TABLE 12.2 PRODUCTS OF THE PHOTOBLEACHING OF M-THPC (**77**) AND M-THPBC (**134**) (PHILIPS MLU 300 LAMP, WATER COOLED REACTOR).

Substrate	Reaction time (h)	Solvent	OH, CO_2Me	O N O H	O N O H	Other products identified
m-THPC	15	$MeOH-H_2O$ (3:2)	1.3%	1.2%	n.d.	Hydroxy m-THPC 2.3%
m-THPBC	5	MeOH	1.4%	n.d.	1%	m-THPC (trace)

12.5 BIOLOGICAL SYSTEMS

We have already seen (section 12.3) that, in the presence of biosystems (eg erythrocyte membranes, thioethers), the rate of photobleaching increases. Not surprisingly, then, photobleaching is found to be important in living tissues. However, certain factors need to be considered when the results are interpreted:

(i) the data often refer to loss of fluorescence. As mentioned earlier, this is an exquisitely sensitive mode of detection, but is particularly subject to quenching, for example by aggregation.
(ii) in some cases the loss of fluorescence has been shown not to parallel loss of absorption or loss of photocytotoxicity. For example, Belichenko and colleages reported in 1998 that, for m-THPC in an aqueous medium containing 2% fetal calf serum, the rate of photobleaching measured by loss of fluorescence was 15 times the rate measured by loss of absorbance.They also observed a lack of correlation between loss of fluorescence and phototoxicity towards human colon adenocarcinoma cells, attributing their results to preferential bleaching of the monomer and the marshalling of aggregated sensitiser reserves.
(iii) the sensitiser may move from one compartment to another on irradiation.

12.5.1 *In vitro* Studies

Protoporphyrin, generated from the pro-drug δ-ALA, is rapidly photobleached in cells: it is estimated that 70–95% of protoporphyrin in cells is degraded at clinically relevant light fluences (40–200 J cm^{-2} at 630 nm). A small amount of photoprotoporphyrin (**132**, **191**), which is also photolabile, is detected. Much of the protoporphyrin is protein bound: during irradiation the binding sites are destroyed, those close to tryptophan residues being the most sensitive (section 4.4.2.iii).

In Chinese hamster lung fibroblasts (V79 cells) the rate of photobleaching of m-THPC is greater than that of m-THPP and of Photofrin. Studies with a human carcinoma *in situ* cell line (NHIK 3025) show that m-THPP and Photofrin photobleach at about the same rate when studied separately. However, when present in the cells together, m-THPP is degraded 3–6 times more efficiently by photons absorbed by the fluorescencent (monomeric) fraction of m-THPP than by photons absorbed by the fluorescent fraction of Photofrin. Moan and Berg have used these experiments to estimate the distance which the reactive intermediate (presumably $^1O_2$) can diffuse: the value derived in this way turns out to be about 0.01–0.02 μm, corresponding to a lifetime of only 0.01–0.04 μs.

12.5.2 *In vivo* Studies

Photobleaching has been observed for Photofrin and protoporphyrin (δ-ALA treatment) in skin tumours in patients, and in experimental animals. For Photofrin photobleaching in mice with mammary carcinoma implants, porphyrin fluorescence was shown to parallel drug levels determined by extraction. The photobleaching of m-THPC and m-THPBC in mice has been reported.

Panel 12.1 Aggregation

1. In the present context *aggregation* is a solution phenomenon involving association between solute molecules, *but not involving covalent bonding*, to give assemblies (*aggregates*) of individual units (*monomers*). These assemblies may be dimers, trimers, oligomers etc, or mixtures of these. At the extreme, the aggregates may become so extensive that the solution becomes opalescent and particulates begin to precipitate.

The phenomenon arises when intermolecular interactions between solute molecules become energetically more favourable than the interactions between solute and solvent molecules. Aggregation may involve interaction between two or more solutes (hetero-aggregation), but most examples which have been studied involve self-aggregation.

2. The molecules are held together by weak intermolecular forces, which may include hydrogen bonding, electrostatic interactions including π-π interactions, van der Waals forces and hydrophobic interactions, depending on the detail of the structure. In simple porphyrins (and other aromatics), π-π interactions make a significant contribution to binding. According to Hunter and Sanders (1990), it is in fact π-σ interactions which cause attraction: these workers have proposed three ground rules:-

(i) π-π repulsion dominates in the face-to-face geometry (a),
(ii) π-σ attraction dominates in edge-on (b) and T-shaped (c) geometries, and
(iii) π-σ attraction dominates in an offset π-stack (d).

(a) **(b)** **(c)** **(d)**

Polarising substituents may stabilise the face-to-face arrangement (a); for example, 2,3,5,6-tetramethylbenzoquinone crystallises with a face-to-face geometry. The latter is also favoured by van der Waals and by hydrophobic interactions, since it maximises both.

In the porphyrin series a common aggregate structure is the offset π-stack (d) with a separation between the macrocyclic planes of 3.5–4Å. Typically in porphyrin structures, an electron-rich pyrrole ring is located above the electron-deficient hole (or metal ion) of the offset but parallel molecule, thus:

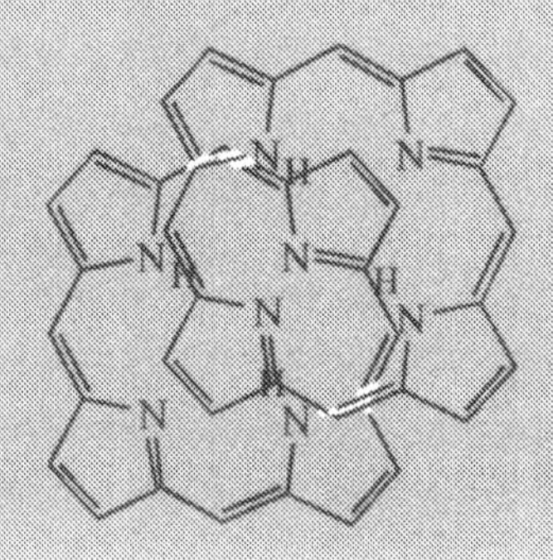

3. The formation of aggregates is a reversible process, and the equilibrium is affected by physical factors as shown below:

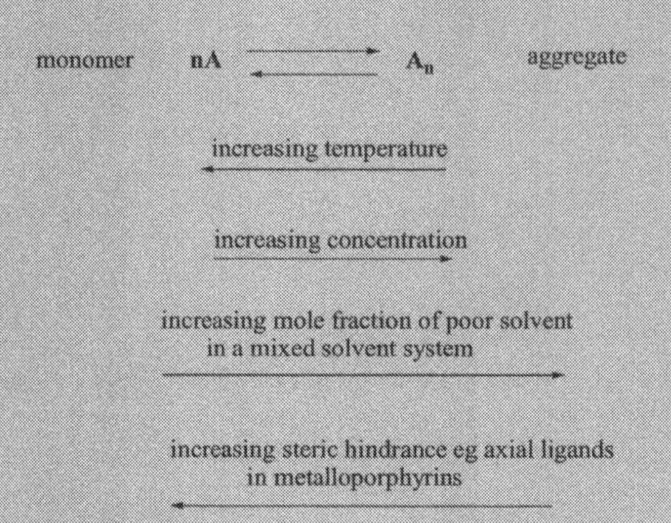

4. The electronic spectra of aggregates show rather complex behaviour, which was first noticed for cyanine dyes in the 1930s. Shifts of λ_{max} are observed and often (but not always — see below) the bands become broader (lower ε_{max}, larger width at half-height, $W_{1/2}$).

Kasha and his colleagues developed a point dipole model for exciton coupling between the transitions of the neighbouring chromophores. This has been subsequently modified, but still offers a useful qualitative approach. If the electronic transition energy in the monomer is represented by Figure 12.8 (a), then (b) represents the predicted situation for in-line transition dipoles in the dimer. Coupling between the dipoles results in two energy levels, and the allowed transition in this dimer is to the lower energy level, the transition to the higher level being forbidden for symmetry reasons. Hence in the dimer the band suffers a red (bathochromic) shift.

Conversely, for the parallel arrangement of transition dipoles, shown at (c), splitting again occurs, but it is the transition to the higher energy state that is allowed, and the dimer absorption shows a blue (hypsochromic) shift. Extensions of the treatment predict a diminution of fluorescence in the aggregate, but enhanced triplet excitation.

Figure 12.9 develops this idea by showing what is predicted to happen as the angle θ (between the coplanar transition dipoles and the axis interconnecting the units of the dimer or aggregate) is varied from 0° (in-line, side-by-side arrangement) to 90° (parallel, face-to-face arrangement). At one specific value of θ, (which turns out to be $54.7^\circ = \text{arc cos } 1/\sqrt{3}$) exciton splitting is zero: systems with lower angles θ than this (side-by-side in the extreme) are referred to as J; those with larger angles (face-to-face in the extreme) are called H.

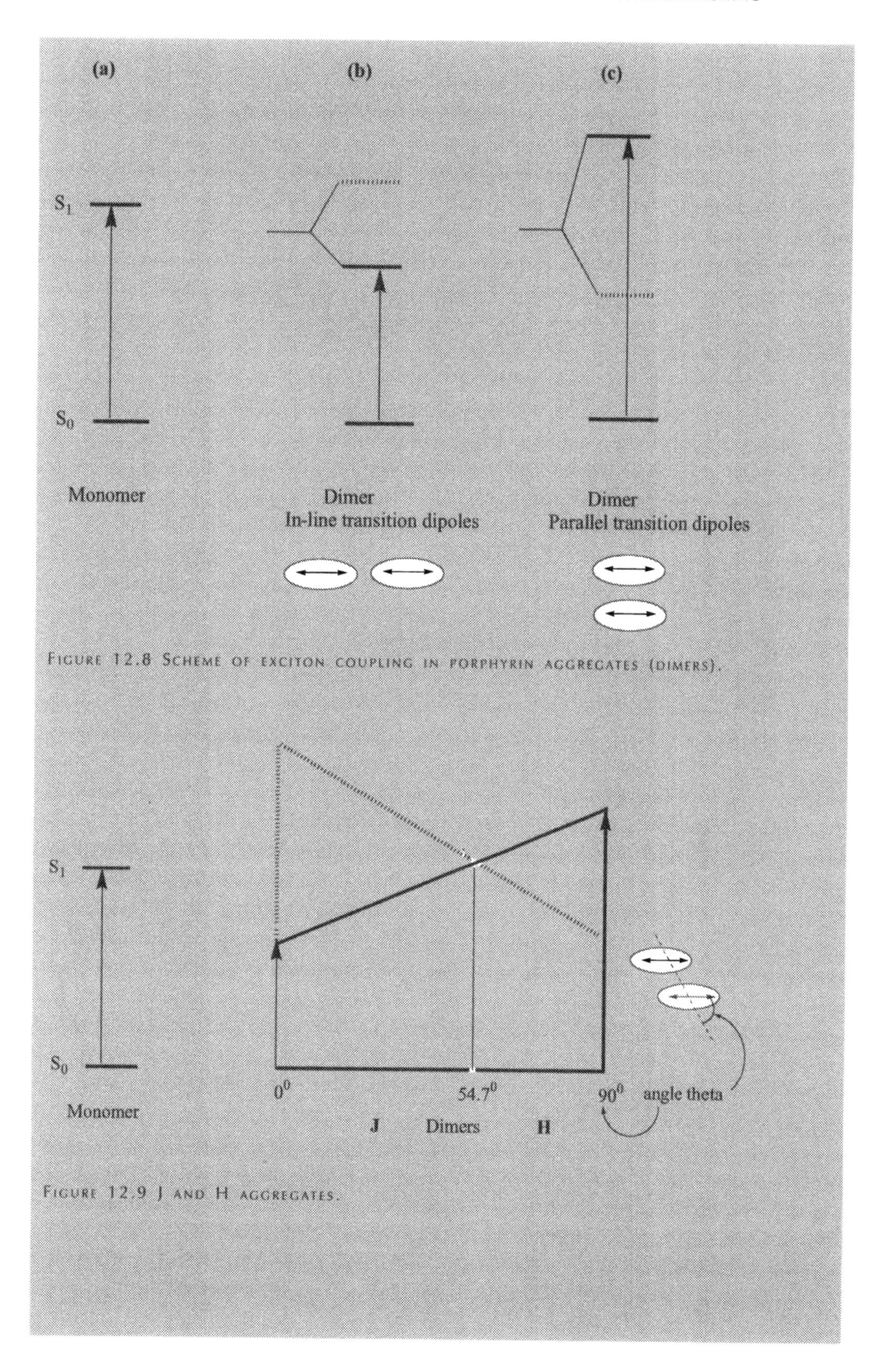

FIGURE 12.8 SCHEME OF EXCITON COUPLING IN PORPHYRIN AGGREGATES (DIMERS).

FIGURE 12.9 J AND H AGGREGATES.

RONIC

SPECTRA.

Aggregate Type	Schematic Structure	Electronic Spectrum with respect to monomer
J	θ Large offset $0^0 < \theta < 54.7^0$ when θ is small structure is ribbon-like	Red shift of Soret and Q bands. Either band sharpening or slight band broadening
H	Small or no offset face-to-face arrangement $90^0 > \theta > 54.7^0$	Blue shift of Soret and Q bands Usually marked broadening with decrease in ε and increase in $W_{1/2}$

Table 12.3 summarises the variety of effects of aggregation on the electronic spectra of porphyrins.

(The J and H designations appear to arise from *J*elley, who with Scheibe, described these phenonmena for cyanine dyes in the 1930s; and from the *h*ypsochromic shift which the H-systems show. E.E. Jelley, *Nature*, 1936, **138**, 1009; G. Scheibe, *Angew. Chem.*, 1936, **49**, 563).

5. Aggregation is important in PDT because it occurs readily between photosensitiser molecules (which are usually large aromatics) in water-rich media (such as cytoplasm), and the aggregate species has different properties from those of the monomer. In particular, the aggregate:

(i) has a different spectrum as described above;
(ii) has a lower quantum yield of fluorescence;
(iii) in spite of predicted triplet enhancement (4, above) is usually found to have lower a ϕ_Δ value. For example, haematoporphyrin is monomeric in methanol, and ϕ_Δ is 0.76: as water is added to the solution, ϕ_Δ decreases, and the value for the dimer is calculated to be 0.12; and

(iv) is less effective as a PDT sensitiser. It can, nevertheless, act as a reservoir of the monomer, with which it is equilibrium.

6. Aggregation may be detected by:

(i) deviations from the Beer-Lambert Law (section 2.3.1) with increasing concentration;
(ii) broadening of absorption peaks and lowering of ε with increasing concentration. This behaviour is characteristic of H aggregates, where it is accompanied by new absorption with a blue shift. J aggregates show new absorption with a red shift, and the new peak may be sharp in this case;
(iii) observing, in mixed solvents, whether or not the changes described in (ii) occur on increasing the proportion of the poorer solvent. It is important to bear in mind that simply changing the solvent will have some effect on the monomer spectrum (solvatochromic effect);
(iv) changes in δ values with concentration. ^{1}H-NMR studies are particularly useful in the study of porphyrin-type dimers because the ring current effect (Panel 8.1.3) is highly directional in the size and sign of the chemical shift changes which it causes. Consequently, specific aggregation geometry can often be deduced. The concentrations required for NMR studies are 2–3 orders of magnitude greater than those used in UV-VIS studies, which usefully extends the range of the observations.

7. Examples.

(i) m-THPC
Example of the effect of solvent composition at constant concentration

TABLE 12.4 SORET BAND PARAMETERS FOR M-THPC (77) IN AQUEOUS METHANOL AT 4.59 × 10^{-5} M, 20°C.

Solvent composition % H_2O in MeOH, v/v	λ_{max} nm	ε_{max} M^{-1} cm^{-1}	$W_{1/2}$ nm
0	414.4	189 000	32
10	415.6	193 000	32
30	415.6	231 000	34
65	421.6	117 000	46
68.75	425.2	57 000	68
73.75	428.8	45 000	80

(A. Nguyen, 1997)

Table 12.4 illustrates para 6(iii) above. In pure methanol m-THPC is monomeric in the concentation range 0.46–36.7 × 10^{-5} M. In aqueous methanol at 4.59 × 10^{-5} M it remains largely monomeric up to about 50% water v/v. Above this

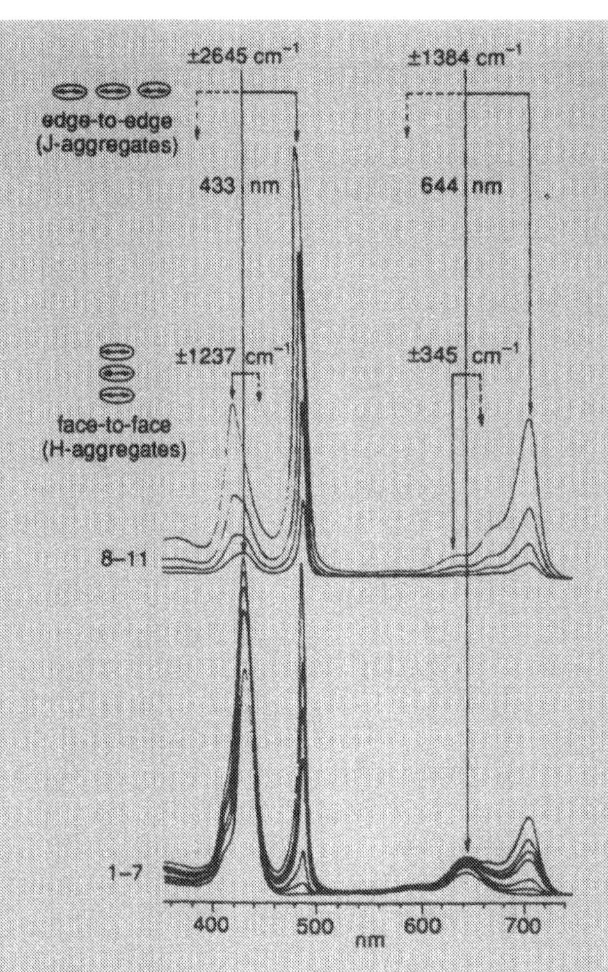

FIGURE 12.10 ELECTRONIC SPECTRA AT pH 3.5 (0.1 M ACETATE BUFFER) OF TETRASODIUM 5,10,15,20-TETRAKIS(*p*-SULPHONATOPHENYL)PORPHYRIN ($H_2TPPS_4^{4-}$) $4Na^+$ AT VARIOUS CONCENTRATIONS. CURVES 1–7 : CELL LENGTH = 0.1 CM; 4,5,6,7,8,9 AND 10 × 10^{-5}M. CURVES 8–11 : CELL LENGTH = 0.01 CM; 1.25,2.5,5 AND 10 × 10^{-4} M. (REPRODUCED FROM J.M. RIBÓ, J. CRUSATS, J.A. FARRERA AND M.L. VALERO, *J. CHEM. SOC., CHEM. COMMUN.*,1994, 681–682, WITH PERMISSION).

proportion the Soret band broadens markedly, λ_{max} decreases, and the band shows a bathochromic (red) shift. The red shift suggests that an aggregate of the J type is formed. At the higher water concentrations, solutions are metastable, and ε_{max} decreases on keeping the solution for some hours.

(ii) $TPPS_4$ (73)

Effect of increasing concentration at constant solvent composition

At pH 3.5 (acetate buffer, 0.1M) this substance exists as the tetrasodium salt of the diprotonated species, and shows a remarkable series of changes as the concentration is increased. Below 5×10^{-5} M the Soret band of the monomer appears at 433 nm. As the concentration is increased a new species appears with red shifted Soret and Q bands (Figure 12.10): in the Soret region a sharp band ($W_{1/2}$ = 8 nm) appears

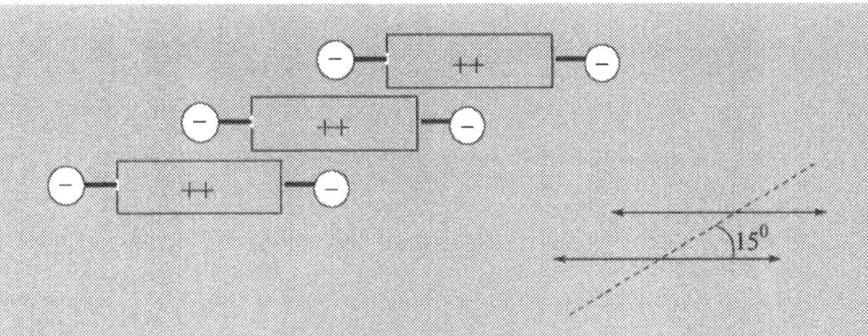

FIGURE 12.11 PROPOSED J AGGREGATE OF H_4^{++} $TPPS_4^-$ AT CONCENTRATIONS $>10^{-5}$ M AT pH 3.5.

at 489 nm, and hence the aggregate is regarded as of the J type. However as concentration is increased above 10^{-4} M, new *blue* shifted absorption appears (Figure 12.10). Thus in the Soret region, the 489 nm band is now accompanied by a broad band at 422 nm.

Ribó and his colleages interpret the J type aggregate as being of the form shown in Figure 12.11, where the zwitterion $H_4^{2+}TPPS_4^{4-}$ is represented in a diagrammatic way, and where $\theta \approx 15°$.

The second change, which has the character of H aggregation, is then interpreted in terms of the face-to-face aggregation of the ribbon-like J aggregates, and is accompanied by the development of gel-like rheological behaviour.

This example is of interest since it shows that aggregation can occur in spite of net ionic repulsions (since each H_4^{2+} $TPPS_4^{4-}$ bears two net negative charges). The aggregation is subject to steric hindrance: it does not occur with the tetrakis(2,6-dimethylphenyl) analogues.

(iii) $AlPcS_n$

Aggregation and species diversity (R. Edrei, V.Gottfried, J.E. van Lier and S. Kimel, *J. Porphyrins Phthalocyanines*, 1998, **2**, 191–199).

In preparations of sulphonated aluminium phthalocyanines, it is reported that materials with the smallest number of isomeric species showed the greatest tendency to aggregation. Complex mixtures, it seems, are more likely to be monomeric in an aqueous environment. Cellular uptake parallels hydrophobicity of the preparation and decreases as aggregation increases. Of the purified disulphonic acids (see **74**), $AlPcS_{2a}$ shows the highest cell uptake, but an even higher cell uptake is observed for the unseparated mixture of disuphonic acids. The interactions occurring in these complex mixtures are difficult to disentangle and are not yet fully understood. It is probably premature to conclude that mixtures will give the most favourable photodynamic result.

BIBLIOGRAPHY

Photobleaching

R.S.Sinclair, 'The light stability and photodegradation of dyes', *Photochem. Photobiol.*, 1980, **31**, 627–629.

J.D. Spikes, 'Quantum yields and kinetics of the photobleaching of haematoporphyrin, Photofrin II, tetra(4-sufonatophenyl)porphyrin and uroporphyrin', *Photochem. Photobiol.*,1992, **55**, 797–808.

H.H.Inhoffen, H. Brockmann and K.M. Bliesener, 'Photoprotoporphyrins and their transformation into spirographis- and isospirographis- porphyrins', *Liebig's Ann. Chem.*, 1969, **730**, 173–185.

G.S. Cox and D.G.Whitten, 'Mechanisms for the photooxidation of protoporphyrin in solution', *J. Am. Chem. Soc.*, 1982, **104**, 516–521.

G.S. Cox, M. Krieg and D.G.Whitten, 'Self-sensitised photooxidation of protoporphyrin IX derivatives in aqueous surfactant solutions: product and mechanistic studies', *J. Am. Chem. Soc.*, 1982, **104**, 6930–6937.

R. Rotomskis, S. Bagdonas and G. Streckyte, 'Spectroscopic studies of photobleaching and photoproduct formation of porphyrins used in tumour phototherapy', *J. Photochem. Photobiol.,B:Biol.*, 1996, **33**, 61–67.

I. Belitchenko, V. Melnikova, L. Bezdetnaya, H. Rezzoug, J.L. Merlin, A. Potapenko and F. Guillemin, 'Characterisation of photodegradation of meta tetra(hydroxyphenyl)chlorin (mTHPC) in solution: biological consequences in human tumour cells', *Photochem. Photobiol.*, 1998, **67**, 584–590.

J. Moan, G.Streckyte, S. Bagdonas, O. Bech and K. Berg, 'Photobleaching of protoporphyrin IX in cells incubated with 5-aminolevulinic acid', *Int. J. Cancer,* 1997, **70**, 90-97.

J. Moan and K. Berg, 'The photodegradation of porphyrins in cells can be used to estimate the lifetime of singlet oxygen', *Photochem. Photobiol.*, 1991, **53**, 549–553.

T.S. Mang, T.J. Dougherty, W.R. Potter, D.G. Boyle, S. Somer and J. Moan, 'Photobleaching of porphyrins used in photodynamic therapy and implications for therapy', *Photochem. Photobiol.*, 1987, **45**, 501–506.

Aggregation

W.I. White, 'Aggregation of porphyrins and metalloporphyrins', in *The Porphyrins* (ed D. Dolphin), Academic Press, New York, 1978. Vol. V, pp.303–339.

C. A. Hunter and J.K.M. Sanders,'The nature of π-π interactions', *J. Am. Chem. Soc.*, 1990, **112**, 5525–5534.

M. Kasha, H.R. Rawls and M.A. El-Bayoumi, 'The exciton model in molecular spectroscopy', *Pure Appl. Chem.*, 1965, **11**, 371–392.

C.A. Hunter, J.K.M. Sanders and A.J. Stone, 'Exciton coupling in porphyrin dimers', *Chem. Phys.*, 1989, **133**, 395–404.

CHAPTER 13

BIOLOGICAL ASPECTS

In earlier chapters reference has been made as occasion merited to a variety of biological procedures and results. In this chapter some important biological aspects will be summarised, and some account given of the localisation of sensitisers in biological systems.

13.1 BIOLOGICAL ASSAY

The overall aim of photosensitiser bioassay in the present context is to arrive at a substance which is suitable for clinical PDT. This is a complex process, and there is no one way of doing it. But it eventually requires assays using a mammalian model, usually a rodent. It can be argued that such results on animal models do not apply closely to the human animal; this is true, but this approach is the closest we can get with the current state of knowledge. Certainly using a model with functioning organs and a vascular system is more likely to be successful in its outcome than experiments in a cell culture, which has none of these things.

Nevertheless, a number of preliminary screens must be used to weed out unpromising photosensitisers, and so reduce animal experiments.

13.1.1 Preliminary Physical Tests

Two measurements on potential PDT photosensitisers can be made: they are not decisive in themselves, but since the experiments are relatively easy to perform, they should be done earlier rather later.

(i) *Partition Coefficient.* (Panel 7.1) The sensitiser should be amphiphilic: lipophilic enough to enter membranes, but with sufficient polar character to aid solubilisation for injection (section 7.2.4 and Figure 7.3).

(ii) *Singlet oxygen quantum yields* (ϕ_Δ) can be measured either from the luminescence at 1270 nm (section 4.4.3.2ii), or from the rate of photobleaching of 1,3-diphenylisobenzofuran (section 3.3.3).

An alternative approach, which does not refer only to Type II processes, is to use a physical method (eg an oxygen electrode) to measure the rate of oxygen depletion in the presence of a photooxidisable substrate.

The λ_{max} value of a sensitiser changes slightly in a biological system: usually a small bathochromic shift is observed. As a preliminary, spectra are usually recorded in foetal calf serum in order to determine the most advantageous wavelength at which to irradiate *in vitro* and *in vivo.*

13.1.2 *In vitro* Bioassays

This refers to assays with cells in culture, the cells usually being derived from a tumour source. Different workers adopt different cell lines and different procedures. It would be a valuable and progressive step if an agreed standard cell culture and irradiation protocol were adopted as a point of reference between laboratories, not to displace the variety of cell lines used (which is valuable, since cancer is not a single disease), but to allow a better correlation of results.

Prior to irradiation, two experiments are needed. *First,* a study of acute dark toxicity, for which normal cells can be used. If all the cells are killed in the dark, the photosensitiser is eliminated, at least for PDT applications. *Second,* a study of photosensitiser uptake by the tumour cells, using extraction followed by fluorescence or absorption measurements.

After irradiation, it is the photochemically-induced cell kill that is being determined against appropriate dark controls. Cell kill may be determined either by the staining of the dead cells with a dyestuff (eg with propidium iodide **200**), followed by counting under a microscope; or by dilution and sub-culture, and counting the new colonies (clonogenic assay).

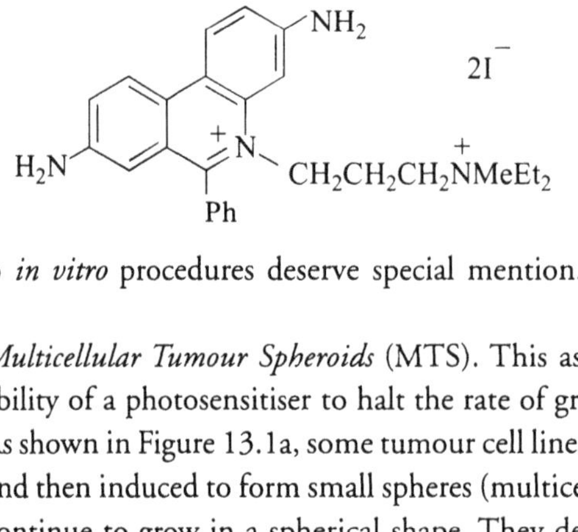

200 Propidium iodide (λ_{max} 493 nm)

Two *in vitro* procedures deserve special mention.

(i) *Multicellular Tumour Spheroids* (MTS). This assay depends on measuring the ability of a photosensitiser to halt the rate of growth of a model microtumour. As shown in Figure 13.1a, some tumour cell lines can be cultured as monolayers, and then induced to form small spheres (multicellular tumour spheroids) which continue to grow in a spherical shape. They develop a necrotic centre, and in this respect are thought to be good models for much larger tumours (Figure 13.1b). Such spheroids are placed in growth medium in multi-well plates, and volume is measured with time in the presence and absence of photosensitiser and light. Activity is indicated by a diminution in the rate of growth (estimated as spherical volume) of the spheroids compared with controls (Figure 13.1c).

(ii) *MTT Assay.* MTT stands for 3-(4,5-dimethylthiazol-2-yl)-2,5-diphenyl-2*H*-tetrazolium bromide (**201**): metabolically active mitochondria reduce MTT, which is yellowish, to the corresponding formazan (**202**), which is blue.

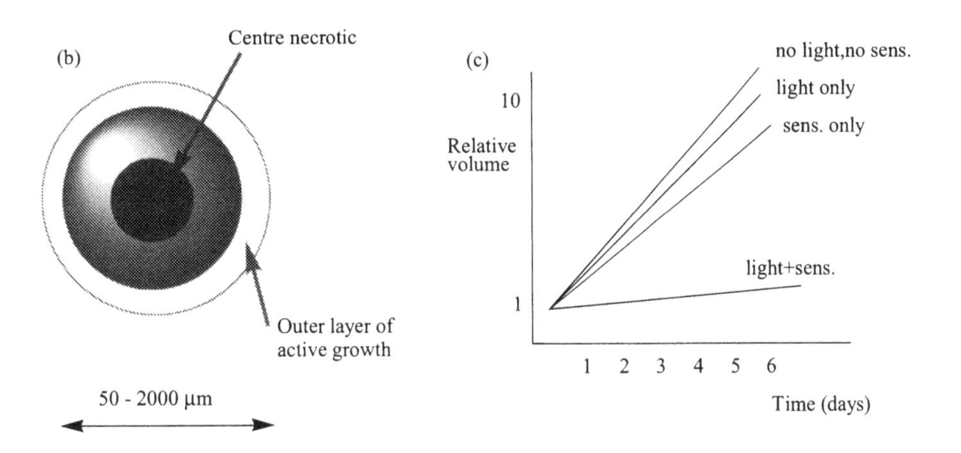

FIGURE 13.1 MULTICELLULAR TUMOUR SPHEROIDS (MTS); (A) PREPARATION, (B) SCHEMATIC CROSS SECTION, (C) SCHEMATIC OF THE EFFECT OF PHOTOSENSITISER AND LIGHT IN HALTING THE RATE OF GROWTH.

$2e, H^+$

201 MTT ⇌ **202**

If it is accepted that metabolically active mitochondria signify a living cell, then this offers a convenient and rapid spectroscopic method for estimating cell death. As more cells are photonecrosed, the amount of formazan decreases.

13.1.3 Intermediate Bioassays

The chick embryo assay, which is still being developed, falls into an intermediate category, since the developing embryo is regarded as non-sentient in regulatory terms up to half-hatching time (10–11 days). Consequently, observations are made 6–10 days after the commencement of incubation.

As shown schematically in Figure 13.2, the tumour cells are placed in an inert retaining ring on the surface of the chorioallantoic membrane (CAM, hence CAM assay). This is highly vascularised (because it is the oxygen supply route from the air sac to the embryo), and the vascularisation quickly invades the developing neoplasm. The sensitiser is administered to the tumour or to the embryo, the tumour is irradiated from above, and photodamage to the tumour is monitored.

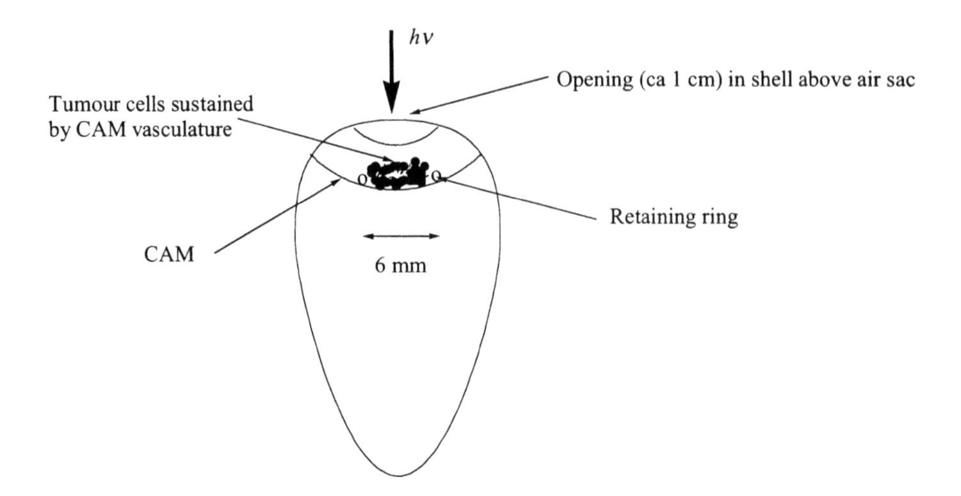

FIGURE 13.2 SCHEMATIC FOR THE CAM ASSAY OF PHOTOSENSITISERS. IN ORDER TO PREVENT THE CAM FROM DRYING OUT, THE OPENING AT THE TOP OF THE AIR SAC IS NORMALLY COVERED WITH PARAFILM.

13.1.4 *In vivo* Bioassays

13.1.4.1 Tumour photonecrotic activity

Assays involving animals are subject to strict regulatory control of both the scientist and the laboratory. Since the results are subject to biological variation (more so than with *in vitro* bioassays) it is necessary to have several animals in each group in order to give a reliable mean value.

The model tumour may be produced by induction (for example, by ionising radiation, by UV-B, or by application of a carcinogen such as dimethylnitrosamine or 2-formamido-4-(5-nitro-2-furyl)thiazole, FANFT, **203**; *induced tumour*); or by subcutaneous or intramuscular transplantation of cells (*ca.* 10^6) from an established tumour cell line (*implanted tumour, xenograft*). Animals are selected with tumours of approximately the same size, and treated with the sensitiser (i.v.,i.p., or for δ-ALA topically or orally). Significant variables are the drug-light interval, the drug dose, and the light fluence (but see also Table 7.1).

O_2N NHCHO

203

N-[4-(5-Nitro-2-furanyl)-2-thiazolyl]formamide, 'FANFT']

NaO_3S NH_2 OH N=N N=N OH NH_2 SO_3Na SO_3Na SO_3Na

204 Evan's blue (λ_{max} 611 nm)

Tumour diminution may be measured in a variety of ways. One way is to measure the decrease in the rate of growth of the tumour, by estimating tumour volume (*eg*

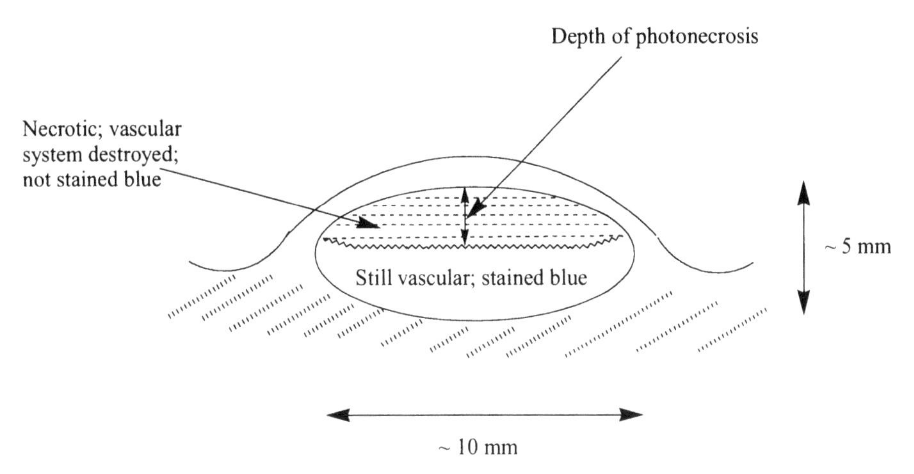

FIGURE 13.3 DIAGRAM OF FIXED SECTION OF EXCISED TUMOUR AFTER PHOTODYNAMIC TREATMENT *IN VIVO* FOLLOWED BY EVAN'S BLUE INJECTION. (BASED ON M.C. BERENBAUM, R. BONNETT AND P.A. SCOURIDES, *BR. J. CANCER,* 1982, **45**, 571–581).

with callipers, by excision and weighing, or by total animal weight) at given times after irradiation. The growth rate changes used in constructing Figure 7.3 were obtained from tumour volume measurements made using electronic callipers. Another method involves injecting Evan's blue (a harmless water soluble azo dye (**204**) which rapidly pervades the vascular system) 24 hours after irradiation. After a further hour to allow dissemination of the dye, the animal is sacrificed, and the depth in mm of unstained tissue (*ie* dead tumour) in a vertical fixed section of the tumour (Figure 13.3) is measured using a stereoscopic microscope fitted with a micrometer eyepiece. The depths of tumour necrosis given in Table 9.1 were obtained in this way.

13.1.4.2 Tumour selectivity

Thus, the first hurdle in these assays is to find *activity*, that is to find a sensitiser of low dark toxicity which efficiently causes tumour photodestruction. The second hurdle is *selectivity:* of such active substances, which ones, if any, selectively photodamage tumour tissue with respect to normal tissue? There are two objectives here: (i) to diminish damage to normal peritumour tissue, since the area surrounding the tumour will inevitably receive some irradiation (*eg* by scattering) during phototherapy, and (ii) to diminish long term general photosensitivity.

Selectivity can be assessed by subjecting normal tissue in a sensitised tumour-bearing animal to the same light conditions (drug-light interval, fluence, fluence-rate) as the tumour. Thus, if the tumour implant is on the right hind leg, the left hind leg of the same animal can be employed for selectivity studies with respect to skin and muscle. For example, the chlorinylglucose (**142,** section 9.5.2) is found to be an excellent tumour photonecrotic agent, but it is not sufficiently selective, since both skin and muscle show strong sensitisation.

It is possible to establish simple indices for damage to normal and tumour tissues, and to plot dose-response curves in the usual way. Figure 13.4 shows some relationships of this sort for m-THPP (**76**).

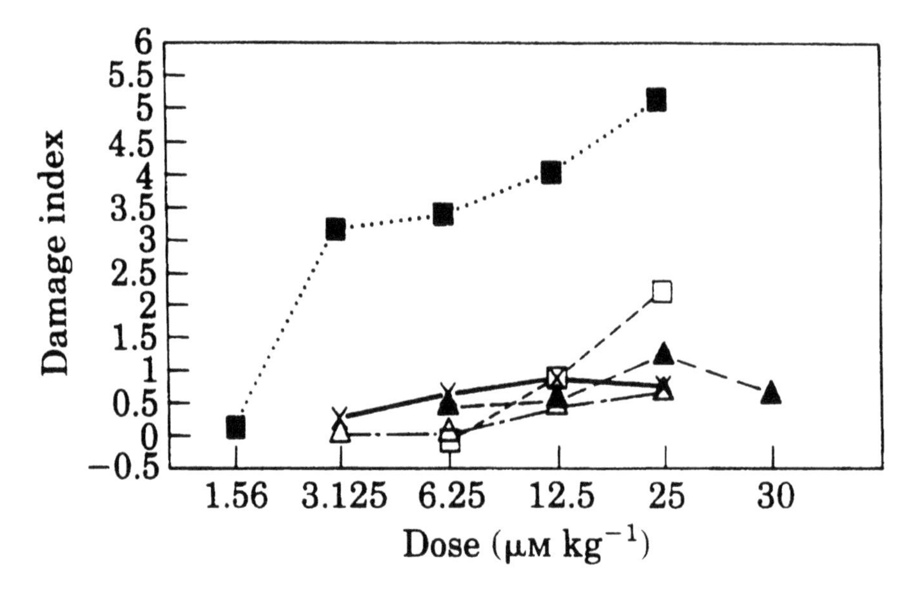

FIGURE 13.4 DOSE-RESPONSE RELATIONSHIPS FOR M-THPP (**76**) AT CONSTANT FLUENCE (10 J CM^{-2}).
KEY: ■, TUMOUR NECROSIS; □, SKIN OEDEMA; Δ, MUSCLE OEDEMA; ×, MUSCLE NECROSIS; AND, ▲, BLADDER OEDEMA. DAMAGE INDICES: TUMOUR, DEPTH OF NECROSIS IN MM; OEDEMA, FRACTIONAL INCREASE IN WEIGHT 4H AFTER ILLUMINATION WITH RESPECT TO UNILLUMINATED CONTROL; MUSCLE NECROSIS, AREA OF NECROTIC TISSUE AS A FRACTION OF THE WHOLE. (REPRINTED WITH PERMISSION FROM M.C. BERENBAUM, R. BONNETT, E.B. CHEVRETTON, S.L. AKANDE-ADEBAKIN AND M. RUSTON, *LASERS MED. SCI.*, 1993, **8**, 235–243).

Berenbaum developed this by applying cost-benefit analysis to an assessment of selectivity. In a general way it can be seen that if cost as ordinate is plotted against benefit as abscissa (Figure 13.5), then desirable outcomes will occupy the bottom right hand portion of the field.

In the present case the benefit is clearly tumour photonecrosis, and the cost is damage to normal tissues, and we have in the damage indices the basis of a quantitative treatment. If, for example, for five sensitisers (Photofrin, m-THPP, p-THPP, m-THPC, p-THPC) skin oedema (cost) is plotted against tumour photonecrosis (benefit), Figure 13.6 emerges, where each point represents cost and benefit at a given sensitiser dose, with the light fluence being kept constant. It is clear that, at certain concentrations, m-THPC, p-THPC and to a marginal extent, m-THPP fall into the favourable bottom right hand box, and hence are superior in this analysis to the other two sensitisers which do not fall into the 'sufficient benefit at reasonable cost" box at all.

13.2 LOCALISATION

The exact location of the photosensitiser is important because the reactive oxygen species is short lived. Singlet oxygen in a biological environment is estimated to have a lifetime of about 0.01 μs and can diffuse about 0.01 μm (section 12.5.1): so reactions can only be initiated quite close to the place where the sensitiser molecule

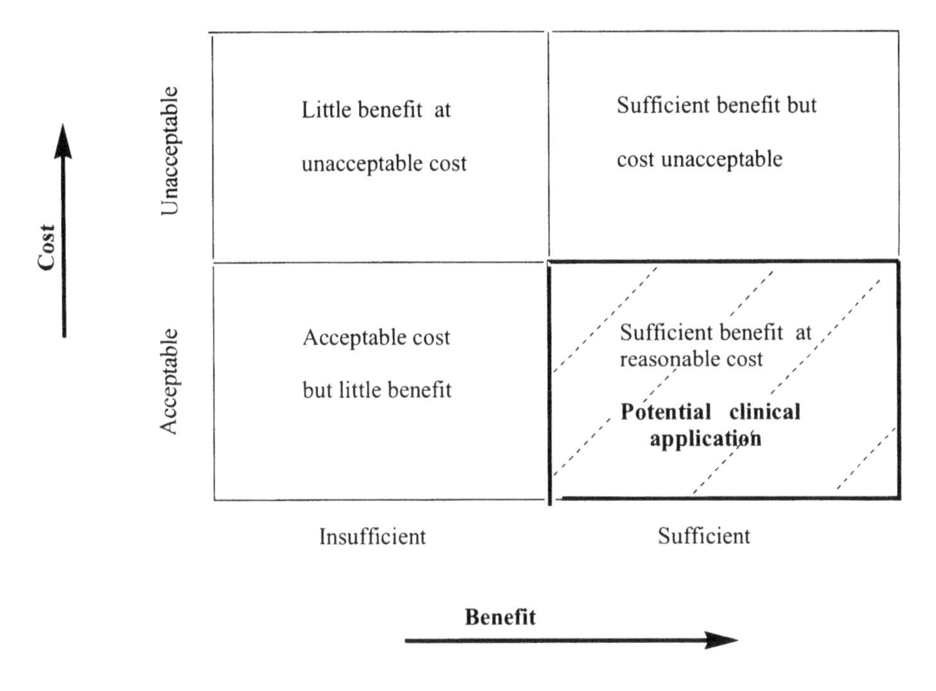

FIGURE 13.5 GENERALISED DIAGRAM OF COST-BENEFIT ANALYSIS.

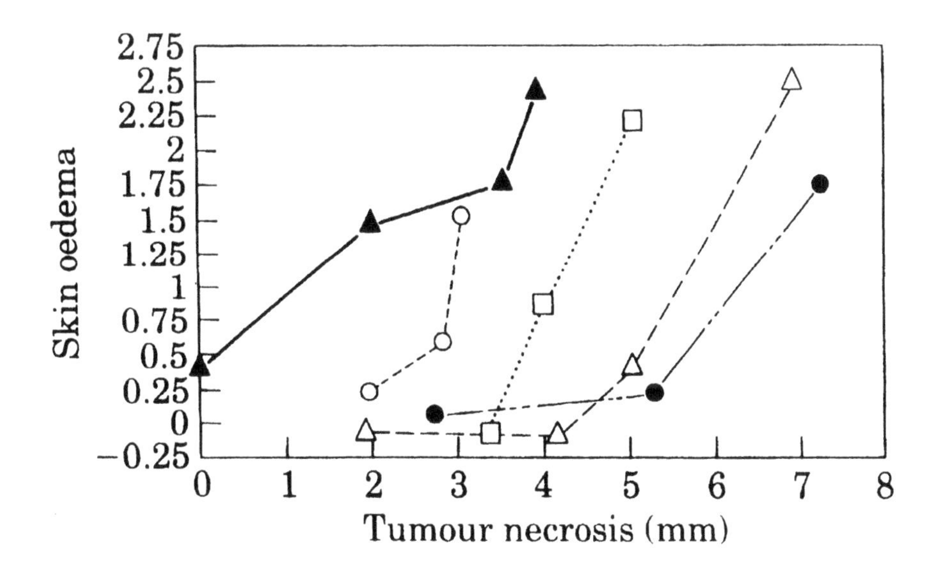

FIGURE 13.6 COST-BENEFIT ANALYSIS SHOWING COST IN SKIN OEDEMA FOR GIVEN DEPTHS OF TUMOUR NECROSIS. FLUENCE 10 J cm^{-2}.
KEY: ▲, PHOTOFRIN; ○, p-THPP; □, m-THPP; Δ, p-THPC; •, m-THPC. (REPRODUCED WITH PERMISSION FROM M.C. BERENBAUM *ET AL.*, *LASERS MED. SCI.*, 1993, **8**, 235–243).

is located. For radical chain reactions the sphere of reaction would be anticipated to be larger, but still within the vicinity of the sensitiser molecule.

The pharmacokinetics of the sensitiser are described by its concentration with respect to time and location in the biological system and are important in another way. What is wanted ideally is a sensitiser that will get in quickly, do its phototherapeutical task, and get out quickly so that residual photosensitisation is minimal. This clearly relates to the concentration of the sensitiser in various tissues ie to localisation.

The presence of a photophysically satisfactory sensitiser in adequate amount is a necessary condition in a biosystem for PDT activity, but it does guarantee it. This arises because different drugs occupy different compartments, or the same drug may aggregate to varying extents in different compartments or in different cell lines. Thus, referring to the mixed naphthalo / phthalo cyanine sulphonic acids described in Panel 10.2.1, experiments in Chinese hamster lung fibroblasts (V-79) show that the disulphonated derivatives are slightly more photoactive than the trisulphonic acid derivatives, whereas the monosulphonic acid is inactive, in spite of a six-fold higher cell uptake.

13.2.1 Protein Binding

Intravenously injected sensitisers are transported by plasma proteins in the blood stream. Many photosensitisers tend to undergo aggregation in aqueous media (Panel 12.1), but, in the presence of protein, monomerisation occurs and it is the (fluorescent) monomer which is generally carried by the protein. Thus, zinc phthalocyanine derivatives substituted with neutral [$SO_2N(C_8H_{17})_2$], cationic ($CH_2NC_5H_5^+$) or anionic ($SO_2NHCH_2CO_2^-$) groups in the benzenoid rings are highly aggregated in aqueous solution but monomeric in the presence of bovine serum albumin (or in methanol). The situation with HpD is more complex because it is a mixture of monomers and covalent oligomers (Chapter 6), each of which may aggregate and disaggregate.

In general, it seems that the more hydrophobic sensitisers (eg ZnPc) associate more with the lipoproteins (HDL, LDL, VLDL: Panel 7.2) while the hydrophilic sensitisers (eg $AlPcS_4$, $TPPS_4$) associate more with serum albumin and other high molecular weight proteins.

However, the distribution between proteins is sensitive to the detail of sensitiser structure in a complex way, and is affected by the use of a delivery agent such as Cremophor or dimethyl sulphoxide. Compounds such as uroporphyrin, which are freely soluble in water, and which have little uptake by serum proteins or membranes, are generally found to be poor tumour photosensitisers.

13.2.2 Intracellular Localisation

Fluorescence microscopy is an important technique here, but results need cautious interpretation.

Membranes have been identified as an important intracellular location, and contain molecules which readily react with singlet oxygen (section 4.4.2): such membranes include the plasma membrane surrounding the cell, the membranes of the endoplasmic reticulum distributed in the cytoplasm, and the membranes of organelles (eg mitrochondria, Golgi apparatus, nucleus). Penetration of the membranes is also expected to occur, depending on the ionicity of the sensitiser.

The primary target in the cell depends on the structure of the sensitiser. As shown in Table 13.1 several of the compounds go initially to the lysosomes, which are vesicles containing hydrolase enzymes. Irradiation causes rupture of the lysosome envelope and redistribution of the photosensitiser. Since the photosensitisers ($AlPcS_4$, $TPPS_4$) tend to aggregate in the lysosome, there is a considerable enhancement of fluorescence when they are diluted by, and disaggregate in, the cytoplasm. At the same time the distribution of hydrolase activity contributes to the destruction of the cells.

This effect has also been observed *in vivo* when mice given $AlPcS_2$ and $AlPcS_4$ i.p. are irradiated at various intervals: after injection the intensity of fluorescence increases twofold during the first few minutes of irradiation. This is interpreted in terms of initial localisation in lysosomes, followed on irradiation by singlet oxygen rupture of the lysosomal membrane, and release of monomeric photosensitiser of enhanced ϕ_f.

However, there is considerable variation. For example tolyporphyrin (**205**), a dioxobacteriochlorin isolated from the blue-green alga (cyanobacterium) *Tolypothrix nodosa*, is a potent sensitiser of the photodestruction of murine mammary tumour cells (EMT-6) both *in vitro* and *in vivo*. Intracellular localisation of the photosensitiser occurs not in the lysosomes or mitochondria, but in the perinuclear region and particularly in the endoplasmic reticulum.

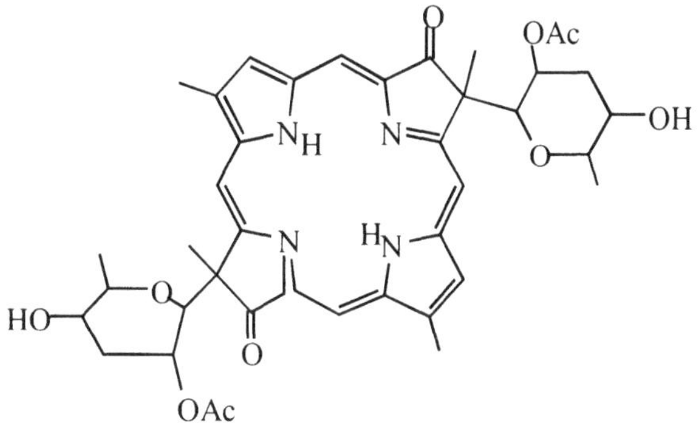

205 Tolyporphin
λ_{max} 676 nm (ε 68 600)

(Stereochemistry: see T.G. Minehan *et al.*, *Angew. Chem. Intl. Edn.*, 1999, **38**, 926)

13.2.3 Localisation *in vivo*

Far more methods are available for examining localisation in whole animals. Here each biological sample can be readily separated by dissection and is large enough for the sensitiser to be extracted. Absorption and fluorescence (on unlabelled

TABLE 13.1 SOME EXAMPLES OF INTRACELLULAR LOCALISATION IN TUMOUR CELL CULTURES.

Photosensitiser	Cell line	Major targets	Comment	Ref
1. HpD / Photofrin	Various	Plasma membrane (Lysosomes) Mitochondria	Most extensively studied system	*a*
2. Protoporphyrin (from δ-ALA)	HeLa	Mitochondria	Strong PD effect	*b*
3. Protoporphyrin (exogenous)	HeLa	Plasma membrane	Weak PD effect	*b*
4. Tolyporphin	EMT-6	Endoplasmic reticulum – Golgi, nuclear membrane	Hydrophilic because of sugar residues	*c*
5. $TPPS_1$	Murine colon carcinoma	Golgi-like complex, nuclear membrane, smaller vesicles	Does not redistribute on irradiation	*d*
6. $TPPS_4$	Murine colon carcinoma; Human cervix carcinoma NHIK 3025.	Lysosomes	On irradiation, some relocation to nucleus, nucleoli, cytoplasm.	*d*
7. ZnPc (cationic substituent)	RIF-1	Lysosomes	Most active of 7-9 in uptake and cell kill; on irradiation, fluorescence moves to cytoplasm and nucleoli	*e*
8. ZnPc (anionic substituent)	RIF-1	Lysosomes	Relocates on irradiation → nucleus → plasma membrane	*e*
9. ZnPc (neutral, non-ionic substituent)	RIF-1	Diffuse: cytoplasm, plasma membrane, Golgi	Least active of 7–9 in uptake and cell kill; no relocation on irradiation; photobleaches	*e*
10. ZnPc	NHIK 3025	Mainly Golgi and mitochondria	Uptake by diffusion; marker enzyme methodology.	*f*
11. Lu texaphyrin	EMT-6	Lysosomes	Irradiation causes break up of lysosomes, cytoplasmic blebbing and cell death	*g*

a. Further examples may be found in K. Berg, 'Mechanisms of cell damage in photodynamic therapy', in *The Fundamental Bases of Phototherapy* (H. Hönigsmann, G. Jori and A.R. Young, eds.), OEMF, Milan, 1996. pp.181–207.
b. K. Tabata, S. Ogura and I. Okura, *Photochem. Photobiol.,* 1997, **66**, 842–846.
c. P. Morlière, J.C. Mazière, R. Santus, C.D. Smith, M.R. Prinsep, C.C. Stobbe, M.C. Fenning, J.L. Golberg and J.D. Chapman , *Cancer Res.,* 1998, **58**, 3571–3577.
d. K. Berg, K. Madslien, J.C. Bommer R. Oftebro, J.W. Winkelman and J. Moan, *Photochem. Photobiol.,* 1991, **53**, 203–210. Z. Malik, I. Amit and C. Rothmann, *Photochem. Photobiol.,* 1997, **65**, 389–396.
e. S.R. Wood, J.A. Holroyd and S.B. Brown, *Photochem. Photobiol.,* 1997, **65**, 397–402.
f. G.H. Rodal, S.K. Rodal, J. Moan and K. Berg, *J. Photochem.Photobiol. B Biol.,*1998, **45**, 150–159.
g. K.W. Woodburn, Q. Fan, D.R. Miles, D. Kessel, Y. Luo and S.W. Young, *Photochem. Photobiol.,* 1997, **65**, 410–415.

photosensitisers) and radiocounting (on labelled photosensitisers) give reliable results, especially in conjunction with HPLC separation .

Within an individual tissue, the localisation of photosensitiser is expected to be different in different compartments. Thus HpD (and other aggregated or water soluble sensitisers) localise in a mouse mammary tumour mainly in the stroma (the matrix in which the cells are embedded), which constitutes a significant compartment in most tumours. Compounds which localise in the tumour cells tend to be more effective photonecrotic agents.

Most photosensitisers are found in high concentrations in the metabolically vigorous tissues (especially liver, spleen, kidney, lung). What is of particular interest is a high ratio of photosensitiser concentration in tumour with respect to concentrations in all other tissues, but particularly peritumour tissues such as muscle and skin. The hope, which is not always realised because of the varied environments of the photosensitiser molecules in different tissues, is that this will translate into selectivity in phototherapy.

Figure 13.7 shows some tissue / tumour ratios for normal and tumour tissue determined by fluoroimetry in mice with a C3H/Tif mouse mammary carcinoma 24 hours after administration of the photosensitisers. Tumour uptake is set at unity for each tissue. It is seen that the ratios are very sensitive to photosensitiser structure, and that for many of the compounds listed, concentration occurs in the lung, stomach, kidney, spleen and liver: there is not consistent uptake, although AlPc appears consistently at a high (*ie* unfavourable) tissue / tumour ratio. However, for the muscle and skin, favourable tumour/tissue ratios are found for all except AlPc.

Photosensitisers labelled with ^{14}C have also been used in this sort of study. Early experiments with labelled HpD are difficult to interpret because of the complex nature of this photosensitiser (Chapter 6), but m-THPC (77) labelled at the *meso*-carbons (starting from *m*-$MeOC_6H_4{}^{14}CHO$; section 8.4.2.ii and Figure 9.7) is a single substance, and has provided useful data. Thus when mice bearing colo26 colorectal carcinoma implants are injected with 10 μg m-THPC (0.35 μCi) via the tail vein, it is found that tumour concentration reaches a maximum at 48 hours. Some tumour selectivity is observed (tumour / underlying muscle =4.0, tumour /

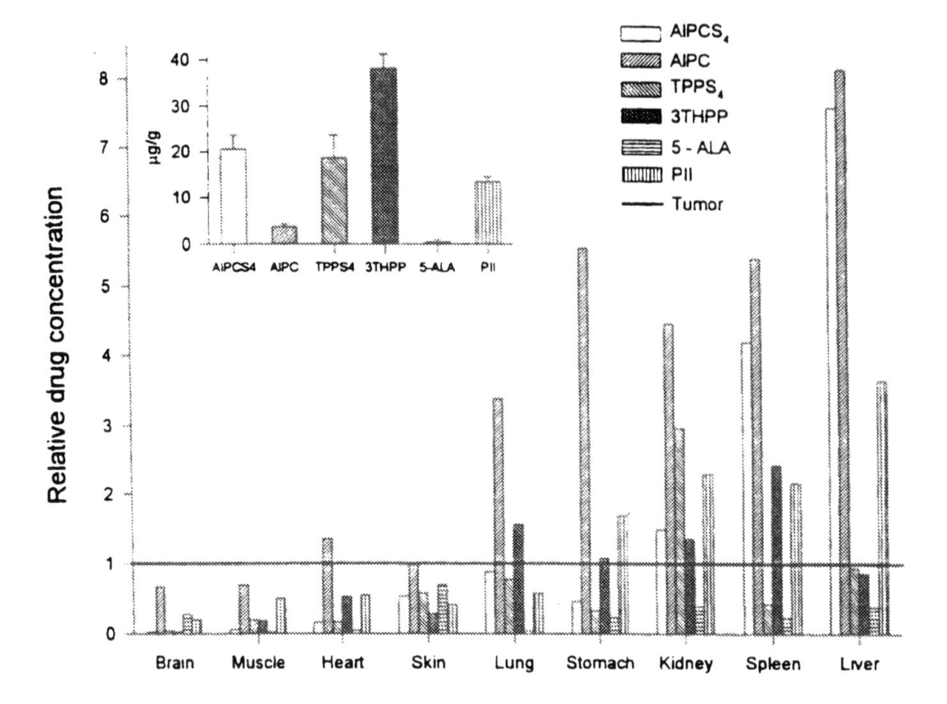

FIGURE 13.7 TISSUE/TUMOUR CONCENTRATION RATIOS FOR $AlPcS_4$ (TETRASULPHONIC ACID OF AL COMPLEX OF **30**), ALPC (**30** AL COMPLEX), $TPPS_4$ (**73**), M-THPP (**76**), ENDOGENEOUS PORPHYRIN (MAINLY PROTOPORPHYRIN, **61**) FROM δ-ALA AND PHOTOFRIN IN NORMAL AND TUMOUR TISSUES OF MICE BEARING MOUSE MAMMARY CARCINOMA C3H/TIF. MEASUREMENTS MADE 24 HOURS AFTER I.P. INJECTION OF 25 MG/KG (δ-ALA, 200 MG/KG). THE INSET FIGURE SHOWS ABSOLUTE CONCENTRATIONS IN THE TUMOUR AT 24 HR. (REPRINTED WITH PERMISSION FROM J. MOAN, Q. PENG, V. IANI, L. MA, R.W. HOROBIN, K. BERG, M. KONGSHAUG AND J.M. NESLAND, ' BIODISTIBUTION, PHARMACOKINETICS AND IN VIVO FLUORESCENCE SPECTROSCOPIC STUDIES OF PHOTOSENSITIZERS', IN *PHOTOCHEMOTHERAPY: PHOTODYNAMIC THERAPY AND OTHER MODALITIES* ISSUED AS *PROC. SOC PHOTO-OPT. INSTRUM. ENG.*, 1996, **2625**, 234–250).

skin=1.6, both at 4 days after injection), and the elimination half-life is 10-12 hours. Figure 13.8 shows changes in radioactivity levels with time in various tissues in this experiment.

13.3 PHOTOSENSITISER EXCRETION

Elimination of photosensitisers from the mammalian (here, mouse) body depends very much on chemical structure. Photosensitisers which are soluble in water, such as $AlPcS_4$, are eliminated mainly in the urine ($AlPcS_4$, 20 mg kg^{-1}, i.p. : 77% in urine, 10% in faeces during 7 days). Photofrin shows a very different distribution under the same conditions (80% in faeces, 3.9% in urine during 7 days). The pattern of elimination of m-THPC is somewhat similar to that of Photofrin: after 7 days, 72% has been excreted in the faeces, but hardly any (< 0.2%) in the urine. The excretion is biphasic, 40% of the radioactivity being excreted in the first 24 hours. The half-life of the initial rapid phase is 12.7–22.8 hours, whereas the second slow phase has a half-life of 5.0–11.7 days.

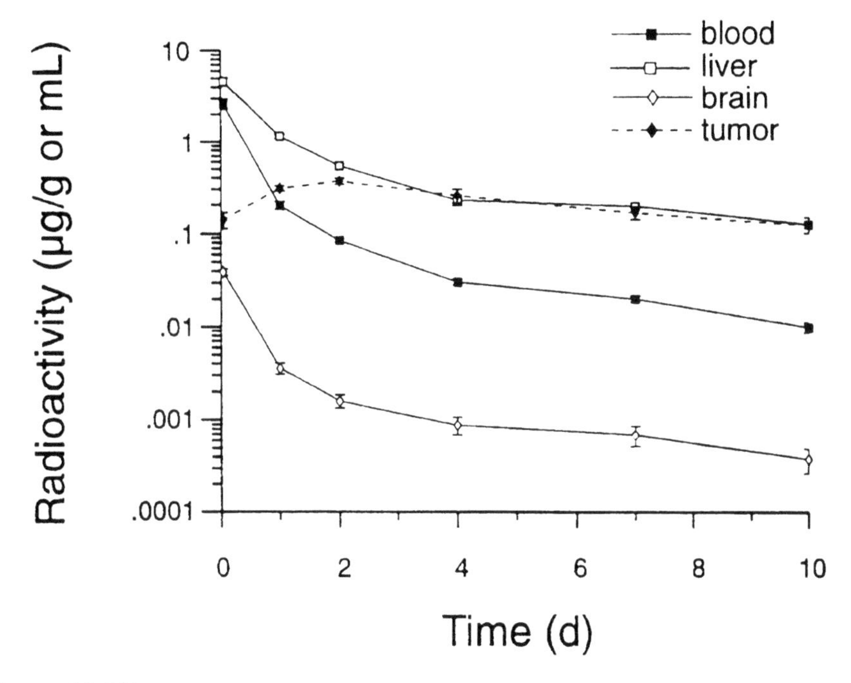

FIGURE 13.8 VARIATION IN RADIOACTIVITY IN TISSUES OF MICE WITH COLORECTAL CARCINOMA IMPLANTS AFTER INJECTION OF 10 μG ^{14}C - LABELLED M-THPC (0.35 μCI). (REPRODUCED WITH PERMISSION FROM R. WHELPTON, A.T. MICHAEL-TITUS, R.P. JAMDAR, K. ABDILLAHI AND M.F. GRAHN, 'DISTRIBUTION AND EXCRETION OF RADIOLABELLED TEMOPORFIN IN A MURINE TUMOR MODEL', *PHOTOCHEM. PHOTOBIOL.*, 1996, **63**, 885–891).

13.4 MECHANISMS OF TUMOUR DESTRUCTION

This topic is currently the subject of increasing interest and speculation. Macroscopic damage to the tumour in PDT appears to occur by at least three pathways: (i) direct damage to tumour cells; (ii) damage to endothelial cells of the vascular system of, and around, the tumour, so shutting off its blood supply; and (iii) macrophage-mediated infiltration of the tumour. For example, according to studies by Moan and his colleagues (*Cancer Res.*, 1995, **55**, 2620–2626), m-THPP (**76**) is mainly localised in the stroma of tumours, and destoys the microvasculature; whereas m-THPC (77) is distributed in the vascular interstitium and neoplastic cells of tumours, and destroys vascular walls and tumour cells.

The death of the tumour cells seems to occur by two distinct processes, the balance between them depending both on the cell line and the photosensitiser structure:

(i) *Necrosis.* This results when membranes (eg, mitochondrial, plasma, lysosomal) or other vital units are so damaged that the cell can no longer sustain essential function, and dies. This type of killing occurs rapidly. Up to this point in this book, the term *necrosis* has been used in the normal (OED) sense of 'death of a circumscribed piece of tissue' (usually tumour), but in the present context necrosis refers to a cause of dying, to make a distinction from apoptosis.

(ii) *Apoptosis.* This is programmed cell death, and it may take 24 hours for the programme to operate and kill the cell. Apoptosis may be recognised by various biochemical markers, including the fragmentation of the DNA in the cell, resulting in a ladder-like appearance of the DNA components on gel electrophoresis. Mitochondrial changes, and the formation of apoptotic bodies, are also observed. With Photofrin, PDT initiated apoptosis has been reported with two out of the three cell lines tested: with merocyanine 450, apoptosis was detected in one leukemic cell line, but was absent in two others. In K562 human chronic myelogenous leukemia cells, 100 minutes of irradiation with haematoporphyrin as the sensitiser causes plasma membrane damage in 20–25% of the cells before any DNA fragmentation is discernible, showing that under these conditions the photo process is causing necrosis. Kessel and his colleagues have reported that two photosensitisers (SnEt2 and a porphycene dimer), which locate in lysosomes, rapidly initiate apoptosis, whereas two cationic sensitisers, which target cell membranes, do not show full apoptotic characteristics: although there was cleavage of DNA to large particles, there was no internucleosomal cleavage. The topic is currently in a state of flux, and overall conclusions are best deferred for the time being. But it will be of considerable interest to discover to what extent, and just how, PDT is able to switch on the apoptotic sequence.

BIBLIOGRAPHY

Localisation

J. Moan, K. Berg, E. Kvam, A. Western, Z. Malik, A. Rück and H. Schneckenburger, 'Intracellular localisation of photosensitisers', in *Photosensitising Compounds: Their Chemistry, Biology and Clinical Use,* (S. Harnett and P. Bock, eds.), *Ciba Foundation Symposium,* 1989, **146**, 95–107.

Q. Peng, J. Moan and J.M.Nesland, 'Correlation of subcellular and intratumoral photo-sensitiser localisation with ultrastructural features after photodynamic therapy', *Ultrastruct Pathol.*, 1996, **20**, 109–29.

J. Moan, Q. Peng, V. Iani, L.W. Ma, R.W. Horobin, K .Berg, M. Kongshaug and J.M. Nesland, 'Biodistribution, pharmacokinetic and *in vivo* fluorescence spectroscopic studies of photosensitisers', *Proc. Soc. Photo-opt. Instrum. Eng.,*, 1996, **2625**, 234–250.

R. Boyle and D. Dolphin, 'Structure and biodistribution relationships of photodynamic sensitisers', *Photochem.Photobiol.*, 1996, **64**, 469–485. Detailed review with useful summarising tables.

Apoptosis

Y.Luo, C.K. Chang and D. Kessel, 'Rapid initiation of apoptosis by photodynamic therapy', *Photochem. Photobiol.*, 1996, **63**, 528–534. Shows rapid internucleosomal DNA cleavage with neutral but not cationic sensitisers. Clear photographs.

V. Carré, C. Jayat, R. Granet, P. Krausz and M. Guilloton, 'Chronology of the apoptotic events induced in the K562 cell line by photodynamic treatment with haematoporphyrin and monoglucosylporphyrin', *Photochem. Photobiol.*, 1999, **69**, 55–60.

CHAPTER 14

CLINICAL AND COMMERCIAL DEVELOPMENTS: PROSPECTS FOR THE FUTURE

"In conclusion, the treatment which I have described seems to have proved its value, and there is every reason to give it the place it deserves in therapeutics, a place which it is at present still far from having obtained, doubtless owing to its strangeness and unintelligibilty. In reality, its scientific basis is much better and more solid than that of many other methods of medical treatment."

Niels Finsen, 'Phototherapy' (J.H. Sequeira, trans.), Arnold, London, 1901.

14.1 INTRODUCTION

After passing a series of hurdles the candidate drug eventually reaches the clinic. Here it is subject to further detailed investigation, often referred to in terms of various phases (see Panel 14.9). From this stage onwards the chemist has less and less influence on what is done, provided the drug moves forward smoothly. If problems arise, then a commercial decision will usually be made depending on the nature of the difficulty, ranging all the way from going back to the bench to modify the structure to abandoning the drug altogether.

Photodynamic therapy for cancer has made considerable headway over the past twenty years, but it still excites suspicion because of its novelty and "high biotech" characteristics. It still has to win over opinion, but this is gradually happening. For example, the first regulatory approval in the UK for a PDT procedure was granted in 1999, six years after the first such approval anywhere (in Canada). Like the Canadian approval, the UK one is carefully limited — it is for Photofrin in the palliation of certain obstructing endobronchial lung cancers and oesophageal cancers — but it represents a step forward in attitude.

PDT is now in a position to join those techniques already in use in tumour treatment and management. Tumour phototherapy has several features which need to be considered when a method of treatment is being selected:-

(i) it can be carried out safely, often on an out-patient surgical basis;
(ii) provided that care is taken to avoid subsequent exposure to strong illumination, it has few side effects. The generalised residual photosensitivity, which lasts for several weeks depending on the drug, is without doubt a significant drawback,

but in the circumstances may be regarded as a bearable one; and the second generation sensitisers cause less generalised photosensitivity, some virtually none at all. In contrast, some forms of cancer chemotherapy are associated with severe side effects;

(iii) because collagen, which is rich in unreactive proline residues, appears to resist photochemical destruction, the general structure of the site is conserved, and the healing and cosmetic outcomes are favourable. This is often not the case for surgical intervention on tumours in delicate thin tissues, such as those of the nose and throat;

(iv) with a non-laser source [such as a tungsten-halogen lamp (section 2.2.2) or a small arc xenon lamp (section 2.2.3)] which is small, manoeuvrable and cheap to operate and maintain, the procedure is inexpensive, although cost will, of course, also depend on the cost of the drug. As an out-patient procedure it would be expected to reduce health costs in the Western world: it can also be seen to be a practical low-cost procedure in Third World countries, as is being demonstrated in current trials in India.

(v) the success rate is satisfactory: if the treatment is unsuccessful, no lasting harm has been done. It does not debar subsequent surgical excision or other treatment should this become advisable. PDT with m-THPC does not induce resistance to chemotherapy, radiotherapy or further PDT in human breast cancer cells studied *in vitro.*

PDT is at its most powerful in the *early detection* (by fluorescence: *diagnostic use,* sometimes referred to as *photodiagnosis*) and *early treatment (therapeutic use ie photodynamic therapy)* of tumours. It is not suitable for all tumours: a tumour has, in the way of things human, often grown to a substantial size before the patient presents, and *PDT does not lend itself to the treatment of large tumours.* However, even then, two variant applications of PDT may be helpful:

(i) *intraoperative,* where after surgical excision of the tumour, the wound is subjected to PDT to kill any remaining patches of tumour tissue;

(ii) *palliative,* where structures, especially tubular structures (eg oesophagus, bronchus), are cleared of tumour to ease congestion.

It is not appropriate in a book of this sort to go into the detail of surgical results, but a short bibliography is provided at the end of this chapter for those who wish to read further on the clinical side. The rest of this final chapter will be devoted to looking at the main second generation drugs which are in the experimental clinical phase; and to cast a glance to the future.

14.2 PHOTOFRIN^{R} (also referred to as porfimer sodium)

Although this preparation has already been discussed in detail (Chapter 6), it is useful here to rehearse some of the clinical detail, since, inevitably, second generation drugs are going to be compared with it. Administration typically involves a drug dose of about 20 mg kg^{-1}, a drug-light interval of 48 hours, with irradiation at 630 nm. Photosensitisation precautions are advised for 30 days. Photofrin is already in the market place, and has been licensed in Canada (Lederle), Europe (Beaufour-Ipsen) and in the United States (Sanofi).

Examples of major approvals for Photofrin are as follows.

(i) *Advanced stage oesophageal cancer* (C.J. Lightdale *et al., Gastrointest. Endoscopy,* 1995, **42**, 507-12). In a multicentre randomised trial for the treatment of partially obstructing oesophageal cancer, comparison was made with thermal ablation (Nd-YAG laser). The trial involved 236 patients. PDT had an advantage over the Nd-YAG laser treatment in that it caused fewer oesophageal perforations, required fewer treatments, and was found to be more comfortable by the patient and, especially in cases of awkward geometry, easier to carry out by the clinician. PDT scored less well than the reference laser ablation in that there were more adverse reactions, including some cases of mild sunburn not found with ablation. Overall, the median survival rate in the two groups was the same.

(ii) *Abdominal non-small-cell lung cancer* (S. Lam *et al., Proc. Soc. Photo-opt Instrum. Eng.,* 1991, **1616**, 20–28).
A comparative trial of radiotherapy combined with PDT against radiotherapy alone (41 patients) showed that adding the PDT component increased the clearance of obstructed airways from 10% to 70%.
A trial of PDT against Nd-YAG laser treatment (211 patients) has shown several advantages for PDT, but generalised photosensitisation with the latter is a problem where patients do not follow instructions.

(iii) *Early Stage Lung Cancer*
Whereas the two proceeding examples are palliative, this is an example of the most promising *early stage* type of treatment, which appears to have high chances of success. Thus, Kato and his colleagues (*J. Clin. Laser Med. Surg.,* 1996, **14,** 235–8) report the use of Photofrin on studies with 75 patients with early lung cancer. The response (see Panel 14.9 for definitions) depends on tumour size, with a complete response of about 96% for tumours less than 5 mm in size, but only about 37% for tumours greater than 2 cm in size.

14.3 SECOND GENERATION DRUGS AT THE CLINICAL STAGE

Thus, at present, only one drug (Photofrin) has regulatory approval. The front-runners amongst the second-generation drugs appear to be benzoporphyrin derivative (Verteporfin) and m-THPC (Foscan). In what follows the photosensitisers are placed in order on the basis of increasing λ_{max} values for Band I.

14.3.1 δ-ALA (104)

The photosensitiser here is essentially endogenous protoporphyrin (see also section 8.5). This is emerging as an excellent photodiagnostic agent, but phototherapeutic applications are essentially limited, it seems, to superficial lesions. However, the response rate to δ-ALA PDT in these superficial neoplasms is striking: in an analysis of 826 superficial basal cell carcinomas treated in hospitals in Canada and Europe, complete responses average 87%, with only 8% of lesions showing no response. Promising results have also been reported for other superficial lesions, including squamous cell carcinoma, Bowen's disease, mycosis fungoides and psoriasis. On the other hand, results for the effect of δ-ALA PDT on *nodular* basal cell carcinomas are less satisfactory.

The dose of δ-ALA is much higher than that which is used with the preformed photosensitisers. When applied topically the dose is difficult to measure, but an oil-water emulsion containing *ca.* 20% δ-ALA is frequently used, followed by a drug light interval of 3–8 hours. Intravenous injection employs about 30 mg kg^{-1}, while the oral dose is about 60 mg kg^{-1}. Long term photosensitivity is minimal. The topical treatment, which is non-invasive, can be repeated without difficulty if this is required. Pain, controllable with local anaesthesia, may be experienced during the irradiation. The cosmetic result is good: this, and the ability to treat disseminated lesions (eg Bowen's disease) appear to be the main advantages over conventional surgical therapies. Panel 14.1 summarises current commercial activity with δ-ALA.

14.3.2 m-THPC (77)

This substance was introduced in section 9.5.1. It appears to be the most potent of the photosensitisers in commercial development, requiring only low drug doses (*ca* 0.1 mg kg^{-1}) at low fluences (*ca* 20 Jcm^{-2}).

The original clinical work (Ris, Berne, 1991) was on malignant mesothelioma (4 cases, intraoperatively, using a drug dose of 0.3 mg kg^{-1}, a drug light interval of 48 hours and a fluence of 10 J cm^{-1} at 650 nm). Subsequent studies, including some with pegylated m-THPC, have extended the early promise of this treatment.

Extensive clinical studies, under the overall coordination of Scotia Pharmaceuticals Ltd., have concentrated on cancer of the upper aerodigestive tract, although there is also activity in prostate, pancreas, gynaeological and palliative applications. For example, in 1993 clinicians in Lausanne described the treatment, mainly of early squamous cell carcinomas of the upper aerodigestive tract of 27 patients. Treatment of 36 early tumours gave complete response (follow up 3–35 months). Complications included perforation of the eosophageal wall (one case) and residual photosensitisation up to one week after drug administration.

The Lausanne group has also studied the pharmacokinetics of m-THPC in plasma in patients undergoing PDT, the drug being injected in water:polyethylene glycol 400: ethanol = 5:3:2. Drug concentrations in plasma fell for about one hour, and then, remarkably, peaked after 10 hours. It is suggested that m-THPC plasma levels monitored just before PDT may be used to adjust PDT light dosimetry.

The related photosensitisers in this series, which are also under commercial development, are (i) *pegylated m-THPC*, which requires a larger mass dose but which is more selective (less generalised photosensitivity); and (ii) *m-THPBC*, which is more active but photobleaches quickly (see sections 12.2 and 12.4.2.1).

Panel 14.2 summarises current commercial activity with m-THPC.

14.3.3 Tin Etiopurpurin — SnEt2 (see section 9.5.3)

This sensitiser is being developed by a company which has changed its name from PDT Inc (which seemed very appropriate) to Miravant Medical Technologies. The substance is hydrophobic, and is injected with a delivery vehicle, eg as an emulsion with Cremophor. Preliminary clinical trials have given encouraging results with cutaneous metastatic breast cancer, BCC and Kaposi's sarcoma.

The drug has a drawback as far as residual photosensitisation is concerned, since photoreaction is reported to persist in some patients for a month or more. Commerical development is summarised in Panel 14.3.

14.3.4 Mono-L-aspartylchlorin e_6 — MACE (125)

As mentioned in section 9.3, this is one of the earliest chlorins with proven PDT application. It is water soluble (so can be administered in PBS), has a short drug light interval, and residual skin photosensitivity is brief.

MACE is believed to be undergoing trials in Japan for the treatment of lung cancer, but little information has been published. Panel 14.4 reveals what little is known on the commercial side.

14.3.5 3-(1-Hexyloxyethyl)-3-devinylpyropheophorbide *a* (75)

This recent introduction (see section 7.2.4.3) has been taken on by QLT Phototherapeutics (Vancouver). Although it is not water soluble, it can be administered in a detergent vehicle. It is an effective sensitiser, and leaves little or no residual skin photosensitivity. Panel 14.5 summarises the little that has been revealed so far.

14.3.6 Sulphonated Aluminium Phthalocyanine

The sulphonation of chloroaluminium phthalocyanine gives a mixture of sulphonic acids (sections 7.2.4.2 and 10.5.3.2). A mixture, essentially of di- and trisulphonic acids, has been employed by Stradnadko in Moscow for extensive clinical trials. Stradnadko reports on the results regularly at international meetings, and a considerable number of patients (several hundred) have now been treated. Residual photosensitivity is thought to be a significant problem with this drug, which is not easily photobleached. The treatment appears to benefit from the use of ascorbic acid or sodium ascorbate (20–50 mg kg^{-1}) as an adjunct (see section 7.2.5). Panel 14.6 outlines some essential features of the clinical use of this photosensitiser.

14.3.7 Benzoporphyrin Derivative (131)

This in a clinical context refers to the mixture of mono acids (**131**) derived from the compound in which the initial Diels-Alder addition has occurred on ring A of protoporphyrin dimethyl ester (section 9.4.2). It may be said to share with m-THPC (77) the lead position amongst the second generation PDT photosensitisers. Although it is reported to have attractive cancer PDT potential, it is currently being advanced by QLT Phototherapeutics Inc., Vancouver as a treatment for age-related macular degeneration (see section 14.5.3.3).

In spite of being a monocarboxylic acid, benzoporphyrin derivative is hydrophobic, and needs a delivery vehicle: usually a liposomal preparation is chosen. The drug light interval is conveniently short, and residual photosensitisation is minimal. See Panel 14.7.

14.3.8 Lutetium Texaphyrin (187)

This substance is not a porphyrin although it is related to the tetrapyrroles (section 11.6). Lutetium texaphyrin (**187**) has built-in hydrophilic chains, has water compatibility and fluorescence, and is an effective photosensitiser which generates little residual photosensitivity after PDT. The phototherapy is painful, however, and this needs to be controlled with local anaesthesia. With absorption at 732 nm, this photosensitiser presents an opportunity to photodegrade melanomas, and this is one of the clinical targets. It appears to be remarkably selective for tumours versus normal skin: cases are described where subcutaneous melanomas show complete response with little damage to overlying skin. Preliminary studies on a range of skin cancers gave 29% complete responses and 17% partial responses.

Current commercial development is indicated in Panel 14.8.

14.3.9 Porphycenes

The porphycenes offer a range of structures which give rise to strong composition-of-matter patents because of their novelty (section 11.4). They are being developed by Cytopharm (Menlo Park, California) and are understood to be in Phase I clinical trials for the treatment of various cancers and for skin diseases. In contrast to the other second generation photosensitisers (other than δ-ALA) which are injected, it is expected that topical application may be employed with the porphycenes as a way of achieving additional targetting.

14.4 COMPARATIVE DATA

Five of the compounds referred to in the panels are chlorins, as is clear from their chemical structures, if not from their names: they are m-THPC, MACE, SnEt2, Photochlor and benzoporphyrin derivative.

Clinical comparisons between the various sensitisers are so far uncommon, and limited to small numbers of patients. For example, a study by Bown and his colleagues (*Neoplasm,* 1998, **45,** 157–161) examined the effect of three different sensitisers in the treatment of various benign and malignant gastrointestinal tumours in the oesophagus, duodenum and rectum in 22 patients who were unsuitable for, or who had refused, surgery. Photofrin (i.v.) was used for 4 patients, oral δ-ALA for 16 patients, and m-THPC (i.v.) for 2 patients. With δ-ALA, photonecrosis was only superficial (up to 1.8 mm), although it was said to improve at higher concentrations or with modified light dosimetry: with Photofrin and m-THPC much deeper tumour photonecrosis was observed. It was found that Photofrin and m-THPC worked better, but (unlike δ-ALA) caused cutaneous photosensitivity (lasting up to 12 and 5 weeks, respectively).

A comparison of the various second generation sensitisers (omitting Photochlor and the porphycenes, for which there is insufficient clinical data) is given in Table 14.1. This illustrates the potency of m-THPC mentioned earlier (section 14.3.2) which shows up in the striking and advantageous brevity of the treatment time.

TABLE 14.1 SOME COMPARISONS OF SECOND GENERATION PHOTOSENSITISERS.

Compound	Approximate irradiation wavelength	ϕ_Δ	Typical drug dose mg kg^{-1}	Typical light dose J cm^{-2}	Typical treatment time (s)
δ-ALA (endogenous protoporphyrin)	635	0.6	60	50–150	500–1500
m-THPC	652	0.43	0.15	5–20	50–200
SnEt2	659	0.6	1.2	150–200	1500–2000
MACE	660	0.8	1.0	25–200	1500
Photosens	675	0.34	1.0	50–200	1500
BPDMA	690	0.7	4	150	1500
Lu Texaphyrin	732	0.11	1.0	150	1500

(Based in part, on L. Milgrom and S. MacRobert, *Chem. Brit.*, 1998, (May), pp.45–50).

14.5 FUTURE PROSPECTS

14.5.1 Synthesis of New Sensitisers

Undoubtedly new photosensitisers will continue to appear, especially as other targets (section 14.5.3) become apparent. It seems likely that the guidelines given in Chapter 7 will become modified as the tendency emerges to tailor different sensitisers to specific cancers and specific diseases. This can already be seen to be happening as the porphycenes (section 11.4) are adapted to topical application, for example for dermatological and peridontal disease. New — and so far clinically untried — photosensitisers have been mentioned earlier in this book, for example, the tetrakis(phenylethynyl)porphyrins (**95, 96**, section 8.4.2 iii) and the squaraines (*eg* **185**, section 11.5). Others will certainly appear.

It also seems likely that further efforts will be made (i) to harness photobleaching capability (Chapter 12) in the design of drugs in order to minimise residual photosensitisation, and (ii) to prepare third generation drugs (section 11.8), that is, those covalently attached to a receptor (or other) targeting system. Combinatorial chemistry using the Lindsey method has been used to prepare libraries of tetraphenylporphyrins, but this approach needs considerable development if it is to prove useful.

14.5.2 Light Sources

Phototherapy is a light-drug package, and on the commercial side it seems likely that manufacturers will produce low or modestly priced packages suitable for use in the local hospital or even in the doctor's surgery. Conventional laser sources are likely to diminish in importance as the cheaper diode lasers and non-coherent sources (section 2.2) become popular.

Light *delivery* is another aspect of development. Interstitial delivery using optical fibres or flexible transparent tubes to deliver the light directly into the tumour at a depth of 2–3 cm below the skin surface have allowed tumour volumes of up to 60 ml to be destroyed in a single treatment.

14.5.3 New Clinical Directions in PDT

What appears to be happening now is that research scientists and clinicians with experience in the PDT of cancer are indulging in some useful lateral thinking and asking the question: what other diseases might be treated with phototherapy? Of course, this is not a new question (see Chapter 5), but it is being approached anew, and from the special viewpoint of the photodynamic effect.

The major prospects for other phototherapies are:

14.5.3.1 Psoriasis (section 5.5)

There have been several preliminary reports on the treatment of psoriasis by PDT. Benzoporphyrin derivative is in clinical trial for this condition (Panel 14.7) and porphycenes are also understood to be in prospective evaluation for it.

14.5.3.2 Arthritis

PDT with benzoporphyrin derivative, as a liposomal preparation, has been found to be effective in the amerlioration of inflammation-related symptoms of rheumatoid arthritis in animal models. The observations suggest that PDT could be used in the management of inflammed joints in autoimmune diseases. If these results are confirmed, there is a fascinating mechanistic problem to be solved here.

14.5.3.3 Age-related Macular Degeneration (AMD)

QLT Phototherapeutics plc and Ciba Vision Corporation are engaged in a joint programme to assess the treatment of the wet-form of age-related macular degeneration. This is a condition of the retina which leads to blindness in sufferers over 50, and which is at present not effectively treatable. Phase III studies using benzoporphyrin derivative (called Visudyne^{TM} in its opthalmological application) with 609 patients at 22 centres in Europe and North America have been completed. These show promising results (after a 12 month follow-up period) and a filing for regulatory approval of the clinical use of this treatment is planned.

14.5.3.4 Other Conditions and Treatments

PDT has also been examined in a preliminary way for the treatment of atheromous plaques. A Phase I clinical trial (30 patients, Pharmacyclics, California) is underway to examine the effect of PDT with lutetium texaphyrin (called ANTRIN^{TM} in this application) for photoangioplasty of peripheral arterial disease.

Endometrial ablation, and other palliative procedures, continue to be studied using PDT.

Photopheresis involves removing blood, irradiating it, or its components, in the presence of a photosensitiser (eg 8-MOP, **48**) and then returning it to the body. It may seem an unlikely technique, but is attracting attention as a treatment for a range of autoimmune diseases, (eg cutaneous T-cell lymphoma, rheumatoid arthritis).

14.5.3.5 Microbicidal Activity

The use of photosensitisers and light to kill microorganisms is promising. Bacteria, yeasts and viruses can be inactivated in this way. Gram-positive and Gram-negative bacteria have different cell-wall chemistry, and the latter are not always killed by neutral and anionic sensitisers. However, cationic sensitisers (eg *meso*-tetrakis (Nalkylpyridinium-4-yl)porphyrin tetratosylate **206**) appear to be able to kill both Gram-positive and Gram-negative bacteria in the presence of light. Cationic porphycenes show similar photomicrobicidal effects.

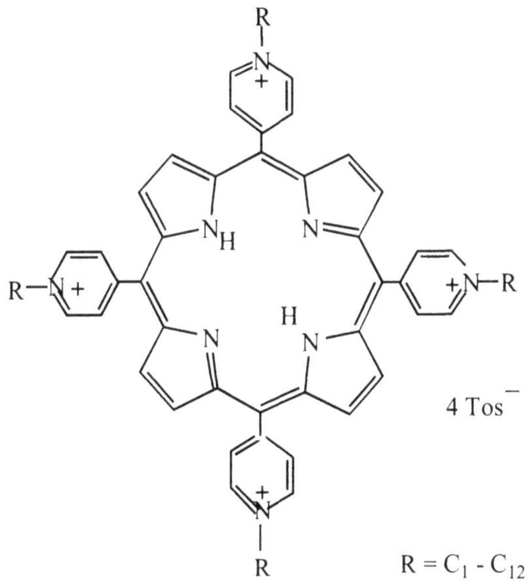

$4\ Tos^-$

$R = C_1 - C_{12}$

206

Photosensitisers have potential application in the killing of microorganisms responsible for peridontal disease.

Sensitisers can also be incorporated (by physical dyeing or by chemical reaction) into polymeric films, and the resulting polymers have photomicrobicidal activity, which may find application.

Photovirucidal activity has been looked at intensively, especially by Ben-Hur and his colleagues. The principle target here, which is related to controlling the spread of HIV disease, is the killing of viruses in blood without disabling the erythrocytes. This has a particular bearing on the spreading of the infection by viruses in transfused blood. Photosensitisers which absorb well into the red (phthalocyanines and naphthalocyanines, Chapter 10), and can therefore be irradiated at wavelengths at which the erythrocytes do not absorb strongly, are being most studied in this regard.

The possibilities for the future are not limitless, of course, but they are extensive, largely because PDT has not been generally appreciated and cultivated in the past. Now that the value of PDT is becoming recognised, the outlook for future development appears to be very encouraging indeed.

Panel 14.1 δ - AMINOLAEVULINIC ACID (δ - ALA) (also called δ-aminolaevulic acid and 5-aminolaevulinic acid, or simply ALA).

$NH_2CH_2COCH_2CH_2CO_2H$ (104), biosynthetic precursor of protoporphyrin (**61**).

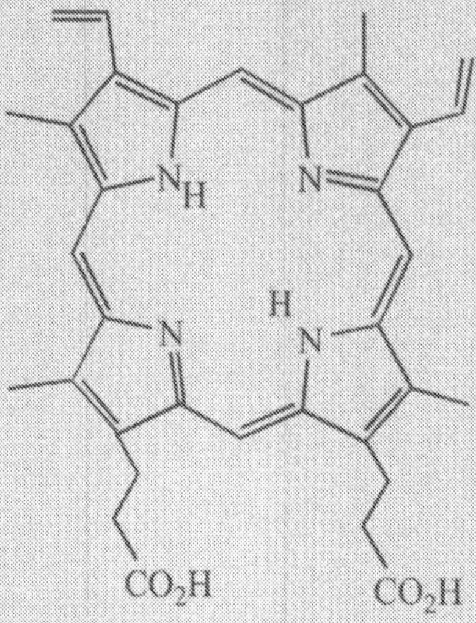

61

Band I (irradiation region): 635 nm (ε 5 000)
ϕ_Δ : 0.6
Trade name: LEVULAN
Company: DUSA Pharmaceuticals Inc, Tarrytown, New York
In Phase III clinical trial for actinic keratoses (sun-induced, potentially cancerous skin lesion), Phase II for bladder cancer photodiagnosis, acne, hair removal.
Current position: *http://www.DUSApharma.com*
Typical protocol: Topical 20% δ-ALA in an oil-water emulsion using a fluence of 150 J cm^{-1}.
δ-ALA esters are under development by PhotoCure AS (Oslo).
Original reference: J.C. Kennedy, R.H. Pottier and D.C. Pross, *J. Photochem. Photobiol. B Biol.*,1990, **6**, 143-148.
Review: Q. Peng, T. Warloe, K. Berg, J. Moan , M. Kongshaug, K.-E. Giercsky and J.M.Nesland, *Cancer*, 1997, **79**, 2282–2308. (225 references).

Panel 14.2 5,10,15,20-TETRAKIS(m-HYDROXYPHENYL)CHLORIN (m-THPC)

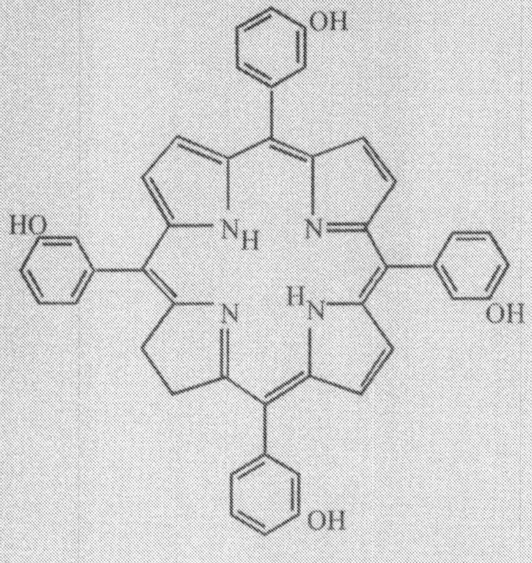

77

Band I (irradiation region): 652 nm (MeOH) (ε 30 000)
ϕ_Δ (air): 0.43
International non-proprietary name (INN): Temoporfin
Proprietary name: FOSCAN
Company: Scotia Pharmaceuticals plc, Stirling, Scotland
In Phase III clinical trials for cancers of upper aerodigestive tract, prostate and others; and palliative procedures. Current position: *http://www.scotia-holdings.com*
Typical protocol: One course, drug dose 0.15 mg kg^{-1}
Delivery : water- polyethylene glycol 400- ethanol or propylene glycol - ethanol
DLI : 96 hr. Wavelength : 652 nm
Fluence : 5–20 J cm^{-2} Fluence rate : 100 mW cm^{-2}
Original reference: R. Bonnett, R.D. White, U.-J. Winfield and M.C. Berenbaum, *Biochem. J.*, 1989, **261,** 277–280.
Clinical reference (head and neck cancer): M.G. Dilkes, M.L. De Jode and A. Rowntree-Taylor, *Lasers Med. Sci.*, 1997, **11**, 23–30.

Panel 14.3 TIN ETIOPURPURIN (SnEt2)

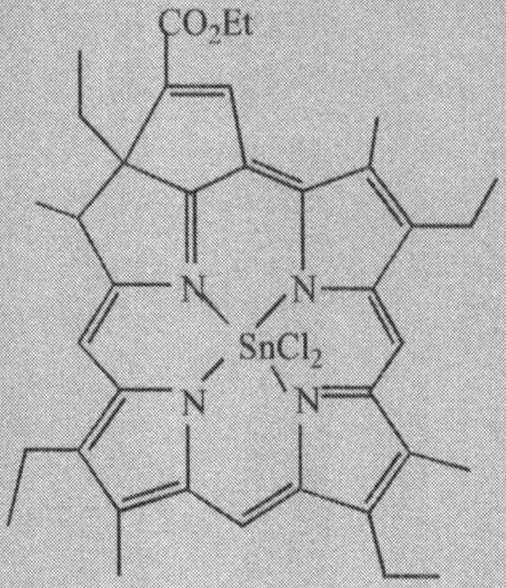

78

Band I (irradiation region): 659 nm (CH_2Cl_2) (ε 30 300)
ϕ_Δ : 0.6
Proprietary name: PURLYTIN
Company: Miravant Medical Technologies, Santa Barbara, California, USA
In Phase III clinical trials for wet AMD (with Pharmacia and Upjohn); Phase I for prostate cancer
Current position : *http://www.miravant.com*
Typical protocol: Drug dose : 1.2 mg kg^{-2}
Delivery : Cremophor emulsion
DLI : 24 hr; Fluence : 200 J cm^{-2}
Original reference: A.R. Morgan, G.M. Garbo, R.W. Keck and S.H. Selman, *Cancer Res.*, 1988, **48,** 194–198.
Clinical reference: N. Razum, A Snyder and D. Dorion, *Proc. Soc. Photo-opt. Instrum. Eng.*,1996, **2675,** 43–46.

Panel 14.4 MONO-L-ASPARTYLCHLORIN e_6 (MACE)

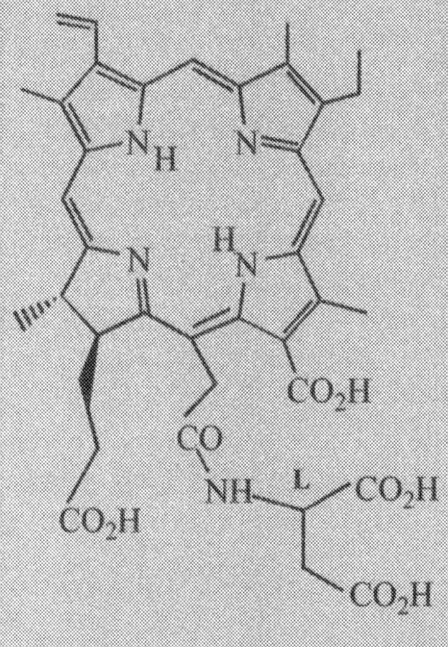

125 MACE

Band I (irradiation region): 660 nm (aq.) (ε 40 000)
ϕ_Δ **: 0.8**
Proprietary designation: NPe6
Companies: Nippon Petrochemicals, Tokyo, Japan & Meiji Seika, Japan
Believed to be in clinical trials for skin cancer and endobronchial lung cancer
Typical protocol: Drug dose : 1 mg kg^{-1}, aqueous delivery
DLI : 3-4 hr
Wavelength : 660 nm
Original reference: J.S. Nelson, W.G. Roberts and J.W. Berns, *Cancer Res.*, 1987, **46,** 4681–4685.
Clinical reference: R.P. Allen, D. Kessel, R.S. Tharratt and W. Volz, in "Photodynamic Therapy and Biomedical Lasers" (P. Spinelli, M. Dal Fante and R. Marchesini, eds.), Elsevier, Amsterdam, 1992. pp.441–445.
Revised structure: S. Gomi *et al., Heterocycles,* 1998, **48,** 2231–2243.

Panel 14.5 3-(1-HEXYLOXYETHYL)-3-DEVINYLPYROPHEOPHORBIDE a

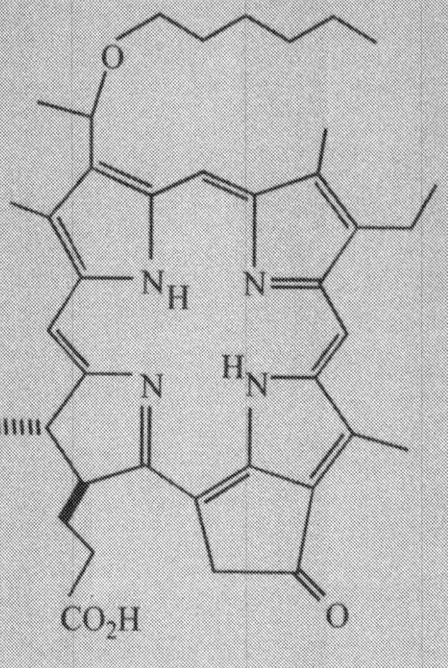

75

Band I (irradiation region): 665 nm (ε 47 500)
ϕ_Δ : 0.48
Proprietary name: PHOTOCHLOR
Company: QLT PhotoTherapeutics plc, Vancouver, Canada
Approved for Phase I clinical trials: two patients with oesophageal cancer treated with encouraging results. Current position : *http://www.qlt-pdt.com*
Possible protocol: Drug dose : 3 mg kg^{-1}
Surfactant delivery system
DLI : 24 hr
Fluence : 135 J cm^{-2}
Original reference: R.K. Pandey, A.B. Sumlin, S. Constantine, M. Aoudia, W.R. Potter, D.A. Bellnier, B.W. Henderson, M.A.Rodgers, K.M. Smith and T.J. Dougherty, *Photochem. Photobiol.*, 1996, **64,** 194–204.
Clinical work: mentioned in R.K. Pandey and C.K. Herman, *Chem. Ind.*, 1998, 739–743.

Panel 14.6 SULPHONATED ALUMINIUM PHTHALOCYANINE ($AlPcS_x$)

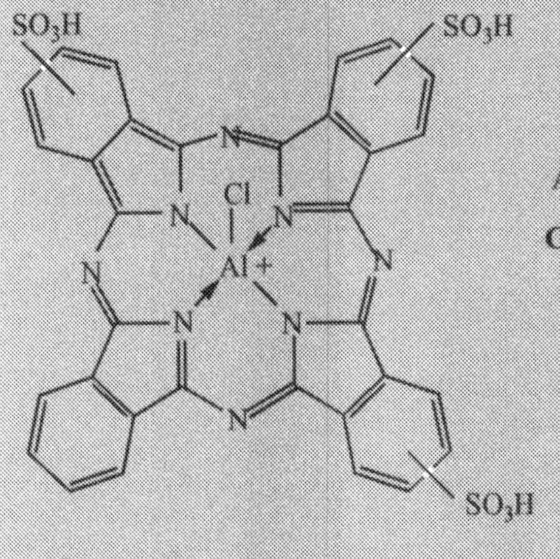

A mixture of di- and tri- sulphonic acids

Chloroaluminium trisulphonic acid shown

Band I (irradiation region): 675 mn (ε 200 000)
ϕ_Δ : 0.34
Proprietary name: PHOTOSENS
Company: State Research Centre for Laser Medicine, Moscow, Russia
Clinical trials started in 1992; mainly skin, breast and oropharyngeal cancers.
Typical protocol: Drug dose : 0.5–1.5 mg kg^{-1}
Aqueous delivery (?)
Fluence : 150–600 J cm^{-2}
Fluence rate : 50–1000 mW cm^{-2}
Clinical references: E. Stranadko, O. Skobelkin, G. Litwin and T. Astrakhankina, *Proc. Soc. Photo-opt. Instrum. Eng.*, 1994, **2325**, 240–246.
E.F. Stranadko, O.K. Skobelkin, G.N. Vorozhtsov, A.F. Mironov, S.E. Beshleul, N.A. Markitchev and M.V. Riabov, *Proc. Soc. Photo-opt. Instrum. Eng.*, 1997, **3191**, 253–262.

Panel 14.7 BENZOPORPHYRIN DERIVATIVE MONOACID RING A (BPDMA)

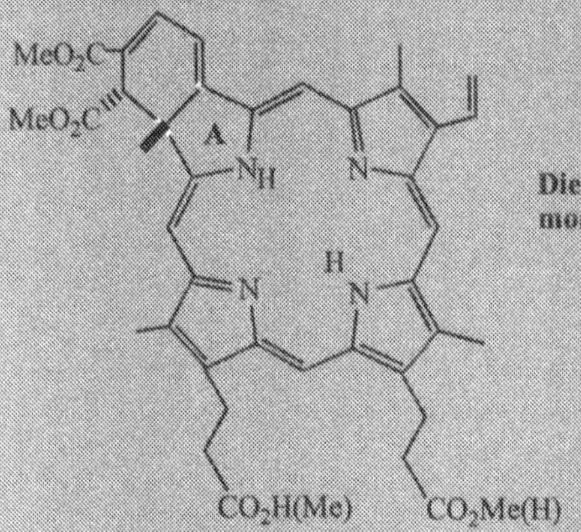

Diels-Alder adduct on ring A: mixture of monoacids at C-13 and C-17(see section 9.4.2)

131

Band I (irradiation region): 690 nm (ε 35 000)
ϕ_Δ **: 0.7**
Proprietary name: VERTEPORFIN
Company: QLT PhotoTherapeutics plc, Vancouver, Canada
Has been in clinical trial for BCC, cutaneous metastases, and psoriasis. Now jointly with CIBA Vision Corporation in trial for age-related macular degeneration under the proprietary name of Visudyne™. Current position : *http://www.qlt-pdt.com* and *http://www.visudyne.com*
Typical protocol (cancer application): Drug dose : 0.3 mg kg^{-1}
Liposomal delivery system; DLI : 3–5 hr
Fluence : 50 J cm^{-2}; Fluence rate : 60 mW cm^{-2}
Original reference: A.M. Richter, B. Kelly, J. Chow, D.J. Liu, G.M.N. Towers, D. Dolphin and J.G. Levy, *J. Natl. Cancer Inst.*, 1987, **79**, 1327–1332.
Clinical reference: J.G. Levy, A. Chan and H.A. Strong, *Proc. Soc. Photo-opt. Instrum. Eng.*, 1996, **2625**, 86–95.

Panel 14.8 LUTETIUM TEXAPHYRIN

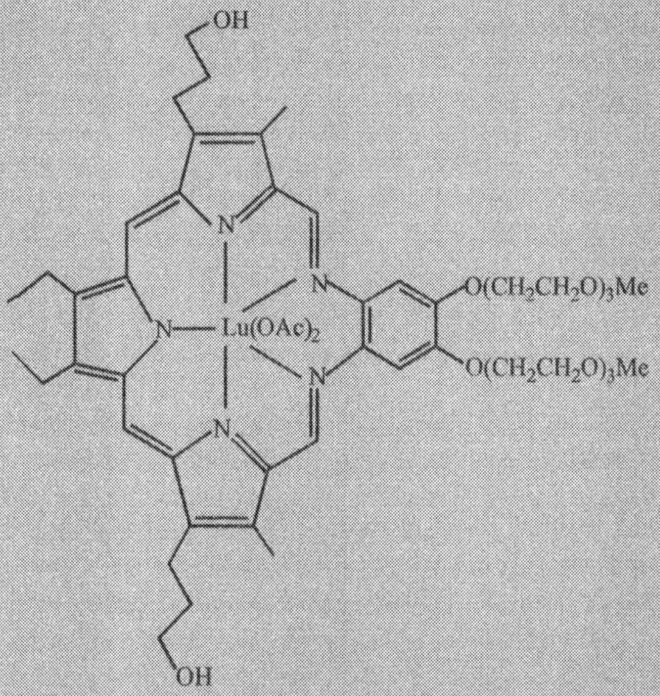

187

Band I (irradiation region): 732 nm (MeOH) (ε 42 000)
ϕ_Δ **: 0.11**
Proprietary name: LUTRIN
Company: Pharmacyclics, Sunnydale, California, USA
In Phase I/II trials for breast cancer, melanoma, Kaposi's sarcoma +
Current position : *http://www.pcyc.com*
Typical protocol: Drug dose : 0.6-7.2 mg kg^{-1}
Aqueous delivery
DLI : 3 hr
Fluence : 150 J cm-2
Initial reference: S.W. Young, K.W. Woodburn, M. Wright, T.D. Mody, Q. Fan, J.L. Sessler, W.C. Dow and R.A. Miller, *Photochem. Photobiol.*, 1996, **63**, 892–897.
Clinical reference: M. Renschler and 13 others, *Photochem. Photobiol.*, 1997, **65**, 47S (abstract).

Panel 14.9 Clinical Evaluation

Phases

The clinical evaluation of drugs is tightly monitored, and falls broadly into four phases, which take place after extensive biological and toxicological studies have been successfully completed.

Phase I establishes the effect on a small number of healthy volunteers, and allows pharmacokinetic studies in humans. (Sometimes patients take part in Phase I trials, depending on the circumstances).

Phase II allows evaluation in patients and numbers are larger. In both phases II and III the drug is administered on an informed patient consent basis, since it is still not, of course, an approved drug. This phase should show whether the drug works against the disease, and whether it produces side effects in patients suffering from the disease.

Phase III is a controlled clinical trial or series of trials involving hundreds or thousands of patients. The new procedure is tried out in comparison with the best available treatment heretofore and is preferably done under 'double blind' conditions. This may not be possible with PDT, where the best alternative is likely to be chemotherapy (but not photochemotherapy). This trial allows the demonstration (or otherwise) of therapeutic advantage, the proof of efficacy and safety, and generates reliable conclusions about dosage, contraindications and precautionary measures.

At about this stage the drug is submitted for regulatory approval. If this is given, it is subject to conditions and usually a specified and limited application. This leads on to

Phase IV where in more routine use post-market surveillance is carried out to provide further information leading to optimisation of the procedure, and to alert clinicians to any further indications or contraindications not recognised earlier.

Cures

Clinicians report results on the experimental treatment of tumours as complete response (or cure), partial response, or no response, importantly giving a time after treatment at which the response was judged. Sometimes the judgement will be made by clinical inspection, sometimes by biopsy, and this should be stated. Often the time period will be a range, because patients may in some cases join programmes which have already started.

These figures need to be interpreted with understanding of the clinician's position.

Complete response is defined as complete absence of tumour at the time of inspection.

Partial response is defined as greater than 50% reduction in tumour volume (or tumour area, if superficial).

No response is any other outcome.

BIBLIOGRAPHY

Popular reviews

R.K. Pandey and C.K. Herman, 'Shedding some light on tumours', *Chem. Ind.,* 1998, 739–743.

L. Milgrom and S. MacRobert, 'Light years ahead', *Chem. Brit.,* 1998, (May), 45–50.

J.Wibley, 'Advances in cancer treatment - photodynamic therapy', *Pharm. J.,* 1997, **258**, 846–848.

Clinical trials — General

R. Stevenson, 'Gold standard for drugs', *Chemistry in Britain,* 1998, **34**, (9), 31–35.

Medical Research Council, *MRC Guidelines for Good Clinical Practice in Clinical Trials*, London, 1998.

Clinical results — PDT

J.J. Schuitmaker, P. Baas, H.L.L.M. van Leengoed, F.W. van der Meulen, W.M. Star and N. van Zandwijk, 'Photodynamic therapy: a promising new modality for the treatment of cancer', *J. Photochem. Photobiol.B: Biol.,* 1996, **34**, 3–12.

T.J. Dougherty, C.J. Gomer, B.W. Henderson, G. Jori, D. Kessel, M. Korbelik, J. Moan and Q. Peng, 'Photodynamic therapy', *J. Natl. Cancer Inst.,*1998, **90**, 889–905. (221references).

J. C. Kennedy, S.L. Marcus and R.H. Pottier, 'Photodynamic therapy (PDT) and photodiagnosis (PD) using endogenous photosensitization induced by 5-aminolevulinic acid (ALA): mechanisms and clinical results', *J. Clin. Laser Med. Surg.,* 1996, **14**, 289–304.

P. Grosjean, J.F. Savary, G. Wagnieres, J. Mizeret, A. Woodtli, J-F. Theumann, C. Fontolliet, H. van den Bergh and P. Monnier, 'Tetra(*m*-hydroxyphenyl)chlorin clinical photodynamic therapy of early bronchial and oesophageal cancers', *Lasers Med. Sci.,* 1996, **11**, 227–235.

T. Glanzmann, C. Hadjur, M. Zellweger, P. Grosjean, M. Forrer, J-P. Ballini, P. Monnier, H. van den Bergh, C.K. Lim and G. Wagnieres, 'Pharmacokinetics of tetra(*m*-hydroxyphenyl)chlorin in human plasma and individualized light dosimetry in photodynamic therapy', *Photochem. Photobiol.,* 1998, **67**, 596–602.

J.G. Levy, A. Chan and H.A. Strong,' The clinical status of benzoporphyrin derivative', *Proc. Soc. Photo-opt. Instrum. Eng.,* 1996, **2625**, 86–95. Refers to cutaneous malignancies, psoriasis, age-related macular degeneration and endometrial ablation.

Future Prospects (for psoriasis and AMD see J.G. Levy et al. above)

K. Berlin, R.K. Jain and C. Richert, 'Are porphyrin mixtures favorable photodynamic anticancer drugs? A model study with combinatorial libraries of tetraphenylporphyrins', *Biotechnol. Bioengineer.,* 1998, **61**, 107–118.

R.K. Chowdhary, L.G. Ratkay, A.J. Canaan, J.D. Waterfield, A.M. Richter and J.G. Levy, 'Uptake of Verteporfin by articular tissues following systemic and intro-articular administration', *Biopharm. Drug Dispos.,* 1998, **19**, 395–400.

H.P. van Iperen and G.M.J. Beijersbergen van Henegouwen, 'Clinical and mechanistic aspects of photopheresis', *J. Photochem. Photobiol. B: Biol.,* 1997, **39**, 99–109.

R. Bonnett, D.G. Buckley, T. Burrow, A.B.B. Galia, B. Saville and S.P. Songca, 'Photobactericidal materials based on porphyrins and phthalocyanines', *J. Mater. Chem.,* 1993, **3**, 323–324.

M. Merchat, G. Bertolini, P. Giacomini, A. Villaneuva and G. Jori, 'Meso-substituted cationic porphyrins as efficient photosensitisers of Gram-positive and Gram-negative bacteria', *J. Photochem. Photobiol., B: Biol.,* 1996, **32**, 153–157.

J. North, H. Neyndorff and J.G. Levy, 'Photosensitizers as virucidal agents', *J. Photochem. Photobiol. B:Biol.,* 1993, **17**, 99–108.

S.J. Wagner, A. Skripchenko, D. Robinette, J.W. Foley and L. Cincotta, 'Factors affecting virus photoinactivation by a series of phenothiazine dyes', *Photochem. Photobiol.,* 1998, **67**, 343–349.

Symposium-in-Print, 'Improving Blood Safety with Light' (eds E. Ben-Hur and B. Horowitz), *Photochem. Photobiol.,* 1997, **65**, 427–470.

Index

A

Abscisic acid, 55
Absorption spectra, 23, 24, 64, 95, 144, 146, 152, 170, 172, 178, 186, 189, 200, 203, 226, 232, 246, 247, 253, 254
  empirical correlation with structure, 153
  Gouterman model, of porphyrin, 155, 156
  on aggregation, 253, 254
  shift in biological medium, 257
Acetophenone, 31
Acifluorfen methyl, 64
Acridine, 34, 57
Acridine orange, 10, 65
Acriflavine, 44, 45, 65
Actinic keratosis, 174, 280
Actinometry, 31
Action spectrum, 35, 92
Age-related macular degeneration, 189, 195, 275, 278, 282, 286
Aggregation, 220, 226, 240, 249–255, 264
  and species diversity, 255
  detection of, 253
  J and H aggregates, 250–252
  of porphyrin tetrasulphonic acids, 254
  reversibility, 250
  spectroscopic changes, 250, 252
3-(1-Alkyloxyethyl)-3-devinylpyropheophorbide *a,* 135, 185, 275
Aluminium(III) phthalocyanine sulphonic acids, 132
  by crossed condensation, 134
  by sulphonation of ClAlPc, 134
  clinical use, 134, 220, 275
  commercial development, 285
  HPLC, 134
  regioisomers of $AlPcS_{2a}$, 134
  tissue relocalisation *in vivo,* 265
Amideacetal, Claisen rearrangement of, 187, 188
Amines, 56
δ-Aminolaevulinic acid (ALA), 61, 173, 248, 277
  as a photodiagnostic agent, 273
  commercial development, 280
  esters, 174
  in phototherapy of superficial lesions, 274
Amphiphilicity, 130, 133, 135, 168, 228
Angelicin, 7, 101
Anhydrides, cyclic, 184, 195
Antioxidants, 78, 80
  effect on photocytotoxicity, 138, 275
'Antrin', 278
Apoptosis, 225, 270
Aromaticity, 151, 178, 201, 227, 229
Arthritis, 189, 278
Ascaridol, 53
Atopic eczema, 7
Atheromatous plaque, 278
ATX-S10 (Na), 190
Axial ligand variation, 138, 221

Azabenzopyrromethene, 216
Azachlorin, 203, 207
Azaporphyrin, 199
5-Azaprotoporphyrin, 203, 204, 207

B

Bacteriochlorin, 177, 178
Bacteriochlorin *a* (bacteriochlorin e_6), 84, 85
Bacteriochlorins, 67, 177, 178
  by reduction of porphyrins, 180
Bacteriochlorin type spectra, 179, 180
Bacteriochlorophyll *a*, 178
  isolation, 183
Bacteriopurpurin derivatives, 185
Baldes, E. J., 115
Barton-Zard pyrrole synthesis, 164
Basal cell carcinoma, 13, 93, 126, 176, 186, 274, 275, 286
Bathochromic, 35, 250, 257
Beer-Lambert law, 23, 258
  deviations from, on aggregation, 253
Benzochlorin iminium salts, 144,195, 197, 198
Benzochlorins, 197
Benzophenone, 34, 47
Benzoporphyrins, 203, 208, 209, 210
Benzoporphyrin derivative, mono ester, ring A (BPDMA), 188, 189, 275, 277, 278
  commercial development, 286
Benzopyrromethene, 208
p-Benzoquinone,
  2,3,5,6-tetramethyl, 249
  2,3-dichloro-5,6-dicyano, *see* DDQ
o-Benzoquinonyl porphyrins, 243
Bergapten, 101
Biladiene-ac synthesis of porphyrins, 162
Biological assay, 257
Bladder carcinoma, 10, 280
Bonar-Law micellar assembly, 165
Bone formation, 94, 99
Boron(III) subphthalocyanine, 215
Bowen's disease, 174, 274
Boyle, Robert, 115
Bilirubin, 7, 55, 83, 104
  E_T, 104
  geometrical isomers, 107, 110
  photocyclisation, 109, 110
  photofragmentation, 104, 107, 108, 113
  photoisomerisation, 104, 107, 113
  photosolubilsation, 104, 107, 113
  positional isomers, 106
  solubility, 107
  synthesis of 5-azaprotoporphyrin from, 203
Bioassay with cancer cell lines, 258
  of chlorin amides, 188
  of $ClGaPcS_n$, 135
  of indocyanine green, 226
  of merocyanine 450 versus m-THPC, 225
  of silicon(IV) naphthalocyanines, 222
  of silicon(IV) phthalocyanines, 221
  of sulphonic acids of Pc/NPc systems, 214, 264
  of m-THPC, effect of 1,3-diphenyliso-benzofuran, 83
  of $TPPS_n$, 132
  of zinc(II) hydroxyphthalocyanines, 221
Bioassay, intermediate, 259
Bioassay *in vivo*, 260
  of bacteriopurpurin derivatives, 185
  of benzochlorins, 197
  of chlorin p_6 derivatives185
  of dimeric systems, 173
  of HpD, 168
  of hypericin, 226
  of lipophilic phenothiazines, 227
  of naphthalocyanine benzamido derivatives, 223
  of porphycene derivatives, 228
  of polyhydroxy chlorins and bacteriochlorins, 193
  of protoporphyrin, 173

of pyropheophorbide *a* derivatives, 135
of m-THPBC, 192
of m-THPC, 192
of m-THPP, 192
of TPP derivatives, 168
of zinc(II) hydroxyphthalocyanines, 221
of zinc(II) octapentylphthalocyanine, 220
of zinc(II) phthalocyanine, 219
Breast cancer, 285, 287

C
Calcitriol, 99
Calcium metabolism, 94, 99
Cancer of upper aerodigestive tract, 273, 274, 281
Carbon-carbon linked dimers in HpD, 121, 125
15-Carboxyrhodoporphyrin anhydride methyl ester, 170
Carcinoma, 13
β-Carotene, quenching by, 82, 83
trans-Carveol, 47
Catalase, 73, 81
Cationic sensitisers, 157, 264, 279
Chiral centres, 126, 130, 157, 189
Chlorins, 67, 177
absorption spectra, 178, 180
by reduction of porphyrins, 190
from protohaem, 186
Chlorin type spectra, 179, 180
Chlorin e_6, 69, 185
trimethyl ester, 184
Chlorin p_6 di and tri methyl esters, 184
Chlorinylglucose, lack of selectivity, 261
Chloroaluminium(III) phthalocyanine sulphonic acid, 34
see also aluminium(III) phthalocyanine sulphonic acids
Chlorophyll *a*, 53, 60, 69, 163, 178, 183, 184
chemical reactions of, 184
isolation from *Spirulina*, 183, 184
Chlorophyll *b*, 183
N-Chlorosuccinimide, 42
Cholecalciferol, 97
Cholesterol, 74, 94
products of Type I and Type II mechanisms, 74
Chorioallantoic membrane (CAM) assay, 259
Chromophore, 35, 97
Clinical applications,
early cancer, 273
intraoperative, 126, 272
new, 278
palliation, 126, 128, 271, 273, 274
photodiagnosis, 272, 273
Clinical results, 174, 192, 271
with d-ALA, 273, 276
with benzoporphyrin derivative, 272, 278
with $ClAlPcS_n$, 275
comparative, 276
with lutetium texaphyrin, 276
with MACE, 275
with Photofrin, 273, 276
with porphycene derivatives, 276
with SnEt2, 275
with m-THPC, 274, 276
Coal tar, 7
Combinatorial chemistry, 277
Congenital porphyria, 61
Conjugated photosensitisers, 233
Conservation of orbital symmetry, 51, 52, 97, 98, 102
Conversions between units, 17
Copper(II) coproporphyrin II tetramethyl ester, 162
Copper(II) octaethylbenzochlorin iminium salt, 144, 145
Copper(II) phthalocyanine, 211
chemistry, 211
Copper(II) phthalocyanine sulphonic acid, 34
Cost-benefit analysis, 262

Cremophor, 138, 140, 221, 275
Crossed condensations, 134, 164, 213, 214, 264
Cyanine dyes, 66, 225
Cycloadditions, 2π + 2π, 51, 75, 96, 108, 244

D

DABCO, in quenching of $^1O_2$, 82
DACE, 186
Definition of terms, 1, 12, 57, 70
  B and Q, 152
  in describing tumours, 12
  in drug administration, 12
  in light intensity, 36
  in photobleaching, 237, 238
  in photochemistry, 35
  J and H aggregation, 252
  solvent/substrate affinities, philicities, phobicities, 133
  Type I and Type II photooxygenation, 70, 74
7-Dehydrocholesterol (7,8-didehydrocholesterol, provitamin D_3), 94, 97
Dehydro-β-ionone (didehydro-β-ionone), 55
Delivery systems, 137, 139, 264, 275
Dermis, 89, 90
DDQ, 161, 164, 165, 172, 210
Detergent, 138, 139, 165
Deuteroporphyrin,
  preparation, 157
  chlorins derived from, 187
opp-Dibenzoporphyrins, 204, 208, 210
Diels-Alder cycloaddition (4π + 2π cycloaddition), 52, 55, 82, 188, 190, 209, 212, 223
1,2-Diethoxyethene (*Z* and *E*), 51
1,2-Diethyl-3-hydroxy-4-pyridone, 174
1α,25-Dihydroxyvitamin D_3, 99
Diiminoisoindolines, 212, 215, 217
Dimeric systems, 123, 173
2,5-Dimethylfuran, 41, 46, 54, 82
1,2-Dimethylindene, 50
Dimethylnitrosamine, 260
5,5-Dimethyl-1-pyrroline-1-oxide, 80, 81, 84
3-(4,5-Dimethylthiazol-2-yl)-2,5-diphenyl-2*H*-tetrazolium bromide (MTT), 258, 259
  derived formazan, 259
Diode lasers, 22, 23, 84
Dipalmitoylphosphatidylcholine (DPPC), 141
9,10-Diphenylanthracene, 42, 45, 53, 82
9,10-Diphenylanthracene-9,10-endoperoxide, 42, 45, 53
1,3-Diphenylisobenzofuran, 54, 82, 83, 257
1,4-Diphenylcyclopentadiene, 52
Diphenylether herbicides, 64
Dipyrrylmethanes, 161
DNA,
  dark repair mechanism, 94
  intercalation of furocoumarins, 103
  photochemistry, 96, 102
  photoreaction in relation to skin cancer, 94
  purine and pyrimidine bases, 95
Dose-response relationships, 262
Dosimetry, 131, 238, 277
Drug administation, 12, 142, 174, 277, 280–287
Dysplasia, 13

E

Electrocyclic reactions, 98, 99
Electromagnetic spectrum, 15
Electronic absorption spectra, 24, 144
  four orbital model, 154, 155
  of porphyrins, 24, 144, 152, 156
  of chlorins, 178
  see also absorption spectra
Electronic structure of dioxygen, 40, 41
Electron-rich alkenes, reactions with $^1O_2$, 49
Electron spin resonance, 80, 85, 86
Electron transfer, 70, 82

El-Sayed rules, 35
Emission spectra, 27, 35
see also fluorescence, phosphorescence, luminescence
Endogenous porphyrin, 173
Energy transfer, 31, 74, 82
Ene addition (ene reaction), 46–48, 74, 75
Endometrial ablation, 278
Eosin Y, 34, 55, 58, 66
synthesis, 233
Epidermis, 89, 90, 94, 174
Ergosterol (provitamin D_2), 94, 97
Erythrocytes (red blood corpuscles), 73
Eschenmoser's reagent, 157
Ester linkage in HpD, evidence, 122
Ester-linked model dimer,123
Estradiol derivative conjugate, 234
Ether linkage in HpD, evidence, 122
Ether-linked model dimer, 124
Etio type spectra, 153
Evan's blue, 260, 261
Exchange transfusion, 8
Excimer, 35
Exciplex, 35,
Excited states, 24
Extended quinones, 64, 226

F

Faraday, Michael, 39
Fenton reaction, 74
Ferrochelatase, 60, 62
Fluence, 19, 36, 192, 280–287
Fluence rate, 19, 36, 280–287
Fluorescein, 231
Fluorescence, 28, 227, 237, 248, 272
in diagnosis, 116, 272, 273
microscopy, 28, 227, 238, 264
photobleaching of, 237, 248
quenching, 28, 248
spectra, 28, 192, 204
Fluoride ion, effect on photocytotoxicity of ClAlPc, 138
Finsen, Niels Rydberg, 3
First generation photosensitisers, 115, 129
advantages, 129
disadvantages, 129
Fischer, Hans, 58, 68, 153
2-Formamido-4-(5-nitro-2-furyl)thiazole (FANFT), 260
N-Formylkynurenine, 75, 79
'Foscan', 192, 281
commercial development, 281
Four orbital model for porphyrin spectra, 154, 155
Free radical mechanisms, 71, 72, 73, 74, 78
experimental tests for, 47, 80
Furocoumarins, 6, 100
DNA intercalation, 103

G

Gadolinium(III) complexes as contrast agents, 230
Gallium(III) phthalocyanine, sulphonation, 132
Geel-dikkop, 60
Germicidal lamps, 4, 19, 95
Glucose,
as an adjunct, 138
covalently attached, to solubilise, 194
Glutathione peroxidase, 73
Golgi apparatus, 265, 266
Gouterman, Martin, 154
Grotthus-Draper law, 1, 3, 32
Group 1 and 2 photobiological processes, 2
Guanidine, 76, 77
Guanine, 76, 77
photochemistry, 95, 96

H

Haematoporphyrin, 10, 24, 34, 58, 67, 115, 119, 142
as a benzylic alcohol, 115, 126
bis-pentyl ether, 159
chirality, 158

derivatives of, 158
dimethyl ester, 123
lipoprotein carrier, 142
photoreactions of, 246, 247
preparation of, 116, 157
sensitisation in man, 58
Haematoporphyrin derivative (HpD), 10, 115, 116, 121, 126, 136
Haematoporphyrin derivative, Stage I, 115, 117
composition, 119
HPLC, 118
Haematoporphyrin derivative, Stage II, 115, 117, 118
capillary electrophoresis (Photofrin), 120
carbon- carbon linked oligomers, 121, 122
chromic acid oxidation, 122
composition, 121, 126
ester-linked oligomers, 120, 122
ether-linked oligomers, 121, 122
HPLC, 120
schematic structure, 121
Haematoporphyrin diacetate, 119
Haematoporphyrin monoacetate isomers, 119
Haematoporphyrin dimers, ether linked, 125
isomerism, 127
Haem catabolism, 7, 104, 105
Haemin (protohaemin), 156
Haemoglobin, 116, 144, 156
H type aggregation, 252
Heavy atom effect, 28, 66, 229
Heliotherapy, 3, 9, 19
'Hematodrex', 117, 126
Hereditary coproporphyria, 62
β-Hexahydroporphyrin, 151, 177, 178
Hexyl ether derivative of pyropheophorbide *a*, 185, 275
commercial development, 284
Histidine, 75, 76, 85, 94
Historical aspects, 3, 39, 57, 115, 116, 177, 199, 237
HOMO/LUMO, 51, 52, 154, 155, 182
HPLC, 118, 120, 134
Hund's rule, 30, 36
Hydrogen peroxide,
enzymatic destruction of, 73
from hydroperoxyl, 73
reaction with *N*-chlorosuccinimide, 42
reaction with hypohalite, 42, 46
Hydroperoxyl radical, 73
Hydrophilicity/hydrophobicity, 130, 132, 133, 135
Hydrophilic substituents, 130, 133
spacing of, in sensitisers, 136
2-Hydroxybenzophenone, 93
Hydroxyethylvinyldeuteroporphyrin (isomers), 119, 120
Hydroxyl radical, 73, 81
Hydroxyphthalocyanines, 216, 220, 221
Hyperchromic, 35, 203
Hypericin, 64, 226
Hypochromic, 35
Hypsochromic, 35, 250

I

Imides and isoimides, cyclic, 185
Inclusion compounds, 138, 142
Indene, 50
Indocyanine green, 226
Interactions, intermolecular,
H-bonding, 249
hydrophobic, 249
π-π and π-σ, 249
van der Waals, 249
Internal conversion, 27, 29
Intermolecular electronic excitation transfer, 31, 33, 74
Intersystem crossing, 27, 29
Intraoperative applications of PDT, 126, 272, 273
Irradiance, 36
Isobacteriochlorin, 177

J
Jablonski diagram, 28, 29, 78
Jaundice, 8
Johnson's biladiene-ac synthesis of porphyrins, 162
J type aggregation, 250–252

K
Kaposi's sarcoma, 287
Kasha, Michael, 39
Kasha's rule, 29, 36
Kasha-Vavilov rule, 36
Kautsky, Hans, 39, 44
Kautsky-Foote mechanism, 43
Keratinocytes, 89
Kinetic studies,
  of chemical reaction of α-amino acids with $^1O_2$, 76
  of irreversible oxidation, 239
  of photobleaching, 238, 240
  of photohaemolysis, 85
  of quenching of $^1O_2$, 76, 239
  on bilirubin, 104

L
Laborant, expert, well admonished, 115
Lasers, 2, 21, 22, 240
  table, 22
Leucomalachite green, 44, 45
Leukemia, 225
'Levulan', 280
Lifetime and half-life, 36
Light, 15
Light absorption, 23
Light delivery, 20, 21, 277
Light-emitting diodes (LEDs), 20, 21
Light fractionation, 18
Light sources,
  carbon arc, 4, 9
  fluorescent tube, 8, 19
  germicidal, 19, 95
  incandescent lamp, 19, 174
  LED, 20, 189
  mercury arc, 19
  Paterson, 20
  see also lasers
Limonene, 46, 48
Lindsey's mild variant of Rothemund-Adler reaction, 165
Linstead, Reginald Patrick, 199, 212
Lipid bilayer, 139
Lipid peroxidation, 72
Lipophilicity/lipophobicity, 130, 133, 135
Liposomes, 138, 139, 141
  effect on protoporphyrin photooxygenation, 242
  in drug administration, 189, 228, 275
  preparation, 141
Lipoproteins, 138, 142, 264
Localisation of sensitisers, 9, 116, 173, 189, 222, 227, 262
  in cellular components, 265
  in vivo, 265, 267, 268
  table, 266
Luminescence, 27, 36, 257
Lumirubin and isolumirubin, 109
Lumisterol$_3$, 97, 98
Lung cancer, 126, 273, 283
Lupus vulgaris, 4
Lutetium texaphyrin derivatives, 231, 277
  absorption spectrum, 232
  clinical applications, 276
  commercial development, 287
'Lutrin', 231, 287
Lysosomes, 265, 266

M
MacDonald synthesis of porphyrins, 163
MACE, 186, 240, 275, 277
  commercial development, 283
Magnesium(II) phthalocyanine, 34
Maleimides, 104, 107, 122, 243
Mannitol, 78, 80, 85
Mechanisms,
  of biladiene-ac cyclisation, 162

of bilirubin photobleaching, 104, 110, 112
of HpD formation, 120, 121
of photodynamic action, 70, 83
of phthalocyanine synthesis, 219
of PUVA, 102
of sunscreen photoisomerisation, 94
of tumour destruction with PDT, 269
of vitamin D_3 photogeneration, 98
Melanin, 91
Melanocytes, 89, 90, 91
Melanoma, 86, 222, 276, 287
Membranes, 133, 140
localisation in, 265, 266
penetration of, 8, 140, 174
Merocyanine 540, 66, 225
Merrifield synthesis, 215
Meso positions, 68
Mesoporphyrin, 58, 123
dimethyl ester, 171
Mesorhodin methyl ester, 171
Mesoverdin, 170
methyl ester, 171
Metalloporphyrins, 150
demetallation of, 151
formation of, 150
Metastasis, 13, 286
Methionine, 75, 76, 241
5-Methoxypsoralen, 101
8-Methoxypsoralen, 7, 33, 34, 101, 278
photoaddition to thymine, 102
synthesis, 103
N-Methylacridone, 51
Methylene blue, 34, 46, 50, 52, 65, 226
4-Methylimidazole, 75
Methyl mesopheophorbide *a*, 184
LAH reduction, 185
Methyl mesopyropheophorbide *a*, 184
LAH reduction, 185
Methyl pheophorbide *a*, 184
Methyl pyropheophorbide *a*, 184
Meyer-Betz, Friedrich, 58
Micelle, 139, 140
critical micellar concentration, 140
effect on photooxygenation of protoporphyrin, 242
Microbicidal activity, 278
Microwave discharge, 43, 46, 50
Molecular orbital interpretations, 40, 51, 52, 98, 112, 150, 155, 181
Monoaspartyl chlorin e_6, 186, 240
commercial development, 283
revised structure, 186, 283
Monoclonal antibodies, 129, 234, 235
MTT assay, 221, 258
Multicellular tumour spheroid (MTS) assay, 258, 259
Multiplicity, 24, 36
Mycosis fungoides, 7, 12, 274

N

Naphthalene, 31, 32
Naphthalocyanines, 70, 141, 202, 222
synthesis, 223
Naphthoporphyrin, 209
Necrosis, 269
Neonatal hyperbilirubinemia, 7, 103
Nickel(II) etioporphyrin, 193
Nitroxide radical, 80
NMR spectra, 152, 253
Nomenclature of porphyrins and relatives, 68, 177, 199, 201
NPe6 (MACE), 186, 283
Nucleophilic cyclisation, in diiminoisoindoline synthesis, 212
Nucleophilic substitution, in phthalonitrile synthesis, 212
Nuclear localisation, 265, 266

O

Octaethylporphyrin, 150, 152
absorption spectrum, 152
Octamethylporphyrin, 150
Octamethyltetrabenzoporphyrin, 204
magnesium(II) complex, 208
Oesophageal cancer, 128, 273, 284
Oligomeric linkages in HpD, 120, 121
Oligonucleotide conjugate, 234

Osmylation, 193, 195
Osteomalacia, 4, 94
13^2-Oxopyropheophorbide *a*,
  by autoxidation of methyl
    pyropheophorbide *a*, 186
  carbene insertion into, 186
Oxo-rhodo type spectra, 153
Oxygen,
  electronic structure, 40
  excited states, first and second, 41
  ground state triplet, 40
  solubility in common solvents, 141
Ozone,
  absorption of light, 90
  complex with triphenyl phosphite, 42, 50

P

Palliation, 126, 272, 273, 281
Parabanic acid, 76, 77
Paramecia, 57
Partition coefficients (*P*, log *P*), 86, 133, 257
  of bilirubin and photobilirubins, 109
  of pyropheophorbide *a* derivatives, 135
  in relation to retention on reverse-phase HPLC, 132, 134, 135, 228
Pegylation, 231, 234, 274
Pentacene, 53
Pentdyopent colour test, 107
Pharmacokinetics, 130, 228, 264, 268, 269, 274
Phases of clinical drug development, 288
Phenols, 56, 93, 167
Phenothiazine dyes, 65, 226
Phlorin, 178
Phospholipids, 14
Phosphorescence, 27, 29
  sensitised, 32
Photoangioplasty, 278
Photobactericidal activity, 278
  of cationic sensitisers, 279
  of sensitised films, 279
Photobleaching, 36, 104, 174, 237, 244, 246
  effect of additives, 240
  effect of solvent, 240
  in biological systems, 248
  in vivo, 248
  kinetic studies, 238–241
  product studies, 241–247
  quantum yields, 241
Photobilirubins I, 107
Photobilirubins II, 109, 112
Photobilirubins III, 112
'Photocarcinorin', 117, 126
Photochemical reaction types, 27
Photochemotherapy, 3, 11, 100
'Photochlor', 275
  commercial development, 284
Photodiagnosis, 116, 143, 173, 189, 272, 280
'Photodyn', 10
Photodynamic action, 57, 58
Photodynamic effect, 9, 57, 58
  sensitiser types, 64–70, 129–235
Photodynamic therapy, 9, 11, 116
  variables in, 131
Photofragmentation, 104, 107, 243
'Photofrin', 116, 117, 126, 272
  cost-benefit analysis, 262
  photobleaching *in vitro* and *in vivo*, 248
'Photofrin I' and 'II', 128
'Photogem' ('Photohem'), 117, 126
Photohaemolysis, 58, 84
Photomicrobicidal activity, 226, 279
Photomodification of sensitisers, 238, 241, 243, 244
Photoneoxanthobilirubinic acid, 111
Photooxygenation of porphyrin systems, 108, 242, 243, 244
Photopheresis, 278
Photoprotoporphyrin A and B, 189, 234, 248
  diaspartyl oximino derivative of A, 190

major products of photomodification of protoporphyrin, 242
Photoreduction, 246
'Photosan', 117, 126
'Photosens', 134, 275, 277
commercial development, 285
Photosensitisers, 43, 64
adjuncts, 138
cellular localisation, 264, 266
delivery systems, 138
excretion, 268
first generation, 115, 129
localisation *in vivo,* 265, 268, 269
photophysical properties, 143
red absorption, 144
second generation, 129
solvents, 137
third generation, 129, 234
Photosynthesis, 2
Phototherapy, 4, 8, 11, 94, 100, 103
direct, 3, 4, 8, 97, 103
indirect, 3, 6, 100
Photovirucidal activity, 42, 221, 226, 235, 279
Phthalocyanines, 7, 69, 141, 199, 201, 202, 203, 205
applications, 211
as photovirucides, 279
lacking D_{4h} symmetry in peripheral substitution, 213–217
photobleaching, 240, 246
solubility, 217
spectra, 202, 205
synthesis, 211, 213, 217
Phthalocyanine sulphonic acids, 132, 136, 137, 211, 214, 220
aggregation, 220
uptake and activity, 220
Phthalonitriles, 212–218
Phylloerythrin, 60
Phyllo type spectra, 153
Phytoporphyrin, 60
Piloty reaction, 159
Pinacol-pinacolone rearrangement, 193–195
α-Pinene, 48
reaction with $^1O_2$ and with $R^\bullet$ / $^3O_2$, 49
Polyethylene glycol, 231, 234
Polyhydroxychlorins, 193, 194
Porphycenes, 227
clinical aspects, 276
spectra, 227, 228
substituted derivatives, 228
synthesis, 227
Porphyrazine, 199
synthesis, 203, 207
Porphyrias, 60, 61, 62
Porphyrin, 34
[26]Porphyrin dication, 145, 146
[34]Porphyrin dication, 146
Porphyrins, 67, 149
aggregate structures, 249, 252
aromatic character, 151
electronic absorption spectra, 152–154
fluorescence, 10, 27, 28, 67, 116
four orbital model, 155
from haemoglobin, 156
metal complexes, 150,
molecular shape and size, 150, 201
NMR spectra, 152
sources, 149
synthesis, 159–172
Porphyrinogens, 61–63, 165, 178
Potentiation of PDT activity by reducing agents, 138
Previtamin D_3, 97
Prostate cancer, 274, 281, 282
Protein binding of sensitisers, 264
Protiodevinylation, 157, 158
Protohaem, 60
Protohaemin, 156, 157, 171
chlorins derived from, 186
Propentdyopent adducts, 55, 104, 107
test for, 107
Propidium iodide, 258
Protoporphyrin, 62, 86, 119, 121, 156
diamide and other derivatives, 86, 87
photobleaching *in vivo,* 248

photooxygenation, 241, 242
Protoporphyrin dimethyl ester, reaction with, acetylene dicarboxylic acid dimethyl ester, 188
Eschenmoser's reagent, 157
singlet oxygen, 189
Protoporphyrinogen, 62
Protoporphyrinogen oxidase, 62, 64
Provitamins D_2 and D_3, 97
Psoralen, 6, 101
Psoriasis, 6, 100, 274, 278, 286
Purines, 95, 96
photochemistry, 76, 77, 96
'Purlytin', 195, 282
commercial development, 282
Purpurins, 178, 196,
photooxygenation, 244
Purpurin 18 methyl ester, 172, 184, 185
Purpurin type spectra, 179
Purpuroporphyrin 18 methyl ester, 172
PUVA, 7, 10
Pyrimidines, 95, 96
photochemistry, 96
Pyrroles, 159-162, 164, 165
Pyrromethenes, 159

Q

Q bands, 152, 153
Quantum yield, 30, 34, 81, 86, 143, 170, 192
of photobleaching of some porphyrins, 241
on aggregation, 252
Quenching,
of luminescence, 28, 31, 32
of singlet oxygen, 82, 104
Quinones,
high potential, use for dehydrogenation, 164, 165, 172, 177, 191, 210
derived from m-THPP, 243

R

Raab, Oscar, 57
Radiance, 37
Radiant power, 37
Reactive dyes, 211
Red absorption, desirability in PDT sensitisers, 144
limitations, 146
Reduction of porphyrins, 190
Reductive cyclotetramerisation of phthalonitrile, 219
Response, clinical, 273, 288
Resorcinol, 157, 158
Retro Diels-Alder reaction, 42, 210
Retro-etio systems, 170
Retro-etio type spectra, 153, 171, 172
Rhodamine 6G, 23
Rhodo type spectra, 153
Rickets, 4, 5, 94
Robertson, James Monteith, 199, 212
Rose bengal, 34, 45, 46, 49, 51, 54, 66, 79, 231
Rothemund-Adler porphyrin synthesis, 164, 167

S

Sarcoma, 13
Schönberg-Schenck mechanism, 43
Schumm reaction, 158
Second generation photosensitisers, 129, 130
comparative data, 277
composition, 130
delivery systems, 137
design criteria, 130
dark toxicity, 130
solution behaviour, 130–143
Selectivity, 130, 227, 228, 238, 261
Sensitisation of phosphorescence, 32
Sensitisers,
in films, 279
intracellular localisation, 264
localisation *in vivo*, 265
new, 277
Serum albumin, 264
Skin, 89, 100
structure, 90

Skin cancer, 93, 94, 283, 285
Skin photosensitisation, 127, 168, 186, 189, 193
Silicon(IV) naphthalocyanines, 222
  in vivo activity, carcinoma implant, 223
  in vitro activity, melanoma cells, 222
  spectrum, 205
  variation of axial ligands, 222
Silicon(IV) phthalocyanines, 221
  variation of axial ligands, 221
Singlet excited state, 24
Singlet oxygen, 33, 39, 43
  diffusion in cells, 248
  formation by chemical methods, 42
  formation by photosensitisation, 32, 34, 43–56, 64, 76, 81
  formation by physical methods, 42, 43, 76
  lifetimes, 39, 41, 248
Singlet oxygen quantum yields, ϕ_{Δ}, 32, 34, 81, 86, 143, 146, 170, 186, 188, 189, 195, 203, 226, 227, 229, 257
Singlet oxygen reactions,
  1,2 cycloaddtion, 49, 50, 75
  Diels-Alder reaction, 42, 46, 52, 75
  ene addition, 46–48, 74
  with amines, 56
  with 5,5-dimethyl-1-pyrroline-1-oxide, 81
  with phenols, 56
  with thioethers, 56, 75
Sodium azide quenching of $^1O_2$, 82
Solubility, 133, 147, 221, 231, 233, 234
  of oxygen, 141, 242
Soret band (B band), 152
Spectral irradiance, 37, 91
Spirographis porphyrin, 241, 242
Squamous cell carcinoma, 13, 93, 126, 174, 186, 274
Squaraines, 229
  acid-base behaviour, spectra, 230
SnEt2, 195, 275, 277, 282
Stark-Einstein law, 1, 30, 143
Stokes shift, 29, 36
Stratum corneum, 89, 90
Stratum malpighii, 89, 90
Structure for HpD, schematic, 121
Sulphones, 56
Sulphoxides, 56, 75
Sulphur compounds, 56
Sunburn, 90
  action spectrum, 92
Sunlight, 19, 89, 91
Sunscreens, 89
  chemical, 92
  physical, 92
Superoxide, 73
Superoxide dismutase, 73, 81
Suprasterols, 99
Synthesis of
  ATX-S10(Na), 190
  benzochlorin derivatives, 197
  benzoporphyrin derivative, (BPDMA), 188
  benzoporphyrins, 208–210
  15-carboxyrhodoporphyrin anhydride methyl ester, 171
  chlorophyll *a*, Woodward, 163, 193, 245
  eosin Y, 233
  MACE, 186
  8-methoxypsoralen, 103
  naphthalocyanines, 223
  phthalocyanines, 212
  phthalocyanines with unsymmetrical substitution pattern, 213–216
  porphycenes, 227
  porphyrazine, 207
  porphyrins, 3+1 approach, 161
  porphyrins, Bonar-Law modification, 165
  porphyrins, classification of ring syntheses, 160, 211
  porphyrins, Johnson's biladiene-ac synthesis, 162
  porphyrins, Lindsey modification, 165, 169

porphyrins, MacDonald synthesis, 163
porphyrins, Rothemund-Adler synthesis, 164
pyrroles, Barton-Zard synthesis, 164
SnEt2, 196
squaraines, 229
tetra(arylethynyl)porphyrins, 169
texaphyrins, 231
m-THPC, 191
m-THPP, 167

T

Tachysterol$_3$, 97, 98
'Temoporfin', 281
α-Terpinene, 53
Tests for Type I mechanism, 80, 84–87
ESR, spin trapping, 80, 85, 86
quenching with antioxidants, 80, 85
quenching with mannitol, 80, 85
superoxide dismutase, 81, 85, 144
Tests for Type II mechanism, 81, 84–87
effect of deuteriated solvents on rate, 39, 83, 85, 104
high ϕ_Δ, 81
luminescence from singlet oxygen, 82
product distribution, 46, 83
quenchers, 82, 85, 104
traps, 82, 85, 104
Tetrabenzoporphyrin, 199, 208
Tetra-t-butylnaphthalocyanine, 70
Tetra-t-butylphthalocyanine, 218
Tetraethylpropentdyopent-methanol adduct, 55
Tetraethylpyrromethenone, 55
5,10,15,20-Tetrakis (*N*-alkylpyridinium-4-yl)porphyrin tetratosylate, 279
5,10,15,20-Tetrakis(arylethynyl) porphyrins, 168
spectra, 170
Tetrakis(3,6-dimethylbenzo)porphyrin, 204, 208
5,10,15,20-Tetrakis(*m*-hydroxyphenyl) bacteriochlorin, *see* m-THPBC
5,10,15,20-Tetrakis(*m*-hydroxyphenyl) chlorin, *see* m-THPC
5,10,15,20-Tetrakis(*p*-hydroxyphenyl) chlorin, p-THPC
cost-benefit analysis, 263
5,10,15,20-Tetrakis(*m*-hydroxyphenyl) porphyrin, *see* m-THPP
5,10,15,20-Tetrakis(*o*-hydroxyphenyl) porphyrin, (o-THPP), 137, 167
skin photosensitisation, 168
5,10,15,20-Tetrakis(*p*-hydroxyphenyl) porphyrin, (p-THPP),137, 166, 167
cost-benefit analysis, 263
Tetramethoxyethene, 50
meso-Tetraphenylbacteriochlorin, 69
Tetraphenylcyclopentadienone, 53
2,7,12,17-Tetraphenylporphycene, 228
meso-Tetraphenylporphyrin, 34, 51, 67, 168
sulphonation, 131, 166
meso-Tetraphenylporphyrin sulphonic acids, 131, 136, 166
aggregation, 254
in vitro activity, 132
Texaphyrins, 229
metal complexes, 230
Thionine, 66, 79
Third generation photosensitisers, 129, 234
m-THPBC,
absorption spectrum, 180, 181
development, 274
photobleaching, kinetics, 240
photobleaching, products, 247
photophysical properties of, 192
synthesis, 191
m-THPC, 137, 277
absorption spectrum, 180, 181
aggregation, effect on absorption spectrum, 253
commercial development, 281
cost-benefit analysis, 263
pegylated, 274
photobleaching, kinetics, 240

photobleaching, products, 247
photobleaching, in cell culture, 248
photomodification, products, 245
photophysical properties, 192
potency, 277
m-THPP, 137, 167
absorption spectrum, 180, 181
cost-benefit analysis, 263
dose response curve, 262
photobleaching in cell culture, 248
photooxygenation products, 242
photophysical properties, 192
photoreduction, 246
reduction, 191
synthesis, 167
Thymine, 95
photochemical reactions, 96, 102
Tin etiopurpurin, (SnEt2), 138, 142, 275
commercial development, 282
Tissue transmittance, 130, 144
Titanyl phthalocyanine, 211
Toluidine blue, 65, 226
Topical administration, 9, 174, 276
Traps for singlet oxygen, 82
1α, 24(*R*), 25-Trihydroxyvitamin D_3, 99
2, 3, 6-Trimethylphenol, 56
4, 5′, 8-Trimethylpsoralen, 101
Triplet energy, 27, 29, 47, 143, 146
Triplet excited state, 24
Triplet lifetimes, 27, 29, 143, 222, 229
Triplet quantum yields, 81, 192, 226
table, 81
Tris(2,2′-bipyridyl)ruthenium chloride, 84
True photobleaching, 238
Tryptophan, 75, 76, 79, 248
Tubulin, 136
Tumour, 12
induced, 260
implanted, 260
Tween 80, 138, 140
Type isomerism, 68, 217-219
Type I mechanisms, 74, 76, 80
in the presence of micelles and liposomes, 242
Type II mechanisms, 74, 76, 81
Type spectra, 153, 179

U

Urocanic acid, 93
Uroporphyrins I and III, 68, 264
Uroporphyrin III octamethyl ester, 163
Uroporphyrinogen III, 61
Uroporphyrinogen I, 63
UV-A, 16, 90, 92
UV-B, 16, 90, 92, 94, 97
UV-C, 16, 90, 92

V

Variegate porphyria, 62,
'Verteporfin', 189
commercial development, 286
Vesicles, 139, 140
Vilsmeier-Haack formylation, 193, 196, 198
Z-Vinylneoxanthobilirubinic acid, photocyclisation, 111
mechanism, 112
Vinylogous Vilsmeier reaction, 195, 198
Virus, 42
'Visudyne', 278, 286
Vitamin D, 4, 6, 94
Vitamin D_2, 94, 97
Vitamin D_3, 97
1α, 25-dihydroxy, 99
1α, 24(*R*), 25-trihydroxy, 99

W

Wagner-Meerwein shift, 196
Weak intermolecular forces, 249
Wigner rule, 32, 33, 36
Wittig reaction, 193, 196
Woodward-Hoffmann rules, 97
Woodward's chlorophyll synthesis, 163, 193, 245

X

Xanthene dyes, 66, 231
Xanthotoxin, 101
Xenograft, 260
Xeroderma pigmentosum, 94
X-radiation, 2, 4, 15
X-ray structure of bilirubin, 109

Z

Z/E isomerism, 93, 97, 110, 111, 113
Zinc(II) dihydroxyphthalocyanine, 216
Zinc(II) monoaspartyl chlorin e_6, photobleaching, 240
Zinc(II) monohydroxyphthalocyanine, 213
Zinc(II) octa(decyl)phthalocyanine, 205
Zinc(II) octaethylbenzochlorin iminium salt, 144
Zinc(II) octaethylporphyrin, absorption spectrum compared with that of free base, 152
Zinc(II) phthalocyanine, 69, 136, 138, 142, 219
  lipoprotein carrier, 142
Zinc(II) phthalocyanine sulphonic acid, 34
  derivatives, 264
Zinc(II) tetrabenzoporphyrin, 203
Zinc(II) tetrakis(*p*-octylphenylethynyl) porphyrin, 169
  spectrum, 170
Zinc(II) tetraneopentyloxyphthalocyanine, 218
Zinc(II) tetraphenylporphyrin, 34, 49, 50
  spectrum, 170
  cleavage by photooxygenation, 244

For Product Safety Concerns and Information please contact our EU
representative GPSR@taylorandfrancis.com
Taylor & Francis Verlag GmbH, Kaufingerstraße 24, 80331 München, Germany

www.ingramcontent.com/pod-product-compliance
Ingram Content Group UK Ltd.
Pitfield, Milton Keynes, MK11 3LW, UK
UKHW051828180425
457613UK00007B/251

* 9 7 8 9 0 5 6 9 9 2 4 8 4 *